Frontiers in Clinical Drug Research - CNS and Neurological Disorders
(Volume 1)

Editor

Atta-ur-Rahman, *FRS*
Honorary Life Fellow
Kings College
University of Cambridge
UK

CONTENTS

PREFACE

Frontiers in Clinical Drug Research - CNS and Neurological Disorders represents comprehensive advances in the development of pharmaceutical agents for the treatment of central nervous system (CNS) and other nerve disorders. It covers a range of topics including the medicinal chemistry, pharmacology, molecular biology and biochemistry of contemporary molecular targets involved in neurological and CNS disorders. The reviews are mainly focused on clinical and therapeutic aspects of novel drugs intended for these targets.

Frontiers in Clinical Drug Research - CNS and Neurological Disorders is a valuable resource for pharmaceutical scientists and postgraduate students seeking updated and critical information for developing clinical trials and devising research plans in the field of neurology. The contributions by leading researchers in the field shed light on a variety of therapeutic areas such as:

- Advances in the treatment of cerebral gliomas, multiple sclerosis and schizophrenia,

- Different uses of antidepressants in treating drug users and adolescents suffering from depression,

- Researches on epilepsy and autism spectrum disorder therapy.

In the first chapter, Peter R. Kufahl *et al.* have discussed the potential use of antidepressants as therapies for drug use disorders. They have provided an overview of preclinical and clinical research and explained the animal models used to obtain information about the treatment potential of various pharmaceutical compounds. Primary neurotransmitter targets and pharmacokinetic aspects of the therapeutic compounds are presented in the chapter. In chapter two, Eluen A. Yeh and Naila Makhani describe the mechanism of action, efficacy data, and side effect profile of multiple sclerosis therapies with a focus on the newly emerging agents. Gerardo Caruso *et al.* discuss antisense oligonucleotides therapy as a new potential tool in the treatment of cerebral gliomas in chapter three.

In chapter four, Shigeto Yamamoto and Shigeru Morinobu focus on D-cycloserine (DCS), a partial agonist of N-methyl-D-aspartate (NMDA) receptors, and on histone deacetylase (HDAC) inhibitors. They have provided an overview of the effects of these agents, the clinical implications and drawbacks associated with their use, and directions for future research. Lactate appears to play an important role in energy production. Takashi Uehara and Tomiki Sumiyoshi provide a comprehensive overview of lactate metabolism as a new target for the treatment of schizophrenia in chapter five.

In chapter six, Francesco Pisani *et al.* review the principles and current issues of antiepileptic drug therapy. Schizophrenia is a devastating mental disorder characterized by positive, negative, affective, and cognitive symptoms. In chapter seven, Seiya Miyamoto provides a review on the effects of currently available antipsychotics and cognitive enhancers under development on cognition in schizophrenia.

In chapter eight Kunio Yui describe autism spectrum disorders (ASD) as a neurodevelopmental disorder with reduced cortical functional connectivity relating to social cognition. The current treatment of ASD and future directions of research in the light of the present understanding of this disease is discussed. In another article the same author presents the current state of development of therapeutics for treatment of adolescent depression.

I hope that the readers will find these reviews valuable and thought provoking so that they trigger further research in the quest for the development of pharmaceutical agents for the treatment of central nervous system (CNS) and other nerve disorders.

I am grateful for the timely efforts made by the editorial personnel, especially Mr. Mahmood Alam (Director Publications) and Mr. Shehzad Naqvi (Senior Manager Publications) at Bentham Science Publishers.

Atta-ur-Rahman, FRS
Kings College
University of Cambridge
Cambridge
UK

List of Contributors

Casey Halstengard

Department of Psychology, Arizona State University, Tempe, Arizona, USA

Eluen A. Yeh

Division of Neurology, Hospital for Sick Children and University of Toronto, ON M5G 1X8, Canada

Francesco Pisani

Department of Neurosciences, University of Messina, Messina, Italy

Gerardo Caruso

Department of Neuroscience, Neurosurgical Clinic, University of Messina, Messina, Italy

Giancarla Oteri

Department of Neurosciences, University of Messina, Messina, Italy

Giuseppe Raudino

Department of Neuroscience, Neurosurgical Clinic, University of Messina, Messina, Italy

Kunio Yui

Research Institute of Pervasive Developmental Disorders, Ashiya University Graduate School of Education, Hyogo, Japan

Laura Rosa Pisani

Department of Neurosciences, University of Messina, Messina, Italy

M. Foster Olive

Department of Psychology, Arizona State University, Tempe, Arizona, USA

Maria Angela Pino

Department of Neuroscience, Neurosurgical Clinic, University of Messina, Messina, Italy

Maria Caffo

Department of Neuroscience, Neurosurgical Clinic, University of Messina, Messina, Italy

Naila Makhani

Division of Neurology, Hospital for Sick Children and University of Toronto, ON M5G 1X8, Canada

Peter R. Kufahl

Department of Psychology, Arizona State University, Tempe, Arizona, USA

Piroska Barabas

Department of Psychology, Arizona State University, Tempe, Arizona, USA

Seiya Miyamoto

Department of Neuropsychiatry, St. Marianna University School of Medicine, Kawasaki, Kanagawa, Japan

Shigeru Morinobu

Department of Psychiatry and Neurosciences, Applied Life Sciences Institute of Biomedical & Health Sciences, Hiroshima University, Hiroshima, Japan

Shigeto Yamamoto

Department of Psychiatry and Neurosciences, Applied Life Sciences Institute of Biomedical & Health Sciences, Hiroshima University, Hiroshima, Japan

Takashi Uehara

Department of Neuropsychiatry, University of Toyama Graduate School of Medicine and Pharmaceutical Sciences, Toyama, Japan

Tomiki Sumiyoshi

Department of Neuropsychiatry, University of Toyama Graduate School of Medicine and Pharmaceutical Sciences, Toyama, Japan

Vincenzo Belcastro

Department of Neurosciences, University of Messina, Messina, Italy

CHAPTER 1

Potential Use of Antidepressants as Therapies for Drug Use Disorders

Peter R. Kufahl[*]**, Piroska Barabas, Casey Halstengard and M. Foster Olive**

Department of Psychology, Arizona State University, Tempe, Arizona, USA

Abstract: The diseases of drug addiction and alcoholism are characterized by a transition from experimental and recreational use to uncontrolled and compulsive intake, often accompanied by chemical dependence. The primary therapeutic target of drug abuse is the persistent craving experienced by abstinent patients that precedes relapse to drug taking. Growing clinical and preclinical evidence indicates negative affect and dysphoria as important contributors to drug and alcohol craving. Due to the high comorbidity between drug use disorders and other psychological disorders, including depression and anxiety, there is an interest in the potential use of readily available antidepressant drugs as preventative treatments against drug and alcohol relapse. This chapter provides an overview of preclinical and clinical research investigating these indications, and explains the animal models used to obtain information about the treatment potential of various pharmaceutical compounds. Overviews of primary neurotransmitter targets and pharmacokinetic aspects of the therapeutic compounds are provided throughout the chapter.

Keywords: Antidepressants, central nervous system, addiction, alcoholism, chemical dependence, neurochemical tolerance, stress, neurotransmitter, norepinephrine, psychostimulants.

1. INTRODUCTION

1.1. Contributing Factors to Drug Relapse

Drug addiction, also known as substance dependence, is a chronic disorder described as the pursuit and consumption of harmful substances, despite the recognized adverse medical, social and legal consequences [1-3]. The definitive characteristics of drug addiction include impaired control over drug use and craving, which leads to a long-lasting vulnerability to relapse [4, 5]. Intense feelings of craving for drugs or alcohol are provoked by a variety of factors,

---

*****Address correspondence to Peter R. Kufahl:** Department of Psychology, Arizona State University, Tempe, Arizona, USA; Tel: 480-965-7598; Fax: 480-965-8544; E-mail: pkufahl@asu.edu

which can be categorized as encounters with drug use-related cues, exposure to stress or the development of altered circuits within the central nervous system [6].

As an individual experiences episodes of drug intoxication and withdrawal, environmental stimuli associated with these experiences become part of drug-related memories, which can trigger feelings of craving at later times [7, 8]. The occurrence of such cue-elicited craving sensations is an important factor in determining the likelihood of relapse to drug taking [9]. However, exposure to drug-related stimuli can also provoke autonomic responses that precede drug relapse without eliciting conditioned memories or explicit feelings of craving [10-12]. Hence, reduction of self-reports of craving, extended maintenance of abstinence from drug use, and prevention of relapse behaviors are all targets in the development of addiction therapies.

Another factor in driving relapse to alcohol and drug use is the presence of *stress* [13, 14], or the process of experiencing and adapting to threatening, harmful, or challenging events [15]. Intermittent exposure to stress can facilitate the transition of regular but recreational use of drugs into a pattern of compulsive use that becomes a maladaptive response to later challenges [16]. As an individual develops chemical dependence and experiences withdrawal, development of neurochemical tolerance initiates adaptations within the brain stress circuits [17, 18]. These changes result in both a heightened sensitivity to drug-related stimuli and dysregulated mood states [19-21].

A third contributing factor to the likelihood of relapse is the ongoing process of neuroadaptations in response to repeated drug or alcohol abuse [22]. Early drug-using experiences in the binge/intoxication stage of addiction are accompanied by adaptations within the dopamine and opioid neurotransmitter systems, known to mediate arousal and reward processing [23, 24]. Sufficient drug experiences eventually leave an individual vulnerable to withdrawal experiences that occur upon cessation of drug use. The negative emotional aspects of withdrawal are longer lasting than the physiological effects, and in fact begin to form the drive to seek further drug experiences, forming the withdrawal/negative affect stage of addiction [25]. Major neuroadaptations in this stage are mediated in large part by corticotrophin-releasing factor, norepinephrine, and dynorphin. When relapse to drug use is the primary method by which the individual relieves the protracted withdrawal state, the disease has entered the final stage of addiction, the preoccupation/anticipation stage. Craving is the predominant phenomenon at this stage, and is mediated by neuroplastic changes in the glutamate, dopamine, and GABA transmission systems [26].

1.2. Comorbidity of Drug Addiction and Depressive Disorders

The co-occurrence of depression and dependence on alcohol and/or illict drugs is extremely prevalent. Drug abuse increases the risk of depression by a factor of nearly five [27], and alcohol dependence increases the same risk by a factor of four [28]. Depressed patients that also have substance abuse disorders are known to have more severe clinical symptoms and are more difficult to effectively treat [29, 30], partially due to the longstanding reticence on the part of clinicians to prescribe drugs to substance-dependent individuals [31]. Approximately 80% of alcohol use disorder (AUD) patients report suffering from a co-morbid psychiatric condition, the most common being symptoms of depression [27, 32]. Smoking cessation is also known to induce clinical depression between 3% (with no history of depression) and 30% (with a history of severe depression) of persons quitting without additional therapy [33]. This prevalence means that antidepressant medications should at least be evaluated for effects on addiction treatment strategies. The clinical studies that have been used to support FDA approval of current antidepressants have largely excluded AUD and substance use disorder (SUD) subjects, due to the perception that these populations do not comply with regular treatment or respond as well to doses with proven therapeutic value for non-SUD/non-AUD participants [34, 35]. However, given the milder side effect profile of SSRIs and other second generation antidepressants, accounting for the added complexities of AUD or SUD comorbidities is arguably more possible. Current psychiatric practice prioritizes the treatment of substance abuse liabilities over that of anxiety and depressive symptoms, since the stress and negative affect characteristics of untreated addiction processes would likely negate efforts to treat most other psychiatric symptoms [36].

2. PRECLINICAL MODELS OF THE PROGRESSIVE STAGES OF DRUG ADDICTION

Much of the progress made in understanding the neurobiological aspects of addiction is due to the accumulation of observations made in animal research [6, 37]. As described above, addiction is comprised of several stages of drug-motivated behavior and medical consequences, making up a complex progression of disease that is not encapsulated by any single animal model. Moreover, experimental animals do not become dependent on drugs or alcohol without the continued assistance from human experimenters. Hence, animal models of distinct stages within the progression of addiction have been developed to create usable laboratory measures and testable hypotheses [22].

2.1. Intracranial Self-Stimulation

Discovered by Olds and Milner [38], the method of intracranial self-stimulation (ICSS) training begins with the surgical implantation of electrodes into the medial forebrain bundle of rodents, a collection of neurons projecting from the brainstem along ventral midline of the brain. These neurons are understood to be dopaminergic and important for the arousal, rewarding, and reinforcing properties of environmental stimuli. Rodents can be trained to rapidly acquire a behavior where a response on a lever is reinforced with a small amount of electrical current, motivating additional lever pressing to receive the same stimulation. After a criterion of responses per time unit is reached, the schedule of reinforcement is increased to require more lever press responses for each subsequent current delivery. This procedure eventually generates a threshold response rate that arguably reflects the physiological state of the brain reward circuit [39]. The frequency (Hz) or intensity (mA of current) of ICSS stimulation can then be manipulated to generate a curve of threshold response rates exhibited by the animal [40].

Importantly, the dopaminergic neurons of the medial forebrain bundle are also primary targets of nearly all known drugs of abuse [41], and they have formed the primary neuroanatomical basis for experiments throughout much of the history of addiction research [37]. Administration of virtually all classes of abused drugs (psychostimulants, opiates, nicotine, and alcohol) shifts the curve of ICSS threshold response rates to the left, implying a more easily stimulated reward circuitry [42, 43]. This method has also been utilized to demonstrate the rewarding effects in rodents of emerging synthetic drugs of abuse [44, 45]. Importantly, withdrawal from chronic exposure to many drugs also modifies ICSS performance, shifting the curve of threshold responses to the right [46]. This effect of diminished reward function is indicative of a state of negative affect that underlies the hypothesized maladaptive processes associated with drug dependence [37].

2.2. Conditioned Place Preference

Another common and simple method for measuring the subjective rewarding effects of abused drugs is the conditioned place preference (CPP) paradigm. In the basic CPP model, rodent subjects are exposed to either side of a conditioning apparatus, which is comprised of two environmentally unique chambers separated by a divider [47]. In repeated sessions, the animals are administered a psychoactive drug prior to placement inside one chamber, the drug-paired side, and in other sessions administered the vehicle (*i.e.*, saline) prior to placement inside the other chamber, the vehicle-paired side. If the drug effects are rewarding, the drug-paired side acquires a rewarding value *via* a Pavlovian conditioning process. This is measured in a test session where the divider is removed and the animal, in a drug free state, exhibits a preference for the drug-paired side as evidenced by increased time spent in that chamber relative to the vehicle-paired side. The expression of CPP for drug-paired sides can be blocked or reversed by a variety of pharmaceutical compounds, thereby indirectly linking drug reward with different types of neurotransmitter receptors. A commonly used variant of the CPP model is the addition of a third "neutral" chamber between the drug-paired and cue-paired chambers, and serves as a starting position in the test session [48]. Whether a neutral chamber is preferable or a two-chamber apparatus with an initial bias expressed by the subjects creates more robust drug-induced effects has been debated in the literature [49, 50], and the most appropriate experimental design may be dependent on the drug being studied [47].

While CPP is useful in determining the emotional valence of drug experiences [51], the preference shifts are generally modest and can make for a cumbersome approach to evaluate the efficacy and potency of pharmaceutical compounds [52]. For example, CPP evaluations of the clinically used anti-craving compounds naloxone and acamprosate have failed to consistently reduce preferences for drug-paired sides in rodents [53, 54]. Furthermore, some drugs of abuse can produce CPP and conditioned place aversion (CPA), where animals avoid the drug-paired side during the probe session, under different experimental circumstances. Peripheral exposure to nicotine has been shown to produce both CPP and CPA [55, 56], and ethanol has induced CPA in rats following drinking [57] but CPP when intragastrically administered, albeit in rats genetically selected to consume alcohol [58]. While these ambiguities can lead to theoretical difficulties in interpreting the effects of pharmaceutical compounds on behavior, the simplicity of the experimental design and capability to simultaneously measure rewarding, aversive and locomotor properties of drugs ensures that the CPP model will remain popular and of high interest in the field of addiction.

2.3. Self-Administration

A more direct method of assessing the reinforcement value of a drug in rodents is the self-administration model [59]. In a series of training sessions, animals are placed in mechanized chambers where a typical operant response (depressing a lever or breaking an infrared beam with a nose-poke into an aperture) results in delivery of drug reinforcement (Fig. **1**). Drugs that are highly reinforcing support increasing amounts of operant behavior throughout training, until a baseline daily rate of drug intake is reached [60]. When the dose of drug reinforcement is reduced by the experimenters, the animals tend to compensate with higher levels of operant responding. To aid the acquisition of self-administration behavior, sensory cues are often presented immediately prior to and during the delivery of drug reinforcement.

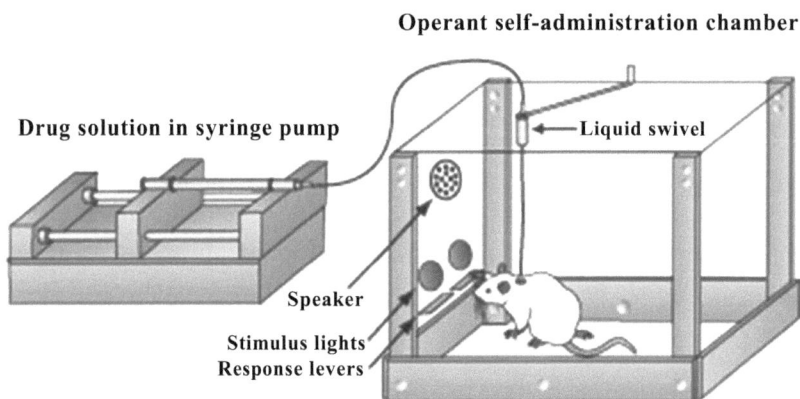

Figure 1: Operant chamber for self-administration training.

For many drugs of abuse, including psychostimulants and opioids, self-administration experiments in rodents require intravenous delivery of reinforcement. Drugs are delivered through a spring tether connected to a vascular port on the animal, leading to a catheter implanted in a major blood vessel. Invasive surgery and a recovery period are therefore required in the preparation of animals before the start of training, but the rapid absorption of drugs into the CNS following drug infusion enables robust timing of sensory cues with rewards. Drugs that are readily self-administered intravenously by laboratory animals are generally thought to have significant abuse potential [61].

Alcohol reinforcement is typically delivered in controlled amounts to a drinking receptacle, creating a self-administration procedure with obvious face validity. The pharmacokinetics of alcohol involves a longer absorption period than intravenous delivery of drugs, complicating the temporal pairing of activity-contingent cues with the drug effects of alcohol intake. Availability of alcohol reinforcement is therefore paired with ambient cues presented throughout self-administration sessions, and these sessions are typically matched by exposure to training sessions without alcohol availability [62]. The non-availability sessions utilize a distinct set of ambient sensory cues, creating a reinforcement contingency signaled by a balanced set of cue conditions [6].

Outbred laboratory rodents generally do not self-administer alcohol to the point of severe intoxication, thereby placing natural limits on the amount of daily intake during self-administration experiments. To create animal models of heavy drinking, several lines of rats and mice have been selectively bred to exhibit high voluntary consumption of alcohol [63, 64]. Of these varieties, one genetic model of alcohol-preferring rats have been known to exhibit physiological withdrawal signs upon termination of free drinking [65, 66].

2.4. Extinction and Reinstatement

The phenomenon of drug craving is most often studied as drug-seeking behavior in rodents using some variant of what has become known as the extinction-reinstatement model [67]. Following a series of self-administration training sessions (Fig. **2a**), animals are then subjected to a series of daily extinction training sessions (Fig. **2b**). During extinction training the animals are exposed to the self-administration environment, but responses are no longer reinforced with drug infusions. Extinction training can take place with or without the environmental cues previously paired with drug reinforcement, and continues until the operant behavior is extinguished, *i.e.*, decreased to a small fraction of self-administration responding levels. Following successful extinction, reinstatement of drug seeking is then triggered in a session by either a small priming dose of the drug delivered by the experimenter, a stressful experience (*i.e.*, footshock), or a cue previously paired with drug reinforcement (provided that this cue was not used during extinction training, Fig. **2c**). Experimental compounds that are administered to the subjects prior to reinstatement testing can then be evaluated for their potential influence on craving and drug relapse.

a) Acquisition and maintenance of drug self-administration

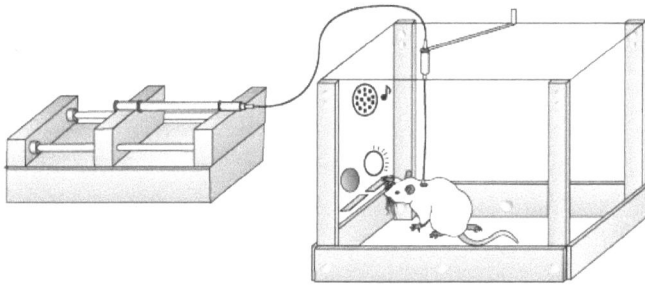

b) Extinction training

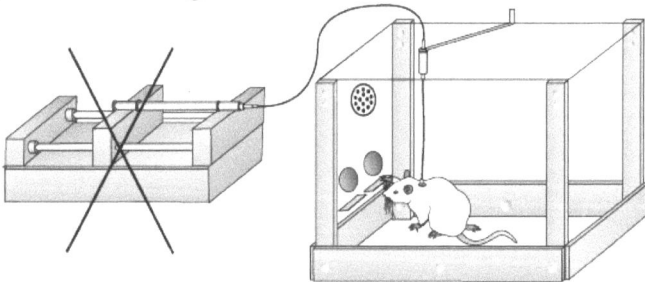

c) Cue-induced reinstatement of drug-seeking behavior

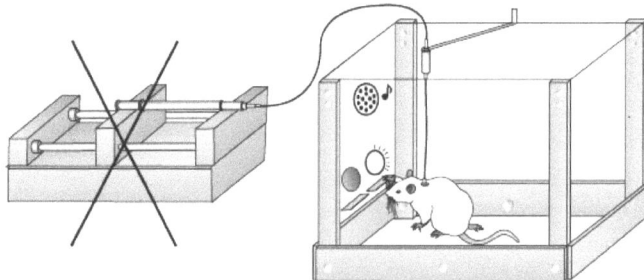

Figure 2: Training sequence for the extinction-reinstatement model of relapse.

3. ANTIDEPRESSANTS

The high comorbidity of substance abuse and clinical depression has presented complications in the treatment strategy of major depressive disorder (MDD). Clinicians commonly require an extended period of abstinence from drugs and alcohol before starting MDD treatment, and symptoms of MDD can overlap those of SUD or side effects of SUD treatment [68]. Consequently, patients with MDD and SUD have been largely excluded from clinical trials of MDD pharmacotherapies, and the interactions between these drugs and substance abuse behaviors are not clearly understood [69]. However, in the recent Combining Medications to Enhance Depression Outcomes (CO-MED) clinical study [70], MDD/SUD patients responded to escitalopram monotherapy, escitalopram + bupropion combination and venlafaxine + mirtazapine combination treatments as well as MDD patients without a comorbid SUD [71]. Additionally, results from the very large Sequenced Treatment Alternatives to Relieve Depression (STAR*D) clinical study [72] included an equivalent success rate of citalopram between MDD/SUD and SUD patients [73]. Improving treatment of MDD symptoms by combining antidepressant treatment with supervision and education resulted in equivalently improved outcomes in patients, regardless of the presence of SUD comorbidity [74]. Hence, the presence or history of an SUD does not appear to require the delay of antidepressant treatment in MDD patients.

The pathophysiology of MDD and other mood disorders characterized by episodes of depression has been linked with dysfunction of the norepinephrine (NE), dopamine (DA) and serotonin (5-HT) neurotransmitter systems [75]. While the classical hypothesis associating MDD with deficiencies in one or more of NE, DA and 5-HT is understood to be outdated, all of the clinically available antidepressants that target these neurotransmitters enhance their release by blocking their corresponding reuptake processes or *via* other indirect mechanisms [36]. Hence, treatment by antidepressant therapy potentially impacts addictive behaviors [37], which are also strongly linked to neurotransmitter dysfunction. DA in particular has been long been a focus of addiction research, due to its role in mediating arousal and reward [76, 77]. However, many drugs of abuse also have immediate and long-lasting effects on NE and 5-HT function [78], and these neurotransmitters have been implicated in the progression from initial drug use to addiction [79-82]. As listed below, several drugs clinically used as antidepressants have also shown potential as treatments for addiction in clinical trials and/or preclinical experiments.

3.1. Selective Serotonin Reuptake Inhibitors

Selective serotonin reuptake inhibitors (SSRIs) are the most prescribed treatments for MDD symptoms. These drugs include fluoxetine (Prozac), paroxetine (Paxil/Seroxat), escitalopram (Lexapro/Cipralex), citalopram (Celexa/Ciprail), sertraline (Zoloft/Lustral) and fluvoxamine (Luvox/Faverin), and target the serotonin transporter (SERT), the protein that clears synaptic serotonin (5-HT) following its release by serotonergic neurons. By inhibiting the activity of SERT, SSRIs are thought to restore depleted synaptic levels of 5-HT to normal levels throughout the brain. SSRIs are known to have a greatly improved side effect profile over earlier antidepressants and anxiolytics. However, like other antidepressants, SSRI drug actions have features not fully explained by current research. First, the maximum benefit of treatment, as assessed by consultations and depression ratings, are not reached until 4 to 6 weeks of regular therapy. This is hypothesized to be reflective of second messenger pathway or gene expression changes that may be critical to obtaining a biochemical state in the brain that improves symptoms of MDD. The consequence of enhanced synaptic 5-HT levels by SSRI treatment is presumably a cascade of cellular responses, including the downregulation of genes associated with receptor synthesis and upregulation of other genes needed for the synthesis of complementary proteins [36]. This process of adaptations could result in decreased 5-HT receptor sensitivity, providing the time-delayed clinical effect. The immediate but usually temporary and moderate side effects reported by patients, including sleep disturbances, sexual dysfunction and anxiety, are usually associated with stimulation of 5-HT_{2A} receptors, an important locus of action for the eventual therapeutic benefits of SSRI treatment [83].

Another incompletely understood feature of SSRI treatment is the source of widely varying reports of efficacy and side effects in individual patients treated with different drugs with presumably the same principal pharmacological action. While differences in the clinical efficacy among the six principal SSRI agents appear to be moderate at best for MDD [84], each of these drugs have pharmacological properties that distinguish it from the others, including side effect profiles, as discussed below. These secondary binding characteristics may have important implications in how MDD/SUD and SUD patients react to treatment.

3.1.1. Fluoxetine

In addition to occupying SERT, fluoxetine is also known to be a 5-HT_{2C} antagonist [85]. Normally, serotonin acting at 5-HT_{2C} receptors inhibits NE and DA release, and fluoxetine may therefore disinhibit to some degree the release of these neurotransmitters. This mechanism may explain the energizing effects of treatment reported by patients, as well as the beneficial effect of fluoxetine and olanzapine (also a 5-HT_{2C} antagonist) in the treatment of bipolar disorder. Since lack of positive affect and impaired responsiveness to non-drug rewards is a recognized characteristic of the withdrawal/negative affect stage of addiction [22], the 5-HT_{2C} antagonistic properties of fluoxetine could arguably be used to remediate these symptoms in abstinent addicts [86]. Early clinical studies demonstrated promising effects of fluoxetine in reducing cocaine usage in abstinent heroin addicts [87, 88], but these results were not confirmed in larger double-blind trials of cocaine addicts [89-91] or cocaine-using heroin addicts on methadone-maintenance [91]. Furthermore, fluoxetine treatment failed to reduce cocaine use or reported craving in comorbid MDD/SUD patients [92, 93].

Despite the lack of convincing evidence that fluoxetine has any benefits beyond placebo treatment in managing craving for cocaine, it has shown some promise in the treatment of alcohol use disorder (AUD). In male AUD hospital inpatients where alcohol was readily available, fluoxetine treatment initially reduced drinking, but levels of consumption returned to pretreatment levels after the first week [94]. Another study found that fluoxetine treatment reliably reduced the self-reported desire to drink in AUD patients, but subsequent reductions in drinking behavior during the outpatient phase were not significantly reduced from placebo-retreated subjects [95]. Fluoxetine treatment also did not effectively reduce drinking behavior or relapse in a 12-week trial of severely dependent AUD patients [96]. A double-blind placebo-controlled trial demonstrated reduced MDD symptoms and lowered drinking in severe MDD/AUD patients after 12 weeks of fluoxetine treatment [97], and these symptoms remained lower one year after treatment [98]. However, adolescent MDD/SUD and MDD/AUD patients continued to exhibit lower drinking and cannabis use five years after fluoxetine treatment, but no significant improvements of MDD symptoms were evident relative to placebo-treated individuals [99]. Additionally, fluoxetine failed to yield results better than placebo treatment when given to MDD/AUD adolescents and combined with cognitive behavioral therapy in a double-blind study [100].

In preclinical experiments, the effectiveness of repeated fluoxetine injections on the reduction of drug self-administration behaviors has been mixed. Fluoxetine has been shown in multiple studies to reduce cocaine intake [101-104], but this effect may not have been drug-specific and unable to persist after cessation of treatment [101]. Fluoxetine treatment also reduced alcohol drinking in alcohol-preferring rats, but this effect was not conclusively established in outbred rats [105]. Repeated fluoxetine (or sertraline or paroxetine, with identical results) injections in mice appeared to initially reduce ethanol consumption, but a behavioral tolerance developed and the effect diminished over time [106]. In a study of cocaine-trained rhesus monkeys, chronic fluoxetine treatment resulted in a desensitization of 5-HT$_{2A}$ receptors and lower DA release in response to a priming injection of cocaine, but presynaptic 5-HT release was unaffected. This study found that fluoxetine treatment did not affect cocaine self-administration but significantly reduced cocaine-primed reinstatement of drug-seeking behavior [107], but a study of outbred rats reported that fluoxetine reduced responding during extinction but failed to significantly attenuate cocaine-primed reinstatement [108]. It remains possible that the monkey and rat extinction-reinstatement models fundamentally differ to the extent that compounds such as the SSRIs may have inconsistent effects on drug-seeking behavior across these species.

Fluoxetine treatment has also demonstrated potential in counteracting the behavioral effects of chronic amphetamine exposure. Normally, withdrawal from repeated injections of amphetamine results in increased ICSS thresholds, indicative of reduced brain reward function [109]. Fluoxetine pretreatment reversed the effect of amphetamine withdrawal and cessation of repeated fluoxetine injections did not produce shifts in the ICSS threshold on its own [110]. In rats trained to self-administer amphetamine, fluoxetine treatment did not reduce responding during a test session (where the amphetamine reinforcement was replaced with saline vehicle) when administered prior to testing [111].

However, when given one day before the test and prior to the final amphetamine training session, fluoxetine appeared to reduce subsequent responding for saline [111]. Fluoxetine also reduced the rate of self-administration for amphetamine across a range of doses, but these same treatments of fluoxetine failed to attenuate self-administration for cocaine [112]. The 5-HT system was also found to be important for the drug effects of methamphetamine: rats trained to press a lever for food pellets after methamphetamine but not saline injections exhibited a reduced ability to discriminate between these conditions when pretreated with fluoxetine [113]. Fluoxetine treatment also interfered with the development and expression of methamphetamine CPP in mice [114]. However, to our knowledge no preclinical research of the effects of fluoxetine, or any other SSRI, on methamphetamine reinstatement has been performed [115]. Given the negative clinical findings reviewed earlier, there presently appears to be a reduced incentive to identify the mechanisms by which fluoxetine modulates psychostimulant-motivated behavior in animal models.

3.1.2. Sertraline

Sertraline is distinguished from other SSRIs by having additional characteristics of binding to and inhibiting the dopamine transporter (DAT) [116] and binding to the sigma$_1$ receptor. The DAT occupancy rate of therapeutic levels of sertraline is small compared to those of methamphetamine or cocaine [117, 118], but even a modest level of DAT inhibition could theoretically counteract some of the reported deficits in energy and concentration associated with SERT inhibition [119, 120]. There is evidence that a subset of MDD patients express fewer adverse side effects with sertraline than other SSRI treatments [121].

Despite the theoretical advantages of sertraline over other SSRIs that lack affinity for DAT, several clinical experiments have yielded mixed results in the treatment of SUD and AUD patients. While sertraline treatment resulted in reduced drinking by alcoholics during active treatment and six months following cessation of treatment, this effect was isolated to late-onset, "lower risk/severity" AUD patients [122, 123]. A later study by the same authors further restricted the therapeutic effect to men and not women [124]. Sertraline also failed to significantly reduce AUD symptoms in MDD/AUD comorbid patients, regardless of the severity of alcohol dependence [125]. Treatment by a combination of sertraline and naltrexone did not extend the measured time to relapse in AUD patients, compared to participants who were treated with naltrexone alone [126]. However, treatment of MDD/AUD patients with this combination resulted in significantly improved abstinence rates and improvements of MDD symptoms than either sertraline or naltrexone alone, or placebo [127]. Finally, the clinical benefits of sertraline in male AUD subjects were evident when the population was comprised of late-onset alcoholics; in early-onset, "higher risk/severity" patients, sertraline treatment outcomes were noticeably worse than placebo [128].

Clinical attempts to establish sertraline as a treatment of psychostimulant dependence have also produced variable results, depending on the nature of the SUD. Large double-blind studies of methamphetamine SUD patients found that drug use was increased in sertraline-treated compared to placebo-treated groups [129, 130]. A later report linked sertraline treatment with increased reported cravings for methamphetamine, conclusively making this SUD a contraindication for sertraline [131]. A double-blind placebo-control experiment found that sertraline treatment decreased self-reports of cocaine craving but not physiological evidence of cocaine use [132]. However, sertraline treatment resulted in significantly delayed drug relapse and reduced depression symptoms in MDD patients who were also abstinent cocaine addicts [133]. While discrepancies in the methodology among the different clinical studies may account for the variance in treatment outcomes for sertraline, it appears more likely that the therapeutic effects of this treatment depend greatly on the nature of the SUD being investigated.

Preclinical studies of sertraline have generally found that this SSRI exerts some effects on addictive behaviors, but only *via* mechanisms that also control feeding. Sertraline was found to reduce cocaine self-administration in rhesus monkeys [134] and alcohol drinking in rats [135, 136], but these effects were not shown to be drug-specific. Further investigation of the rat drinking model suggested that the effects of chronic sertraline treatment were anorectic and the reduced alcohol intake was a secondary phenomenon [137, 138]. Additionally, sertraline treatment did not reliably reduce nicotine self-administration in rhesus monkeys [139], but in rats it reduced the hyperphagia and weight gain that are typical consequences of nicotine withdrawal [140].

3.1.3. Paroxetine

In addition to its SSRI properties, paroxetine is a moderate inhibitor of muscarinic cholinergic receptors and a weak blocker of the norepinephrine transporter (NET). These secondary characteristics have made paroxetine a more popular treatment for MDD patients with anxiety symptoms. It also has a relatively short half-life and metabolic effects that result in more severe cessation consequences when treatment is abruptly discontinued. These include akathisia, restlessness, and tingling sensations; there are also symptoms of sudden discontinuation in other SSRIs, but perhaps more prevalent with paroxetine due to rebound effects within the cholinergic system [141].

Due to its activity at cholinergic receptors, paroxetine has been a drug of interest in the treatment of nicotine addiction. A small double-blind clinical trial of paroxetine treatment and nicotine patch found the combination effective in reducing cravings for nicotine as well as depressive symptoms [142]. Since then paroxetine has been a treatment of interest for MDD and smoking comorbidity [143], but is generally seen as less effective than NE-modulating drugs, such as bupropion in bringing about and maintaining smoking cessation outcomes [144].

An exploratory experiment (albeit with a high rate of subject dropout) indicated potential for paroxetine treatment to reduce craving for methamphetamine in regular methamphetamine users [145]. Administration of paroxetine also appeared to abolish the expression of methamphetamine CPP in mice [146]. However, paroxetine failed to outperform placebo treatment in reducing drug use in a more complete study of cocaine addicts [147].

SSRIs are also a first-line treatment for social anxiety disorder (SAD), a disease highly comorbid with alcohol dependence [148]. A small study of paroxetine treatment in SAD/AUD patients showed reduction of anxiety symptoms as well as some improvement in measures of drinking [149]. Follow-up double-blind experiments established that comorbidity with AUD did not reduce the anxiolytic effects of paroxetine treatment in SAD patients [150], but reductions in drinking behavior were not evident compared to placebo treatment [151]. However, paroxetine-treated subjects reported less of a reliance on alcohol to engage in social situations [151], suggesting that treating SAD symptoms with SSRIs may promote participation in addiction treatment programs [152].

3.1.4. Fluvoxamine

Like sertraline, fluvoxamine has sigma$_1$ receptor binding properties, only with a greater affinity. Though developed and marketed for depression, fluvoxamine has long been approved in the United States for treatment of anxiety disorders but not MDD. Sigma$_1$ receptor pharmacology has recently been a focus of addiction research [153], where specific sigma$_1$ antagonists have been shown to reduce cocaine CPP [154] and reinstatement of cocaine-seeking behavior following self-administration training [155]. Furthermore, self-administration of methamphetamine resulted in upregulation of sigma$_1$ receptors in rats [156], and sigma$_1$ function was shown to have a role in alcohol-induced CPP in mice [157].

Preclinical studies of fluvoxamine and drugs of abuse are limited, but fluvoxamine treatment appears to be more effective in reducing voluntary intake of alcohol than of cocaine or nicotine in rats [158]. Fluvoxamine also failed to reduce the expression of methamphetamine-induced CPP in mice, whereas paroxetine and fluoxetine had been effective [146]. However, fluvoxamine and fluoxetine reduced alcohol consumption in alcohol-preferring rats, whereas paroxetine failed to do so [159]. Further studies of alcohol self-administration using responding for food pellets as a control behavior demonstrated that acute treatment of fluvoxamine resulted in reduction of operant responding for alcohol but not food [160, 161]. However, chronic fluvoxamine treatment resulted in inconclusive effects in rats, where some animals experienced transient reductions in alcohol intake relative to food intake, while others exhibited opposing effects [162]. The limited preclinical data therefore suggests that fluvoxamine may be more effective in treating AUD than non-alcohol SUD behaviors, but the effectiveness of this drug would be highly variable among individuals.

Clinical evaluation of fluvoxamine in alcoholic patients demonstrated equivocal treatment effects. Fluvoxamine, like other SSRIs, reduced drinking behavior in some AUD patients but the results were highly variable, with baseline alcohol consumption, gender and AUD subtype (low *versus* high risk) identified as significant contributors to the treatment efficacy [163]. A double-blind placebo-controlled trial of prevention of alcohol relapse largely failed to show effectiveness of fluvoxamine treatment over placebo [164].

3.1.5. Citalopram and Escitalopram

Citalopram is comprised of two enantiomers, R-citalopram and S-citalopram. The racemic mixture of these two enantiomers has moderate antihistaminergic properties and metabolic effects. The R enantiomer may act to interfere with the SSRI mechanism of the active S enantiomer, making the efficacy of citalopram inconsistent at low doses [165]. Escitalopram was developed as containing only the S enantiomer, and is associated with a higher potency, more consistent efficacy and an extremely limited side-effect profile [36, 166]. Escitalopram has been shown in a series of double-blind placebo-controlled trials to be effective for a spectrum of anxiety disorders, including social anxiety disorder, panic disorder, generalized anxiety disorder and anxiety symptoms present in MDD patients [167].

Citalopram treatment has been shown to reduce ethanol consumption in rats, but this effect was accompanied by reduced water and food intake [159, 168, 169]. Furthermore, behavioral tolerance developed in response to repeated citalopram treatments, and rats with different histories of alcohol exposure generally returned to pretreatment levels of voluntary alcohol consumption [170].

Like fluoxetine, treatment with citalopram resulted in reduced drinking *versus* placebo in AUD patients over the first week of observation [171], but after 12 weeks the reduction of alcohol consumption in citalopram and placebo groups were similar [172]. Male AUD patients appeared to respond more to citalopram intervention than female AUD patients [173]. However, in a double-blind placebo-controlled study solely reliant on testimonials of the patients and their families, citalopram treatment resulted in a long-term reduction of AUD behaviors (significant in the data from reports of the families, but not from self-reports of the participants) [174]. A more promising clinical study found that citalopram combined with behavioral therapy extended abstinence and reduced long-term drug consumption in cocaine SUD/MDD patients [175]. A small open-label study found that escitalopram improved the therapeutic effects of SUD medications naltrexone and GHB, when given simultaneously [176]. In a double-blind study, escitalopram treatment reduced drinking and craving for alcohol in severe AUD/MDD patients over a 6 month observation period, but the dropout proportions were nearly 50% of participants [177]. These results indicate a possible use for citalopram and escitalopram in treating severely addicted individuals, but further studies will require more subjects and complementary methods of assessing drug or alcohol intake.

3.2. Dual Serotonin/Norepinephrine Reuptake Inhibitors - Venlafaxine and Duloxetine

SSRIs were developed as an improvement over the generation of tricyclic antidepressants that were known to have a variety of troublesome side effects. The high selectivity of SSRIs also consequently lack the multiple therapeutic mechanisms needed to generate an adequate therapeutic effect in many patients, particularly those who experience chronic pain [178]. This created a need for the development of antidepressants with additional selective targets, such as the serotonin norepinephrine reuptake inhibitors (SNRIs). These drugs, which include venlafaxine (Effexor) and duloxetine (Cymbalta), inhibit the activity of SERT and the reuptake transporter protein of norepinephrine (NET). An important secondary action of SNRIs is that due to the relative scarcity of DAT in the prefrontal cortex, inhibition of SERT and NET results in the enhanced synaptic levels of dopamine in this part of the brain [179], as a result of high affinity of prefrontal NET for dopamine.

The first SNRI to be developed, venlafaxineis widely prescribed for a variety of depression and anxiety disorders. Withdrawal symptoms and other immediate side effects have been addressed with the extended release formulation of venlafaxine XR, which is taken once per day by patients. However, venlafaxine appears to have an increased risk of fatal overdose, although frequency of overdose has not reached the levels observed with tricyclic antidepressants [180]. A small clinical study reported outstanding responses to venlafaxine treatment by comorbid cocaine SUD/MDD patients, including a 75% reduction in cocaine use during participation [181]. However, an eight-week double-blind study did not confirm this early result, indicating no significant treatment effect of venlafaxine on cocaine use over the length of the study, compared to placebo [147]. Neither the SSRIs discussed above nor venlafaxine have been shown to substantially improve rates of smoking cessation [182]. Venlafaxine was also found to be ineffective in managing anxiety symptoms arising from alcohol detoxification [183].

Duloxetine is an SNRI that is prescribed for MDD, anxiety and chronic pain associated with diabetes. Compared to venlafaxine, duloxetine has appreciably greater potency in blocking SERT and NET [184]. Duloxetine was described as facilitating smoking cessation in a case report [185], but to our knowledge this observation has not be followed by a controlled study. Also intriguing is a recent report that duloxetine reduced alcohol drinking in rats trained to self-administer to achieve blood alcohol levels consistent with human binge drinking [186]. Unlike SSRI effects on rat alcohol intake, which had been accompanied by anorexic effects [187], duloxetine treatment did not reduce consumption of a high palatable and caloric solution [186].

3.3. Dual Norepinephrine/Dopamine Reuptake Inhibitors - Bupropion

Bupropion (Wellbutrin/Zyban) is the only available antidepressant that is a norepinephrine and dopamine reuptake inhibitor (NDRI), containing relatively weak blocking actions for DAT and NET [188]. It is prescribed as a supplementary treatment to counteract undesirable side effects of SSRI treatment [189, 190], or as temporary relief from acute anxiety associated with grief [191]. Bupropion is frequently described as stimulating and is a common treatment strategy for MDD patients undergoing antidepressant therapy but reporting a lack of positive affect.

Importantly, bupropion is widely used as an anti-craving drug in smokers using nicotine replacement therapy [192, 193]. It counteracts the depression symptoms that often accompany nicotine abstinence and essentially doubles the success rate of smoking cessation over the option of no treatment [194]. These effects are confirmed in preclinical studies, where bupropion has been shown to attenuate nicotine self-administration behavior [195, 196]. However, bupropion treatment enhanced reinstatement of responding for nicotine-paired cues after self-administration and extinction training, indicating that its therapeutic effects may be limited to achieving smoking cessation but not the prevention of relapse [197].

The moderate occupancy of DAT by bupropion has been associated with its stimulating effect, but its effect on DAT is far less than that of many abused drugs [198, 199], possibly substituting for these drugs at the receptor level and reducing their intake [113, 200]. Bupropion treatment reduced methamphetamine self-administration in rats, both during [201] and after this behavior was acquired [202]. The presumptive therapeutic use of bupropion in treating psychostimulant

SUDs was then analogous to the reinforcement substitution strategy that underlies the maintenance of heroin abstinence by methadone [203].

The success of bupropion in aiding smoking cessation inspired investigation of the NDRI as an anti-craving treatment for other SUDs. Initial pilot experiments indicated the bupropion was capable of reducing cocaine or methamphetamine intake in treatment-seeking SUD patients [204, 205]. However, in a double-blind placebo-controlled trial of cocaine addicts, bupropion treatment failed to reduce cocaine use beyond placebo [206]. In a double-blind placebo-controlled study of methamphetamine addicts, bupropion treatment did not outperform placebo across all groups considered, but did show a significant reduction in methamphetamine use among patients with a low baseline rate of drug use [207]. Bupropion combined with citalopram also failed to have a significant treatment effect in methadone-maintained opiate SUD patients [208].

3.4. Alpha$_2$ Receptor Antagonists - Mirtazapine

Unlike all of the previously discussed therapeutics that increased levels of 5-HT and/or NE by blocking their reuptake by transporter proteins, the antidepressant mirtazapine (Remeron) is an antagonist of the alpha$_2$ adrenergic receptor. NE normally activates presynaptic alpha$_2$ autoreceptors on noradrenergic neurons and presynaptic alpha$_2$ heteroreceptors on serotonergic neurons to regulate NE and 5-HT release; alpha$_2$ receptor antagonism therefore results in 5-HT and NE disinhibition (SNDI) [188]. Additionally, the NE neurons of the locus coeruleus project to the 5-HT neurons of the raphe nuclei, and the disinhibition of this pathway could act to accelerate the release of 5-HT throughout the brain [209]. Mirtazapine is also an potent blocker of 5-HT$_{2A}$, 5-HT$_{2C}$ and 5-HT$_3$ receptors, leaving 5-HT to stimulate 5-HT$_{1A}$ receptors and consequently increasing the release of dopamine [210]. While mirtazapine does not have the serotonin-related side effects of restlessness, nausea and sexual dysfunction, likely because it inhibits the 5-HT$_{2A}$, 5-HT$_{2C}$ receptors, it also blocks histamine H$_1$ receptors and is known to cause drowsiness [211].

Due to its multifaceted pharmacology and rapid drug action, mirtazapine has been successfully used in an off-label manner to manage a variety of clinical drug abuse problems. Its capability to reduce anxiety has been shown to aid the detoxification of dependent patients from alcohol [212] and amphetamines [213].

Its actions on the 5-HT$_{2A}$ receptors have been associated with the restoring of healthy sleep patterns and have reduced sleep disturbances during withdrawal from methamphetamine [214]. More critically, a double-blind placebo controlled evaluation of MDD/AUD patients demonstrated reduced alcohol cravings and MDD symptoms [215]. Another, open-label study similarly found improvements in mood and reductions of craving for alcohol in MDD/AUD patients [216]. However, mirtazapine treatment did not result in improved retention in a placebo-controlled study of methamphetamine withdrawal and abstinence [217].

Preclinical investigations of mirtazapine have also encompassed behaviors associated with multiple drugs of abuse. Mirtazapine treatment reduced expression of morphine CPP [218] and prototypical behaviors associated with morphine withdrawal [219]. Mirtazapine was found to interfere with the expression of methamphetamine-induced CPP [220, 221]. Moreover, mirtazapine attenuated reinstatement of methamphetamine-seeking behavior in the presence of drug-paired cues [222]. Surprisingly, little is currently known about the effects of mirtazapine in the behavioral models of alcohol and cocaine addiction.

4. CONCLUSIONS

Although the research reviewed here and in other meta-analyses suggest that SSRIs are not conclusively effective in reducing illicit drug or alcohol intake or preventing relapse to substance abuse [223, 224], these drugs are currently prescribed to SUD and AUD patients more often than drugs explicitly developed to reduce SUD and AUD symptoms [225]. The reasons for this include familiarity of SSRIs and the presentation of MDD and anxiety symptoms in these patients, the difficulty of patient compliance with SUD-specific treatment regimens, and the favorable safety profile of SSRIs [225]. The effects of chronically administered SSRIs and other antidepressants on SUD behaviors must therefore continue to be studied, as they are very likely to be combined with supplementary anti-craving and anti-relapse treatments.

Preclinical investigation of antidepressants in the field of addiction has been limited, with virtually no studies making use of animals with a history of drug intake sufficient to elicit physiological signs of chemical dependence. This is a critical gap because the mood disturbances treated by antidepressants are often initiated or exacerbated by the development of the addiction cycle to the point where the addict is trapped in a sustained period of negative affect. The motivation to procure and use illicit drugs and alcohol is changed after the development of chemical dependence, and this transition may have critical implications for the utility of animal models in investigating severe emotional disturbances related to addiction [22]. Chemical dependence to alcohol can be induced in rodents by a variety of methods, including several consecutive weeks of exposure to vaporized ethanol. Rats self-administer elevated amounts of cocaine, methamphetamine and heroin when allowed extended access to the drug, often escalating daily drug intake to putative uncontrolled levels [226-229]. These experimental manipulations subject the animals to episodes of withdrawal to generate "post-dependent" versions of the rodent models. The investigation of chronically administered antidepressants in post-dependent animals is warranted, given the extremely high comorbidity of AUD or SUD and MDD in patients, as discussed above.

The experience of chemical dependence and withdrawal has been shown to change the relevant potency and efficacy characteristics of several therapeutic compounds. For example, naltrexone, to date one of the few approved medications against relapse as result of preclinical research on drug self-administration and reinstatement, has been shown to be less effective in attenuating alcohol seeking in post-dependent rats with histories of alcohol intoxication and withdrawal [230]. Subsequent experiments have also demonstrated altered potency of glutamatergic agents in attenuating alcohol consumption or alcohol seeking in post-dependent rats [231, 232]. Rats allowed to attain histories of escalated methamphetamine consumption have also demonstrated changed sensitivities to glutamatergic agents with anxiolytic actions [233].

Finally, the prevalence of SUD/MDD and AUD/MDD comorbidities appear to make necessary the study of first-line antidepressants in combination with compounds that have shown the capability to reduce drug-motivated behaviors. The overview of SSRIs and other current antidepressants has highlighted main and secondary pharmacological targets that may interact with the actions with novel compounds that target drug relapse. Successful AUD and SUD treatment will likely require coexistence with one or more concurrent MDD treatments.

ACKNOWLEDGEMENTS

The authors would like to gratefully acknowledge Lucas R. Watterson and Kaveish Sewalia of Arizona State University for assistance in the preparation of this manuscript, and the support of the grants DA025606, DA024355 and AA013852 from the National Institute of Health. The authors declare no conflicts of interest.

CONFLICT OF INTEREST

The author(s) confirm that this chapter content has no conflict of interest.

REFERENCES

[1]   A.I. Leshner. Addiction is a brain disease, and it matters. Science 278 (1997) 45-7.

[2]   A.T. McLellan, D.C. Lewis, C.P. O'Brien, H.D. Kleber. Drug dependence, a chronic medical illness: implications for treatment, insurance, and outcomes evaluation. JAMA 284 (2000) 1689-95.

[3]   C.P. O'Brien, and A.T. McLellan. Myths about the treatment of addiction. Lancet 347 (1996) 237-40.

[4]   F. Weiss, R. Ciccocioppo, L.H. Parsons, *et al.* Compulsive drug-seeking behavior and relapse. Neuroadaptation, stress, and conditioning factors. Ann NY Acad Sci 937 (2001) 1-26.

[5]   G.F. Koob, M. Le Moal. Drug abuse: hedonic homeostatic dysregulation. Science 278 (1997) 52-8.

[6]   F. Weiss, Advances in animal models of relapse for addiction research, Advances in the Neuroscience of Addiction, CRC Press, Boca Raton, FL, 2010, pp. 1-25.

[7]   R.N. Ehrman, S.J. Robbins, A.R. Childress, C.P. O'Brien. Conditioned responses to cocaine-related stimuli in cocaine abuse patients. Psychopharmacology (Berl) 107 (1992) 523-9.

[8]   C.P. O'Brien, T.J. O'Brien, J. Mintz, J.P. Brady. Conditioning of narcotic abstinence symptoms in human subjects. Drug Alcohol Depend 1 (1975) 115-23.

[9]   A.R. Childress, A.T. McLellan, R. Ehrman, C.P. O'Brien. Classically conditioned responses in opioid and cocaine dependence: a role in relapse? NIDA Res Monogr 84 (1988) 25-43.

[10]   J.T. Ingjaldsson, J.F. Thayer, J.C. Laberg. Craving for alcohol and pre-attentive processing of alcohol stimuli. Int J Psychophysiol 49 (2003) 29-39.

[11]   N.S. Miller, M.S. Gold. Dissociation of "conscious desire" (craving) from and relapse in alcohol and cocaine dependence. Ann Clin Psychiatry 6 (1994) 99-106.

[12]   S.T. Tiffany, B.L. Carter. Is craving the source of compulsive drug use? J Psychopharmacol 12 (1998) 23-30.

[13]   T.R. Kosten, B.J. Rounsaville, H.D. Kleber. A 2.5-year follow-up of depression, life crises, and treatment effects on abstinence among opioid addicts. Arch Gen Psychiatry 43 (1986) 733-8.

[14]   S.A. Brown, P.W. Vik, T.L. Patterson, I. Grant, M.A. Schuckit. Stress, vulnerability and adult alcohol relapse. J Stud Alcohol 56 (1995) 538-45.

[15]   R.S. Lazarus, S. Folkman. Stress, appraisal and coping., Springer, New York, 1984.

[16]   R. Sinha. How does stress increase risk of drug abuse and relapse? Psychopharmacology (Berl) 158 (2001) 343-59.

[17]   C.J. McDougle, J.E. Black, R.T. Malison, *et al.* Noradrenergic dysregulation during discontinuation of cocaine use in addicts. Arch Gen Psychiatry 51 (1994) 713-9.

[18]   B. Adinoff, P.R. Martin, G.H. Bone, *et al.* Hypothalamic-pituitary-adrenal axis functioning and cerebrospinal fluid corticotropin releasing hormone and corticotropin levels in alcoholics after recent and long-term abstinence. Arch Gen Psychiatry 47 (1990) 325-30.

[19]   G.R. Breese, K. Chu, C.V. Dayas, *et al.* Stress enhancement of craving during sobriety: a risk for relapse. Alcohol Clin Exp Res 29 (2005) 185-95.

[20]   R. Sinha, T. Fuse, L.R. Aubin, S.S. O'Malley. Psychological stress, drug-related cues and cocaine craving. Psychopharmacology (Berl) 152 (2000) 140-8.

[21]   R. Sinha, D. Catapano, S. O'Malley. Stress-induced craving and stress response in cocaine dependent individuals. Psychopharmacology (Berl) 142 (1999) 343-51.

[22]   G.F. Koob, N.D. Volkow. Neurocircuitry of addiction. Neuropsychopharmacology 35 (2010) 217-38.

[23]   C. Kornetsky, R.U. Esposito. Euphorigenic drugs: effects on the reward pathways of the brain. Fed Proc 38 (1979) 2473-6.

[24]   M. Le Moal, H. Simon. Mesocorticolimbic dopaminergic network: functional and regulatory roles. Physiol Rev 71 (1991) 155-234.

[25]   G. Koob, M.J. Kreek. Stress, dysregulation of drug reward pathways, and the transition to drug dependence. Am J Psychiatry 164 (2007) 1149-59.

[26]   P.W. Kalivas, C. O'Brien. Drug addiction as a pathology of staged neuroplasticity. Neuropsychopharmacology 33 (2008) 166-80.

[27]   D.A. Regier, M.E. Farmer, D.S. Rae, *et al.* Comorbidity of mental disorders with alcohol and other drug abuse. Results from the Epidemiologic Catchment Area (ECA) Study. JAMA 264 (1990) 2511-8.

[28]   B.F. Grant, T.C. Harford. Comorbidity between DSM-IV alcohol use disorders and major depression: results of a national survey. Drug Alcohol Depend 39 (1995) 197-206.

[29]   S.E. Gilman, H.D. Abraham. A longitudinal study of the order of onset of alcohol dependence and major depression. Drug Alcohol Depend 63 (2001) 277-86.

[30]   M.E. Thase, I.M. Salloum, J.D. Cornelius. Comorbid alcoholism and depression: treatment issues. J Clin Psychiatry 62 Suppl 20 (2001) 32-41.

[31]   H.M. Pettinati. Antidepressant treatment of co-occurring depression and alcohol dependence. Biol Psychiatry 56 (2004) 785-92.

[32]   M.A. Schuckit, J.E. Tipp, M. Bergman, W. Reich, V.M. Hesselbrock, T.L. Smith. Comparison of induced and independent major depressive disorders in 2,945 alcoholics. Am J Psychiatry 154 (1997) 948-57.

[33]   L.S. Covey, A.H. Glassman, F. Stetner. Major depression following smoking cessation. Am J Psychiatry 154 (1997) 263-5.

[34]   J.A. Shaw, P. Donley, D.W. Morgan, J.A. Robinson. Treatment of depression in alcoholics. Am J Psychiatry 132 (1975) 641-4.

[35]   D.A. Ciraulo, J.G. Barnhill, J.H. Jaffe. Clinical pharmacokinetics of imipramine and desipramine in alcoholics and normal volunteers. Clin Pharmacol Ther 43 (1988) 509-18.

[36]   S. Stahl. Stahl's Essential Psychopharmacology, Cambridge, New York, 2008.

[37]   G.F. Koob, M. Le Moal. Neurobiology of Addiction, Academic Press, San Diego, CA, 2006.

[38]   J. Olds, P. Milner. Positive reinforcement produced by electrical stimulation of septal area and other regions of rat brain. J Comp Physiol Psychol 47 (1954) 419-27.

[39]   K.A. Campbell, G. Evans, C.R. Gallistel. A microcomputer-based method for physiologically interpretable measurement of the rewarding efficacy of brain stimulation. Physiol Behav 35 (1985) 395-403.

[40]   A. Markou, G.F. Koob. Construct validity of a self-stimulation threshold paradigm: effects of reward and performance manipulations. Physiol Behav 51 (1992) 111-9.

[41]   R.A. Wise. Addictive drugs and brain stimulation reward. Annu Rev Neurosci 19 (1996) 319-40.

[42]   P. Bauco, R.A. Wise. Synergistic effects of cocaine with lateral hypothalamic brain stimulation reward: lack of tolerance or sensitization. J Pharmacol Exp Ther 283 (1997) 1160-7.

[43]   E.F. Domino, M.E. Olds. Effects of d-amphetamine, scopolamine, chlordiazepoxide and diphenylhydantoin on self-stimulation behavior and brain acetylcholine. Psychopharmacologia 23 (1972) 1-16.

[44]   J.E. Robinson, A.E. Agoglia, E.W. Fish, M.C. Krouse, C.J. Malanga. Mephedrone (4-methylmethcathinone) and intracranial self-stimulation in C57BL/6J mice: Comparison to cocaine. Behav Brain Res 234 (2012) 76-81.

[45]   L.R. Watterson, P.R. Kufahl, N.E. Nemirovsky, *et al.* Potent rewarding and reinforcing effects of the synthetic cathinone 3,4-methylenedioxypyrovalerone (MDPV). Addict Biol (2012).

[46]   G.F. Koob. Neuroadaptive mechanisms of addiction: studies on the extended amygdala. Eur Neuropsychopharmacol 13 (2003) 442-52.

[47]   C. Cunningham, P. Groblewski, C. Voorhees. Place Conditioning. in: M. Olmstead, (Ed.), Animal Models of Drug Addiction, Springer, New York, 2011, pp. 167-189.

[48]   E. Carboni, C. Vacca. Conditioned place preference: a simple method for investigating reinforcing properties in laboratory animals. in: J. Wang, (Ed.), Drugs of Abuse: Neurological Reviews and Protocols, Humana Press, Totowa, NJ, 2003, pp. 481-498.

[49]   C.L. Cunningham, N.K. Ferree, M.A. Howard. Apparatus bias and place conditioning with ethanol in mice. Psychopharmacology (Berl) 170 (2003) 409-22.

[50]   B. Le Foll, S.R. Goldberg. Nicotine induces conditioned place preferences over a large range of doses in rats. Psychopharmacology (Berl) 178 (2005) 481-92.

[51]   M.T. Bardo, R.A. Bevins. Conditioned place preference: what does it add to our preclinical understanding of drug reward? Psychopharmacology (Berl) 153 (2000) 31-43.

[52]   C. Sanchis-Segura, R. Spanagel. Behavioural assessment of drug reinforcement and addictive features in rodents: an overview. Addict Biol 11 (2006) 2-38.

[53]   A.J. McGeehan, M.F. Olive. The anti-relapse compound acamprosate inhibits the development of a conditioned place preference to ethanol and cocaine but not morphine. Br J Pharmacol 138 (2003) 9-12.

[54]   V. Herzig, W.J. Schmidt. Anti-craving drugs acamprosate and naloxone do not reduce expression of morphine conditioned place preference in isolated and group-housed rats. Neurosci Lett 374 (2005) 119-23.

[55]   P.J. Fudala, K.W. Teoh, E.T. Iwamoto. Pharmacologic characterization of nicotine-induced conditioned place preference. Pharmacol Biochem Behav 22 (1985) 237-41.

[56]   D.E. Jorenby, R.E. Steinpreis, J.E. Sherman, T.B. Baker. Aversion instead of preference learning indicated by nicotine place conditioning in rats. Psychopharmacology (Berl) 101 (1990) 533-8.

[57]   R.B. Stewart, L.A. Grupp. Conditioned place aversion mediated by orally self-administered ethanol in the rat. Pharmacol Biochem Behav 24 (1986) 1369-75.

[58]   R. Ciccocioppo, I. Panocka, R. Froldi, E. Quitadamo, M. Massi. Ethanol induces conditioned place preference in genetically selected alcohol-preferring rats. Psychopharmacology (Berl) 141 (1999) 235-41.

[59]   G.F. Koob, N.E. Goeders. Neuroanatomical susbstrates of drug self-administration. in: J.M. Liebman, S.J. Cooper, (Eds.), The Neuropharmacological Basis of Reward, Clarendon Press, Oxford, 1989, pp. 214-263.

[60]   S.B. Caine, R. Lintz, G.F. Koob. Intravenous drug self-administration techniques in animals. in: A. Sahgal, (Ed.), Behavoural Neuroscience: A Practical Approach, IRL Press, Oxford, 1993, pp. 117-143.

[61]   R.J. Collins, J.R. Weeks, M.M. Cooper, P.I. Good, R.R. Russell. Prediction of abuse liability of drugs using IV self-administration by rats. Psychopharmacology (Berl) 82 (1984) 6-13.

[62]   S.N. Katner, J.G. Magalong, F. Weiss. Reinstatement of alcohol-seeking behavior by drug-associated discriminative stimuli after prolonged extinction in the rat. Neuropsychopharmacology 20 (1999) 471-9.

[63]   J.M. Murphy, R.B. Stewart, R.L. Bell, *et al.* Phenotypic and genotypic characterization of the Indiana University rat lines selectively bred for high and low alcohol preference. Behav Genet 32 (2002) 363-88.

[64]   R. Ciccocioppo, P. Hyytia. The genetic of alcoholism: learning from 50 years of research. Addict Biol 11 (2006) 193-4.

[65]   M.B. Waller, W.J. McBride, L. Lumeng, T.K. Li. Induction of dependence on ethanol by free-choice drinking in alcohol-preferring rats. Pharmacol Biochem Behav 16 (1982) 501-7.

[66]   A.B. Kampov-Polevoy, D.B. Matthews, L. Gause, A.L. Morrow, D.H. Overstreet. P rats develop physical dependence on alcohol *via* voluntary drinking: changes in seizure thresholds, anxiety, and patterns of alcohol drinking. Alcohol Clin Exp Res 24 (2000) 278-84.

[67]   Y. Shaham, U. Shalev, L. Lu, H. De Wit, J. Stewart. The reinstatement model of drug relapse: history, methodology and major findings. Psychopharmacology (Berl) 168 (2003) 3-20.

[68]   L. Davis, A. Uezato, J.M. Newell, E. Frazier. Major depression and comorbid substance use disorders. Curr Opin Psychiatry 21 (2008) 14-8.

[69]   M.J. Ostacher. Comorbid alcohol and substance abuse dependence in depression: impact on the outcome of antidepressant treatment. Psychiatr Clin North Am 30 (2007) 69-76.

[70]   A.J. Rush, M.H. Trivedi, J.W. Stewart, *et al.* Combining medications to enhance depression outcomes (CO-MED): acute and long-term outcomes of a single-blind randomized study. Am J Psychiatry 168 (2011) 689-701.

[71]   L.L. Davis, P. Pilkinton, S.R. Wisniewski, *et al.* Effect of concurrent substance use disorder on the effectiveness of single and combination antidepressant medications for the treatment of major depression: an exploratory analysis of a single-blind randomized trial. Depress Anxiety 29 (2012) 111-22.

[72] E.V. Nunes, F.R. Levin. Treatment of depression in patients with alcohol or other drug dependence: a meta-analysis. JAMA 291 (2004) 1887-96.

[73] L.L. Davis, S.R. Wisniewski, R.H. Howland, *et al.* Does comorbid substance use disorder impair recovery from major depression with SSRI treatment? An analysis of the STAR*D level one treatment outcomes. Drug Alcohol Depend 107 (2010) 161-70.

[74] K.E. Watkins, S.M. Paddock, L. Zhang, K.B. Wells. Improving care for depression in patients with comorbid substance misuse. Am J Psychiatry 163 (2006) 125-32.

[75] E.J. Nestler, M. Barrot, R.J. DiLeone, A.J. Eisch, S.J. Gold, L.M. Monteggia. Neurobiology of depression. Neuron 34 (2002) 13-25.

[76] G.F. Koob, P.P. Sanna, F.E. Bloom. Neuroscience of addiction. Neuron 21 (1998) 467-76.

[77] R.A. Wise, Drug-activation of brain reward pathways. Drug Alcohol Depend 51 (1998) 13-22.

[78] R.B. Rothman, M.H. Baumann, C.M. Dersch, *et al.* Amphetamine-type central nervous system stimulants release norepinephrine more potently than they release dopamine and serotonin. Synapse 39 (2001) 32-41.

[79] C.P. Muller, R.J. Carey, J.P. Huston, M.A. De Souza Silva. Serotonin and psychostimulant addiction: focus on 5-HT1A-receptors. Prog Neurobiol 81 (2007) 133-78.

[80] S.J. Kish, P.S. Fitzmaurice, I. Boileau, *et al.* Brain serotonin transporter in human methamphetamine users. Psychopharmacology (Berl) 202 (2009) 649-61.

[81] D. Weinshenker, J.P. Schroeder. There and back again: a tale of norepinephrine and drug addiction. Neuropsychopharmacology 32 (2007) 1433-51.

[82] L.J. Nonkes, I.P. van Bussel, M.M. Verheij, J.R. Homberg. The interplay between brain 5-hydroxytryptamine levels and cocaine addiction. Behav Pharmacol 22 (2011) 723-38.

[83] P. Celada, M. Puig, M. Amargos-Bosch, A. Adell, F. Artigas. The therapeutic role of 5-HT1A and 5-HT2A receptors in depression. J Psychiatry Neurosci 29 (2004) 252-65.

[84] K. Kroenke, S.L. West, R. Swindle, *et al.* Similar effectiveness of paroxetine, fluoxetine, and sertraline in primary care: a randomized trial. JAMA 286 (2001) 2947-55.

[85] Y.G. Ni, R. Miledi. Blockage of 5HT2C serotonin receptors by fluoxetine (Prozac). Proc Natl Acad Sci USA 94 (1997) 2036-40.

[86] D.A. Gorelick. Pharmacologic therapies for cocaine and other stimulant addiction. in: A.W. Graham, A.W. Schultz, (Eds.), Principles of Addiction Medicine, American Society of Addiction Medicine (ASAM), 1998, pp. 531-544.

[87] M.H. Pollack, J.F. Rosenbaum. Fluoxetine treatment of cocaine abuse in heroin addicts. J Clin Psychiatry 52 (1991) 31-3.

[88] S.L. Batki, L.B. Manfredi, P. Jacob, 3rd, R.T. Jones. Fluoxetine for cocaine dependence in methadone maintenance: quantitative plasma and urine cocaine/benzoylecgonine concentrations. J Clin Psychopharmacol 13 (1993) 243-50.

[89] S.L. Batki, A.M. Washburn, K. Delucchi, R.T. Jones. A controlled trial of fluoxetine in crack cocaine dependence. Drug Alcohol Depend 41 (1996) 137-42.

[90] L. Covi, J.M. Hess, N.A. Kreiter, C.A. Haertzen. Effects of combined fluoxetine and counseling in the outpatient treatment of cocaine abusers. Am J Drug Alcohol Abuse 21 (1995) 327-44.

[91] J. Grabowski, H. Rhoades, R. Elk, *et al.* Fluoxetine is ineffective for treatment of cocaine dependence or concurrent opiate and cocaine dependence: two placebo-controlled double-blind trials. J Clin Psychopharmacol 15 (1995) 163-74.

[92] J.M. Schmitz, P. Averill, A.L. Stotts, *et al.* Fluoxetine treatment of cocaine-dependent patients with major depressive disorder. Drug Alcohol Depend 63 (2001) 207-14.

[93] J.R. Cornelius, I.M. Salloum, M.E. Thase, *et al.* Fluoxetine *versus* placebo in depressed alcoholic cocaine abusers. Psychopharmacol Bull 34 (1998) 117-21.

[94] D.A. Gorelick, A. Paredes. Effect of fluoxetine on alcohol consumption in male alcoholics. Alcohol Clin Exp Res 16 (1992) 261-5.

[95] C.A. Naranjo, C.X. Poulos, K.E. Bremner, K.L. Lanctot. Fluoxetine attenuates alcohol intake and desire to drink. Int Clin Psychopharmacol 9 (1994) 163-72.

[96] D.I. Kabel, F. Petty. A placebo-controlled, double-blind study of fluoxetine in severe alcohol dependence: adjunctive pharmacotherapy during and after inpatient treatment. Alcohol Clin Exp Res 20 (1996) 780-4.

[97] J.R. Cornelius, I.M. Salloum, J.G. Ehler, *et al.* Fluoxetine in depressed alcoholics. A double-blind, placebo-controlled trial. Arch Gen Psychiatry 54 (1997) 700-5.

[98] J.R. Cornelius, I.M. Salloum, R.F. Haskett, *et al.* Fluoxetine *versus* placebo in depressed alcoholics: a 1-year follow-up study. Addict Behav 25 (2000) 307-10.

[99] J.R. Cornelius, D.B. Clark, O.G. Bukstein, B. Birmaher, I.M. Salloum, S.A. Brown. Acute phase and five-year follow-up study of fluoxetine in adolescents with major depression and a comorbid substance use disorder: a review. Addict Behav 30 (2005) 1824-33.

[100] J.R. Cornelius, O.G. Bukstein, D.S. Wood, L. Kirisci, A. Douaihy, D.B. Clark. Double-blind placebo-controlled trial of fluoxetine in adolescents with comorbid major depression and an alcohol use disorder. Addict Behav 34 (2009) 905-9.

[101] M.E. Carroll, S.T. Lac, M. Asencio, R. Kragh. Fluoxetine reduces intravenous cocaine self-administration in rats. Pharmacol Biochem Behav 35 (1990) 237-44.

[102] R. Peltier, S. Schenk. Effects of serotonergic manipulations on cocaine self-administration in rats. Psychopharmacology (Berl) 110 (1993) 390-4.

[103] N.R. Richardson, D.C. Roberts. Fluoxetine pretreatment reduces breaking points on a progressive ratio schedule reinforced by intravenous cocaine self-administration in the rat. Life Sci 49 (1991) 833-40.

[104] A.C. Glatz, M. Ehrlich, R.S. Bae, *et al.* Inhibition of cocaine self-administration by fluoxetine or D-fenfluramine combined with phentermine. Pharmacol Biochem Behav 71 (2002) 197-204.

[105] W.J. McBride, J.M. Murphy, L. Lumeng, T.K. Li. Effects of Ro 15-4513, fluoxetine and desipramine on the intake of ethanol, water and food by the alcohol-preferring (P) and - nonpreferring (NP) lines of rats. Pharmacol Biochem Behav 30 (1988) 1045-50.

[106] J.M. Gulley, C. McNamara, T.J. Barbera, M.C. Ritz, F.R. George. Selective serotonin reuptake inhibitors: effects of chronic treatment on ethanol-reinforced behavior in mice. Alcohol 12 (1995) 177-81.

[107] E.K. Sawyer, J. Mun, J.A. Nye, *et al.* Neurobiological changes mediating the effects of chronic fluoxetine on cocaine use. Neuropsychopharmacology 37 (2012) 1816-24.

[108] J.J. Burmeister, E.M. Lungren, J.L. Neisewander. Effects of fluoxetine and d-fenfluramine on cocaine-seeking behavior in rats. Psychopharmacology (Berl) 168 (2003) 146-54.

[109] N.E. Paterson, C. Myers, A. Markou. Effects of repeated withdrawal from continuous amphetamine administration on brain reward function in rats. Psychopharmacology (Berl) 152 (2000) 440-6.

[110] D. Lin, G.F. Koob, A. Markou. Differential effects of withdrawal from chronic amphetamine or fluoxetine administration on brain stimulation reward in the rat-- interactions between the two drugs. Psychopharmacology (Berl) 145 (1999) 283-94.

[111] D.S. Yu, F.L. Smith, D.G. Smith, W.H. Lyness. Fluoxetine-induced attenuation of amphetamine self-administration in rats. Life Sci 39 (1986) 1383-8.

[112] L.J. Porrino, M.C. Ritz, N.L. Goodman, L.G. Sharpe, M.J. Kuhar, S.R. Goldberg. Differential effects of the pharmacological manipulation of serotonin systems on cocaine and amphetamine self-administration in rats. Life Sci 45 (1989) 1529-35.

[113] P. Munzar, M.H. Baumann, M. Shoaib, S.R. Goldberg. Effects of dopamine and serotonin-releasing agents on methamphetamine discrimination and self-administration in rats. Psychopharmacology (Berl) 141 (1999) 287-96.

[114] Y. Takamatsu, H. Yamamoto, Y. Ogai, Y. Hagino, A. Markou, K. Ikeda. Fluoxetine as a potential pharmacotherapy for methamphetamine dependence: studies in mice. Ann N Y Acad Sci 1074 (2006) 295-302.

[115] P.R. Kufahl, M.F. Olive. Investigating Methamphetamine Craving Using the Extinction-Reinstatement Model in the Rat. J Addict Res Ther S1 (2011).

[116] M.J. Owens, D.L. Knight, C.B. Nemeroff. Second-generation SSRIs: human monoamine transporter binding profile of escitalopram and R-fluoxetine. Biol Psychiatry 50 (2001) 345-50.

[117] N.D. Volkow, G.J. Wang, M.W. Fischman, *et al.* Relationship between subjective effects of cocaine and dopamine transporter occupancy. Nature 386 (1997) 827-30.

[118] N.D. Volkow, G.J. Wang, J.S. Fowler, *et al.* Relationship between psychostimulant-induced "high" and dopamine transporter occupancy. Proc Natl Acad Sci USA 93 (1996) 10388-92.

[119] D. Nutt, K. Demyttenaere, Z. Janka, *et al.* The other face of depression, reduced positive affect: the role of catecholamines in causation and cure. J Psychopharmacol 21 (2007) 461-71.

[120] R.B. Rothman, B.E. Blough, M.H. Baumann. Dual dopamine/serotonin releasers: potential treatment agents for stimulant addiction. Exp Clin Psychopharmacol 16 (2008) 458-74.

[121] E. Aguglia, M. Casacchia, G.B. Cassano, *et al.* Double-blind study of the efficacy and safety of sertraline *versus* fluoxetine in major depression. Int Clin Psychopharmacol 8 (1993) 197-202.

[122] H.M. Pettinati, J.R. Volpicelli, H.R. Kranzler, G. Luck, M.R. Rukstalis, A. Cnaan. Sertraline treatment for alcohol dependence: interactive effects of medication and alcoholic subtype. Alcohol Clin Exp Res 24 (2000) 1041-9.

[123] W. Dundon, K.G. Lynch, H.M. Pettinati, C. Lipkin. Treatment outcomes in type A and B alcohol dependence 6 months after serotonergic pharmacotherapy. Alcohol Clin Exp Res 28 (2004) 1065-73.

[124] H.M. Pettinati, W. Dundon, C. Lipkin. Gender differences in response to sertraline pharmacotherapy in Type A alcohol dependence. Am J Addict 13 (2004) 236-47.

[125] H.R. Kranzler, T. Mueller, J. Cornelius, *et al.* Sertraline treatment of co-occurring alcohol dependence and major depression. J Clin Psychopharmacol 26 (2006) 13-20.

[126] C.K. Farren, M. Scimeca, R. Wu, S.O. Malley. A double-blind, placebo-controlled study of sertraline with naltrexone for alcohol dependence. Drug Alcohol Depend 99 (2009) 317-21.

[127] H.M. Pettinati, D.W. Oslin, K.M. Kampman, *et al.* A double-blind, placebo-controlled trial combining sertraline and naltrexone for treating co-occurring depression and alcohol dependence. Am J Psychiatry 167 (2010) 668-75.

[128] H.R. Kranzler, S. Armeli, H. Tennen, *et al.* A double-blind, randomized trial of sertraline for alcohol dependence: moderation by age of onset [corrected] and 5-hydroxytryptamine transporter-linked promoter region genotype. J Clin Psychopharmacol 31 (2011) 22-30.

[129] R.A. Rawson, P. Marinelli-Casey, M.D. Anglin, *et al.* A multi-site comparison of psychosocial approaches for the treatment of methamphetamine dependence. Addiction 99 (2004) 708-17.

[130] S. Shoptaw, A. Huber, J. Peck, *et al.* Randomized, placebo-controlled trial of sertraline and contingency management for the treatment of methamphetamine dependence. Drug Alcohol Depend 85 (2006) 12-8.

[131] T. Zorick, C.A. Sugar, G. Hellemann, S. Shoptaw, E.D. London. Poor response to sertraline in methamphetamine dependence is associated with sustained craving for methamphetamine. Drug Alcohol Depend 118 (2011) 500-3.

[132] T.M. Winhusen, E.C. Somoza, J.M. Harrer, *et al.* A placebo-controlled screening trial of tiagabine, sertraline and donepezil as cocaine dependence treatments. Addiction 100 Suppl 1 (2005) 68-77.

[133] A. Oliveto, J. Poling, M.J. Mancino, *et al.* Sertraline delays relapse in recently abstinent cocaine-dependent patients with depressive symptoms. Addiction 107 (2012) 131-41.

[134] M.S. Kleven, W.L. Woolverton. Effects of three monoamine uptake inhibitors on behavior maintained by cocaine or food presentation in rhesus monkeys. Drug Alcohol Depend 31 (1993) 149-58.

[135] K. Gill, Z. Amit, B.K. Koe. Treatment with sertraline, a new serotonin uptake inhibitor, reduces voluntary ethanol consumption in rats. Alcohol 5 (1988) 349-54.

[136] R.D. Myers, S.D. Quarfordt. Alcohol drinking attenuated by sertraline in rats with 6-OHDA or 5,7-DHT lesions of N. accumbens: a caloric response? Pharmacol Biochem Behav 40 (1991) 923-8.

[137] K. Gill, Y. Filion, Z. Amit. A further examination of the effects of sertraline on voluntary ethanol consumption. Alcohol 5 (1988) 355-8.

[138] G.A. Higgins, D.M. Tomkins, P.J. Fletcher, E.M. Sellers. Effect of drugs influencing 5-HT function on ethanol drinking and feeding behaviour in rats: studies using a drinkometer system. Neurosci Biobehav Rev 16 (1992) 535-52.

[139] C.A. Sannerud, J. Prada, D.M. Goldberg, S.R. Goldberg. The effects of sertraline on nicotine self-administration and food-maintained responding in squirrel monkeys. Eur J Pharmacol 271 (1994) 461-9.

[140] E.D. Levin, S.J. Briggs, N.C. Christopher, J.E. Rose. Sertraline attenuates hyperphagia in rats following nicotine withdrawal. Pharmacol Biochem Behav 44 (1993) 51-61.

[141] R. Judge, M.G. Parry, D. Quail, J.G. Jacobson. Discontinuation symptoms: comparison of brief interruption in fluoxetine and paroxetine treatment. Int Clin Psychopharmacol 17 (2002) 217-25.

[142] J.D. Killen, S.P. Fortmann, A.F. Schatzberg, *et al.* Nicotine patch and paroxetine for smoking cessation. J Consult Clin Psychol 68 (2000) 883-9.

[143] K. Miyamoto, R. Yoshimura, N. Ueda, *et al.* Effects of acute paroxetine treatment on the consumption of cigarette smoking and caffeine in depressed patients. Hum Psychopharmacol 22 (2007) 483-90.

[144] M. Kotlyar, M. Golding, D.K. Hatsukami, B.D. Jamerson. Effect of nonnicotine pharmacotherapy on smoking behavior. Pharmacotherapy 21 (2001) 1530-48.

[145] M.P. Piasecki, G.M. Steinagel, O.J. Thienhaus, B.S. Kohlenberg. An exploratory study: the use of paroxetine for methamphetamine craving. J Psychoactive Drugs 34 (2002) 301-4.

[146] Y. Takamatsu, H. Yamamoto, Y. Hagino, A. Markou, K. Ikeda. The Selective Serotonin Reuptake Inhibitor Paroxetine, but not Fluvoxamine, Decreases Methamphetamine Conditioned Place Preference in Mice. Curr Neuropharmacol 9 (2011) 68-72.

[147] D.A. Ciraulo, O. Sarid-Segal, C.M. Knapp, *et al.* Efficacy screening trials of paroxetine, pentoxifylline, riluzole, pramipexole and venlafaxine in cocaine dependence. Addiction 100 Suppl 1 (2005) 12-22.

[148] E.P. Morris, S.H. Stewart, L.S. Ham. The relationship between social anxiety disorder and alcohol use disorders: a critical review. Clin Psychol Rev 25 (2005) 734-60.

[149] C.L. Randall, M.R. Johnson, A.K. Thevos, *et al.* Paroxetine for social anxiety and alcohol use in dual-diagnosed patients. Depress Anxiety 14 (2001) 255-62.

[150] S.W. Book, S.E. Thomas, P.K. Randall, C.L. Randall. Paroxetine reduces social anxiety in individuals with a co-occurring alcohol use disorder. J Anxiety Disord 22 (2008) 310-8.

[151] S.E. Thomas, P.K. Randall, S.W. Book, C.L. Randall. A complex relationship between co-occurring social anxiety and alcohol use disorders: what effect does treating social anxiety have on drinking? Alcohol Clin Exp Res 32 (2008) 77-84.

[152] S.W. Book, S.E. Thomas, J.P. Dempsey, P.K. Randall, C.L. Randall. Social anxiety impacts willingness to participate in addiction treatment. Addict Behav 34 (2009) 474-6.

[153] E.J. Cobos, J.M. Entrena, F.R. Nieto, C.M. Cendan, E. Del Pozo. Pharmacology and therapeutic potential of sigma(1) receptor ligands. Curr Neuropharmacol 6 (2008) 344-66.

[154] P. Romieu, V.L. Phan, R. Martin-Fardon, T. Maurice. Involvement of the sigma(1) receptor in cocaine-induced conditioned place preference: possible dependence on dopamine uptake blockade. Neuropsychopharmacology 26 (2002) 444-55.

[155] R. Martin-Fardon, T. Maurice, H. Aujla, W.D. Bowen, F. Weiss. Differential effects of sigma1 receptor blockade on self-administration and conditioned reinstatement motivated by cocaine *vs.* natural reward. Neuropsychopharmacology 32 (2007) 1967-73.

[156] R. Stefanski, Z. Justinova, T. Hayashi, M. Takebayashi, S.R. Goldberg, T.P. Su. Sigma1 receptor upregulation after chronic methamphetamine self-administration in rats: a study with yoked controls. Psychopharmacology (Berl) 175 (2004) 68-75.

[157] T. Maurice, M. Casalino, M. Lacroix, P. Romieu. Involvement of the sigma 1 receptor in the motivational effects of ethanol in mice. Pharmacol Biochem Behav 74 (2003) 869-76.

[158] A. Mosner, G. Kuhlman, C. Roehm, W.H. Vogel. Serotonergic receptors modify the voluntary intake of alcohol and morphine but not of cocaine and nicotine by rats. Pharmacology 54 (1997) 186-92.

[159] S. Maurel, J. De Vry, R. Schreiber. Comparison of the effects of the selective serotonin-reuptake inhibitors fluoxetine, paroxetine, citalopram and fluvoxamine in alcohol-preferring cAA rats. Alcohol 17 (1999) 195-201.

[160] B.C. Ginsburg, W. Koek, M.A. Javors, R.J. Lamb. Effects of fluvoxamine on a multiple schedule of ethanol- and food-maintained behavior in two rat strains. Psychopharmacology (Berl) 180 (2005) 249-57.

[161] R.J. Lamb, T.U. Jarbe. Effects of fluvoxamine on ethanol-reinforced behavior in the rat. J Pharmacol Exp Ther 297 (2001) 1001-9.

[162] B.C. Ginsburg, R.J. Lamb. Effects of chronic fluvoxamine on ethanol- and food-maintained behaviors. Life Sci 79 (2006) 1228-33.

[163] C.A. Naranjo, D.M. Knoke. The role of selective serotonin reuptake inhibitors in reducing alcohol consumption. J Clin Psychiatry 62 Suppl 20 (2001) 18-25.

[164] J. Chick, H. Aschauer, K. Hornik. Efficacy of fluvoxamine in preventing relapse in alcohol dependence: a one-year, double-blind, placebo-controlled multicentre study with analysis by typology. Drug Alcohol Depend 74 (2004) 61-70.

[165] J. Hyttel, K.P. Bogeso, J. Perregaard, C. Sanchez. The pharmacological effect of citalopram residues in the (S)-(+)-enantiomer. J Neural Transm Gen Sect 88 (1992) 157-60.

[166] R.M. Julien, C.D. Advokat, J.E. Comaty. A Primer of Drug Action, Worth Publishers, New York, 2007.

[167]   S.M. Stahl, I. Gergel, D. Li. Escitalopram in the treatment of panic disorder: a randomized, double-blind, placebo-controlled trial. J Clin Psychiatry 64 (2003) 1322-7.

[168]   R. Alvarado, S. Contreras, N. Segovia-Riquelme, J. Mardones. Effects of serotonin uptake blockers and of 5-hydroxytryptophan on the voluntary consumption of ethanol, water and solid food by UChA and UChB rats. Alcohol 7 (1990) 315-9.

[169]   T.F. Meert. Effects of various serotonergic agents on alcohol intake and alcohol preference in Wistar rats selected at two different levels of alcohol preference. Alcohol Alcohol 28 (1993) 157-70.

[170]   L. Hedlund, G. Wahlstrom. Citalopram as an inhibitor of voluntary ethanol intake in the male rat. Alcohol 16 (1998) 295-303.

[171]   C.A. Naranjo, C.X. Poulos, K.E. Bremner, K.L. Lanctot. Citalopram decreases desirability, liking, and consumption of alcohol in alcohol-dependent drinkers. Clin Pharmacol Ther 51 (1992) 729-39.

[172]   C.A. Naranjo, K.E. Bremner, K.L. Lanctot. Effects of citalopram and a brief psycho-social intervention on alcohol intake, dependence and problems. Addiction 90 (1995) 87-99.

[173]   C.A. Naranjo, D.M. Knoke, K.E. Bremner. Variations in response to citalopram in men and women with alcohol dependence. J Psychiatry Neurosci 25 (2000) 269-75.

[174]   J. Tiihonen, O.P. Ryynanen, J. Kauhanen, H.P. Hakola, M. Salaspuro. Citalopram in the treatment of alcoholism: a double-blind placebo-controlled study. Pharmacopsychiatry 29 (1996) 27-9.

[175]   F.G. Moeller, J.M. Schmitz, J.L. Steinberg, *et al.* Citalopram combined with behavioral therapy reduces cocaine use: a double-blind, placebo-controlled trial. Am J Drug Alcohol Abuse 33 (2007) 367-78.

[176]   L. Stella, G. Addolorato, B. Rinaldi, *et al.* An open randomized study of the treatment of escitalopram alone and combined with gamma-hydroxybutyric acid and naltrexone in alcoholic patients. Pharmacol Res 57 (2008) 312-7.

[177]   L.H. Muhonen, J. Lahti, D. Sinclair, J. Lonnqvist, H. Alho. Treatment of alcohol dependence in patients with co-morbid major depressive disorder--predictors for the outcomes with memantine and escitalopram medication. Subst Abuse Treat Prev Policy 3 (2008) 20.

[178]   S.M. Stahl, M.M. Grady, C. Moret, M. Briley. SNRIs: their pharmacology, clinical efficacy, and tolerability in comparison with other classes of antidepressants. CNS Spectr 10 (2005) 732-47.

[179]   P. Weikop, J. Kehr, J. Scheel-Kruger. The role of alpha1- and alpha2-adrenoreceptors on venlafaxine-induced elevation of extracellular serotonin, noradrenaline and dopamine levels in the rat prefrontal cortex and hippocampus. J Psychopharmacol 18 (2004) 395-403.

[180]   D. Deshaue., Venlafaxine (Effexor): concerns about increased risk of fatal outcomes in overdose. CMAJ 176 (2007) 39-40.

[181]   D.M. McDowell F.R. Levin, A.M. Seracini, E.V. Nunes. Venlafaxine treatment of cocaine abusers with depressive disorders. Am J Drug Alcohol Abuse 26 (2000) 25-31.

[182]   J. Hughes, L. Stead, T. Lancaster. Antidepressants for smoking cessation. Cochrane Database Syst Rev (2004) CD000031.

[183]   J. Liappas, T. Paparrigopoulos, E. Tzavellas, A. Rabavilas. Mirtazapine and venlafaxine in the management of collateral psychopathology during alcohol detoxification. Prog Neuropsychopharmacol Biol Psychiatry 29 (2005) 55-60.

[184]   F.P. Bymaster, L.J. Dreshfield-Ahmad, P.G. Threlkeld, J *et al.* Comparative affinity of duloxetine and venlafaxine for serotonin and norepinephrine transporters *in vitro* and *in*

vivo, human serotonin receptor subtypes, and other neuronal receptors. Neuropsychopharmacology 25 (2001) 871-80.

[185] W.A. Marcil, F. Petty. Duloxetine associated with smoking cessation. Ann Pharmacother 39 (2005) 1578-9.

[186] D. Ji, N.W. Gilpin, H.N. Richardson, C.L. Rivier, G.F. Koob. Effects of naltrexone, duloxetine, and a corticotropin-releasing factor type 1 receptor antagonist on binge-like alcohol drinking in rats. Behav Pharmacol 19 (2008) 1-12.

[187] K. Gill, Z. Amit. Serotonin uptake blockers and voluntary alcohol consumption. A review of recent studies. Recent Dev Alcohol 7 (1989) 225-48.

[188] S.M. Stahl. Basic psychopharmacology of antidepressants, part 1: Antidepressants have seven distinct mechanisms of action. J Clin Psychiatry 59 Suppl 4 (1998) 5-14.

[189] S. Zisook, A.J. Rush, B.R. Haight, D.C. Clines, C.B. Rockett. Use of bupropion in combination with serotonin reuptake inhibitors. Biol Psychiatry 59 (2006) 203-10.

[190] A.H. Clayton, E.L. McGarvey, A.I. Abouesh, R.C. Pinkerton. Substitution of an SSRI with bupropion sustained release following SSRI-induced sexual dysfunction. J Clin Psychiatry 62 (2001) 185-90.

[191] S. Zisook, S.R. Shuchter, P. Pedrelli, J. Sable, S.C. Deaciuc. Bupropion sustained release for bereavement: results of an open trial. J Clin Psychiatry 62 (2001) 227-30.

[192] L. Ferry, J.A. Johnston. Efficacy and safety of bupropion SR for smoking cessation: data from clinical trials and five years of postmarketing experience. Int J Clin Pract 57 (2003) 224-30.

[193] R. West. Bupropion SR for smoking cessation. Expert Opin Pharmacother 4 (2003) 533-40.

[194] J.R. Hughes, L.F. Stead, T. Lancaster. Antidepressants for smoking cessation. Cochrane Database Syst Rev (2007) CD000031.

[195] A.W. Bruijnzeel, A. Markou. Characterization of the effects of bupropion on the reinforcing properties of nicotine and food in rats. Synapse 50 (2003) 20-8.

[196] A.S. Rauhut, N. Neugebauer, L.P. Dwoskin, M.T. Bardo. Effect of bupropion on nicotine self-administration in rats. Psychopharmacology (Berl) 169 (2003) 1-9.

[197] X. Liu, A.R. Caggiula, M.I. Palmatier, E.C. Donny, A.F. Sved. Cue-induced reinstatement of nicotine-seeking behavior in rats: effect of bupropion, persistence over repeated tests, and its dependence on training dose. Psychopharmacology (Berl) 196 (2008) 365-75.

[198] G.G. Nomikos, G. Damsma, D. Wenkstern, H.C. Fibiger. Effects of chronic bupropion on interstitial concentrations of dopamine in rat nucleus accumbens and striatum. Neuropsychopharmacology 7 (1992) 7-14.

[199] G.G. Nomikos, G. Damsma, D. Wenkstern, H.C. Fibiger. Acute effects of bupropion on extracellular dopamine concentrations in rat striatum and nucleus accumbens studied by *in vivo* microdialysis. Neuropsychopharmacology 2 (1989) 273-9.

[200] P. Munzar, M.D. Laufert, S.W. Kutkat, J. Novakova, S.R. Goldberg. Effects of various serotonin agonists, antagonists, and uptake inhibitors on the discriminative stimulus effects of methamphetamine in rats. J Pharmacol Exp Ther 291 (1999) 239-50.

[201] C.M. Reichel, J.D. Linkugel, R.A. Bevins. Bupropion differentially impacts acquisition of methamphetamine self-administration and sucrose-maintained behavior. Pharmacol Biochem Behav 89 (2008) 463-72.

[202] C.M. Reichel, J.E. Murray, K.M. Grant, R.A. Bevins. Bupropion attenuates methamphetamine self-administration in adult male rats. Drug Alcohol Depend 100 (2009) 54-62.

[203] J. Grabowski, J. Shearer, J. Merrill, S.S. Negus. Agonist-like, replacement pharmacotherapy for stimulant abuse and dependence. Addict Behav 29 (2004) 1439-64.

[204] A. Margolin, T. Kosten, I. Petrakis, S.K. Avants. An open pilot study of bupropion and psychotherapy for the treatment of cocaine abuse in methadone-maintained patients. NIDA Res Monogr 105 (1990) 367-8.

[205] T.F. Newton, J.D. Roache, R. De La Garza, 2nd, *et al.* Bupropion reduces methamphetamine-induced subjective effects and cue-induced craving. Neuropsychopharmacology 31 (2006) 1537-44.

[206] A. Margolin, T.R. Kosten, S.K. Avants, *et al.* A multicenter trial of bupropion for cocaine dependence in methadone-maintained patients. Drug Alcohol Depend 40 (1995) 125-31.

[207] A.M. Elkashef, R.A. Rawson, A.L. Anderson, *et al.* Bupropion for the treatment of methamphetamine dependence. Neuropsychopharmacology 33 (2008) 1162-70.

[208] J. Poling, R. Pruzinsky, T.R. Kosten, *et al.* Clinical efficacy of citalopram alone or augmented with bupropion in methadone-stabilized patients. Am J Addict 16 (2007) 187-94.

[209] M.J. Millan, A. Gobert, J.M. Rivet, *et al.* Mirtazapine enhances frontocortical dopaminergic and corticolimbic adrenergic, but not serotonergic, transmission by blockade of alpha2-adrenergic and serotonin2C receptors: a comparison with citalopram. Eur J Neurosci 12 (2000) 1079-95.

[210] G.L. Stimmel, J.A. Dopheide, S.M. Stahl. Mirtazapine: an antidepressant with noradrenergic and specific serotonergic effects. Pharmacotherapy 17 (1997) 10-21.

[211] H.G. Westenberg. Pharmacology of antidepressants: selectivity or multiplicity? J Clin Psychiatry 60 Suppl 17 (1999) 4-8; discussion 46-8.

[212] J. Liappas, T. Paparrigopoulos, P. Malitas, E. Tzavellas, G. Christodoulou. Mirtazapine improves alcohol detoxification. J Psychopharmacol 18 (2004) 88-93.

[213] R. Kongsakon, K.I. Papadopoulos, R. Saguansiritham. Mirtazapine in amphetamine detoxification: a placebo-controlled pilot study. Int Clin Psychopharmacol 20 (2005) 253-6.

[214] C. McGregor, M. Srisurapanont, A. Mitchell, W. Wickes, J.M. White. Symptoms and sleep patterns during inpatient treatment of methamphetamine withdrawal: a comparison of mirtazapine and modafinil with treatment as usual. J Subst Abuse Treat 35 (2008) 334-42.

[215] A.E. Altintoprak, N. Zorlu, H. Coskunol, F. Akdeniz, G. Kitapcioglu. Effectiveness and tolerability of mirtazapine and amitriptyline in alcoholic patients with co-morbid depressive disorder: a randomized, double-blind study. Hum Psychopharmacol 23 (2008) 313-9.

[216] S.J. Yoon, C.U. Pae, D.J. Kim, *et al.* Mirtazapine for patients with alcohol dependence and comorbid depressive disorders: a multicentre, open label study. Prog Neuropsychopharmacol Biol Psychiatry 30 (2006) 1196-201.

[217] C.C. Cruickshank, M.E. Montebello, K.R. Dyer, *et al.* A placebo-controlled trial of mirtazapine for the management of methamphetamine withdrawal. Drug Alcohol Rev 27 (2008) 326-33.

[218] S.M. Graves, A.L. Persons, J.L. Riddle, T. Celeste Napier. The atypical antidepressant mirtazapine attenuates expression of morphine-induced place preference and motor sensitization. Brain Res (2012).

[219] L. Kang, D. Wang, B. Li, M. Hu, P. Zhang, J. Li. Mirtazapine, a noradrenergic and specific serotonergic antidepressant, attenuates morphine dependence and withdrawal in Sprague-Dawley rats. Am J Drug Alcohol Abuse 34 (2008) 541-52.

[220] A.A. Herrold, F. Shen, M.P. Graham, *et al.* Mirtazapine treatment after conditioning with methamphetamine alters subsequent expression of place preference. Drug Alcohol Depend 99 (2009) 231-9.

[221] R.M. Voigt, T.C. Napier. Context-dependent effects of a single administration of mirtazapine on the expression of methamphetamine-induced conditioned place preference. Front Behav Neurosci 5 (2011) 92.

[222] S.M. Graves, T.C. Napier. Mirtazapine alters cue-associated methamphetamine seeking in rats. Biol Psychiatry 69 (2011) 275-81.

[223] M. Torrens, F. Fonseca, G. Mateu, M. Farre. Efficacy of antidepressants in substance use disorders with and without comorbid depression. A systematic review and meta-analysis. Drug Alcohol Depend 78 (2005) 1-22.

[224] N. Iovieno, E. Tedeschini, K.H. Bentley, A.E. Evins, G.I. Papakostas. Antidepressants for major depressive disorder and dysthymic disorder in patients with comorbid alcohol use disorders: a meta-analysis of placebo-controlled randomized trials. J Clin Psychiatry 72 (2011) 1144-51.

[225] T.L. Mark, H.R. Kranzler, X. Song, P. Bransberger, V.H. Poole, S. Crosse. Physicians' opinions about medications to treat alcoholism. Addiction 98 (2003) 617-26.

[226] S.H. Ahmed, G.F. Koob. Transition from moderate to excessive drug intake: change in hedonic set point. Science 282 (1998) 298-300.

[227] S.H. Ahmed, J.R. Walker, G.F. Koob. Persistent increase in the motivation to take heroin in rats with a history of drug escalation. Neuropsychopharmacology 22 (2000) 413-21.

[228] J.L. Rogers, S. De Santis, R.E. See. Extended methamphetamine self-administration enhances reinstatement of drug seeking and impairs novel object recognition in rats. Psychopharmacology (Berl) 199 (2008) 615-24.

[229] O. Kitamura, S. Wee, S.E. Specio, G.F. Koob, L. Pulvirenti. Escalation of methamphetamine self-administration in rats: a dose-effect function. Psychopharmacology (Berl) 186 (2006) 48-53.

[230] R. Ciccocioppo, D. Lin, R. Martin-Fardon, F. Weiss. Reinstatement of ethanol-seeking behavior by drug cues following single *versus* multiple ethanol intoxication in the rat: effects of naltrexone. Psychopharmacology (Berl) 168 (2003) 208-15.

[231] P.R. Kufahl, R. Martin-Fardon, F. Weiss. Enhanced sensitivity to attenuation of conditioned reinstatement by the mGluR 2/3 agonist LY379268 and increased functional activity of mGluR 2/3 in rats with a history of ethanol dependence. Neuropsychopharmacology 36 (2011) 2762-73.

[232] Y. Hao, R. Martin-Fardon, F. Weiss. Behavioral and functional evidence of metabotropic glutamate receptor 2/3 and metabotropic glutamate receptor 5 dysregulation in cocaine-escalated rats: factor in the transition to dependence. Biol Psychiatry 68 (2010) 240-8.

[233] P.R. Kufahl, L.R. Watterson, N.E. Nemirovsky, *et al.* Attenuation of methamphetamine seeking by the mGluR2/3 agonist LY379268 in rats with histories of restricted and escalated self-administration. Neuropharmacology (2012).

CHAPTER 2

Advances in Therapies for Multiple Sclerosis

Naila Makhani and Eluen A. Yeh[*]

Division of Neurology, Hospital for Sick Children and University of Toronto, ON M5G 1X8, Canada

Abstract: Several therapeutic agents are currently used as first-line agents in the management of multiple sclerosis (MS) including interferon-beta 1a, interferon-beta 1b, and glatiramer acetate. Currently approved second-line agents include natalizumab, cyclophoshamide, mitoxantrone and the oral agent fingolimod. In addition, there are several emerging therapeutic agents including laquinimod, alemtuzumab, teriflunomide, ocrelizumab, and fumaric acid that are presently being evaluated in clinical trials. While these new treatments have the potential to have a powerful effect on MS disease activity, many also carry significant risks of adverse events or tolerability concerns. With the expanding repertoire of available MS therapies clinicians will have to carefully consider efficacy, safety, and tolerability when engaging with patients to make treatment decisions. In this article, we describe the mechanism of action, efficacy data, and side effect profile of MS therapies with a focus on the newly emerging agents.

Keywords: Multiple sclerosis, therapy, treatment, immunotherapy, interferon, glatiramer acetate, natalizumab, rituximab, fingolimod, alemtuzumab, ocrelizumab, laquinimod, cyclophosphamide, daclizumab, teriflunamide.

INTRODUCTION

First-line treatment for relapsing-remitting multiple sclerosis (RRMS) currently centers on the use of interferon beta (1a or 1b) or glatiramer acetate (Table **1**). An oral agent, fingolimod, has recently been approved in the US for first-line use. The safety and efficacy of injectable medications is fairly well established. For patients who experience breakthrough relapses or accrue new brain MRI lesions while undergoing treatment with one of these agents, therapy is typically escalated to a second-line agent. There are, in addition, several new agents, including oral agents that are currently in later-phase clinical trials (Table **2**). These emerging agents have the potential to have a powerful effect on clinical

*Address correspondence to Eluen A. Yeh: University of Toronto, Director, Pediatric MS and Demyelinating Disorders Program, Hospital for Sick Children, Department of Neurology, 555 University Avenue, Toronto, ON M5G 1X8, Canada; Tel: 416.813.6332; Fax: 416.813.7096; Email: ann.yeh@sickkids.ca

course and MRI measures of disease activity. Many, however, carry with them significant safety or tolerability concerns. With this ever-expanding repertoire of available MS therapies, clinicians will have to balance treatment efficacy with tolerability and side effect concerns, and make individualized treatment decisions. In this chapter, we review the mechanism of action, efficacy data, and safety profile of MS therapies, focusing on new treatments that have either recently been approved or are currently in Phase III clinical trials.

Table 1: Currently approved disease modifying MS therapies

Drug Name	Dosing	Reported Side Effects	Potential Laboratory Abnormalities
Interferon beta 1a (Avonex)	30 mcg IM weekly	Flu-like myalgia Headache Depression	Transaminitis Rarely fulminant liver failure
Interferon beta 1a (Rebif)	22 or 44 µg SC three times a week	Flu-like myalgia Headache Depression Injection-site reaction	Transaminitis Rarely fulminant liver failure
Interferon beta 1b (Betaseron/Extavia)	250 µg SC every other day	Flu-like myalgia Headache Depression Injection-site reaction	Transaminitis Rarely fulminant liver failure
Glatiramer acetate (Copaxone)	20 mg SC daily	Injection-site reaction Immediate post-injection reaction	None reported
Mitoxantrone	12 mg/m2 IV over 30 minutes every 3 months	Alopecia, cardiotoxicity, leukemia infection, infertility	Thrombocytopenia, anemia, leukopenia
Natalizumab	300 mg IV every 4 weeks	Headache, fatigue, allergic hypersensitivity reaction, progressive multifocal leukoencephalopathy	Elevated liver transaminases, anemia, thrombocytopenia
Fingolimod	0.5 mg po once daily	Bradycardia, AV conduction block, macular edema, herpes simplex virus and varicella zoster virus infections	Lymphopenia

Table 2: Emerging multiple sclerosis therapies

Agent	Suggested Dosing	Proposed Mechanism of Action
BG-12	240 mg po BID to TID	Activation of Nrf2-dependent antioxidant pathways
Laquinimod	0.6 mg po once daily	Altered Th1:Th2 balance; reduced leukocyte adhesion and CNS entry
Teriflunamide	7 mg po to 14 mg po daily	Dihydroorate dehydrogenase inhibition leading to reduced pyrimidine synthesis and lymphocyte proliferation; enhanced Th2 response; proinflammatory cytokine generation
Alemtuzumab	12-24 mg/day for 5 days and again for 3 days at 12 months	Anti-CD52 leads to T and B cell depletion;
Daclizumab	150 and 300 mg SC	Anti-CD25 leads to reduced activated T cell proliferation: NK cell activation; altered antigen presentation
Rituximab	1000 mg IV on Day 1 and 15	Anti-CD20 leads to profound B cell depletion
Ocrelizumab	600 mg or 2000 mg IV over two infusions on Day 1 and 15	Anti-CD20 leads to profound B cell depletion

ESTABLISHED THERAPIES FOR MULTIPLE SCLEROSIS

Interferon Beta (1a and 1b)

There are now several medications in the interferon class approved for use in the US, Canada, and Europe for use in RRMS. They vary by route and frequency of administration (Table **1**).

Mechanism of Action

The mechanisms of action attributed to interferon beta are multiple. Interferon beta reduces blood levels of pro-inflammatory cytokines including IL-17, osteopontin, TNF-alpha and IFN-gamma while increasing levels of the anti-inflammatory cytokines IL-4 and IL-10 [1-3]. Interferon beta may reduce lymphocyte entry into the central nervous system (CNS) by affecting adhesion molecule expression on endothelial cells [4-7]. Interferon beta may also promote neuronal survival and repair *via* increased neuronal growth factor (NGF) expression [8, 9].

Efficacy

The major study evaluating the efficacy of subcutaneous interferon beta 1b (Betaseron) looked at two doses of interferon beta 1b (1.6 and 8 million IU) administered every other day as compared to placebo [10]. Over the course of the study annualized relapse rate (ARR) was 1.12 in the placebo arm, 0.96 for the 1.6 MIU group, and 0.8 for the 8 MIU group (p=0.0057 for placebo *vs.* 1.6 MIU and p=0.0006 for placebo *vs.* 8 MIU), which translates into an approximately 30% relative reduction in relapse rate in the highest dose group as compared to placebo. MRI T2 lesion area increased approximately three times more in the placebo treated participants than in those treated with either interferon dose over the course of the study (median annual change in lesion area of -0.1% in 8 MIU group, 3.0% in 1.6 MIU group and 4.6% in placebo group).

Interferon beta 1b has also been shown to improve clinical outcomes in individuals with clinically isolated syndrome (CIS). In the BENEFIT trial every other day interferon beta 1b was evaluated against placebo in individuals with CIS and at least two T2 weighted lesions on brain MRI imaging [11]. Both time to clinically definite MS (CDMS) and time to MS as defined by McDonald 2001 diagnostic criteria for MS were significantly shorter in the treatment group as compared to placebo (Hazard ratios of 0.15 and 0.54 respectively, p<0.0001 for time to CDMS and p<0.00001 for time to McDonald positive MS).

The major Phase III trial for intramuscular interferon beta 1a (Avonex) showed that as compared to placebo, weekly interferon beta 1a (30 μg) increased time to disability progression (p=0.02) [12]. Annualized relapse rate for all patients was significantly lower in the treatment as compared to the placebo group (0.82 *vs.* 0.67, p=0.04). Interferon beta 1a treated patients also had fewer mean number of T2 weighted brain MRI lesions as compared to placebo (1.65 *vs.* 0.8, p=0.05).

The other formulation of interferon beta 1a (Rebif) at two doses (22 and 44 mcg three times a week) was evaluated in a phase III trial in which 560 patients with RRMS were randomized to either 22 or 44 mcg interferon beta-1a three times a week or placebo for 24 months [13]. Mean number of relapses was significantly lower in both treatment dose groups as compared to placebo (1.82 in 22 mcg group, 1.73 in 44 mcg group *vs.* 2.56 for placebo, p<0.005 for both comparisons). Time to sustained disability progression as measured by the EDSS score was significantly longer in both treatment groups as compared to placebo (p<0.05). MRI lesion area on proton density T2-weighted scans showed a median increase of 10·9% in the placebo group, a median decrease of 1·2% in the 22 mcg group and median decrease of 3·8% in the 44 mcg group (p<0·0001 compared with placebo for both comparisons).

Similar to interferon beta 1b, both formulations of interferon beta 1a have shown benefit in prolonged time to relapse in individuals with CIS [14, 15].

Safety

In the clinical trials cited above, the most frequent side effects associated with interferon therapy were injection site reactions and flu-like symptoms. Laboratory abnormalities most frequently observed were an elevations in liver transaminases. These liver enzyme elevations were usually transient and asymptomatic.

Up to 1/3 of patients treated with interferon beta may develop neutralizing antibodies, the presence of which may affect clinical efficacy [16]. Relapse rates and MRI measures of disease activity seem to vary inversely with neutralizing antibody titers. A recent consensus paper reviews the clinical relevance of these antibodies and suggests an algorithm for therapy selection based on antibody titers [17].

Glatiramer Acetate

Glatiramer acetate (GA) is a unique copolymer that was approved for the treatment of relapsing-remitting multiple sclerosis in 1996. It was originally formulated to try to *induce* experimental autoimmune encephalitis (EAE) in mice, but instead treated animals were found to be resistant to EAE and the compound was re-formulated into a pharmaceutical for MS treatment [18].

Mechanism of Action

Glatiramer acetate is a random polymer of glutamic acid, lysine, alanine and tyrosine. There are several proposed mechanisms of action for glatiramer acetate [19]. These include: (1) Modulating regulatory B cell function; (2) Providing neurotrophic support; (3) Altering the ability of antigen presenting cells to promote T cell differentiation; (4) Enhancing regulatory T cell function; (5) Enhancing suppressor T cell proliferation; and (6) Induction of peripheral immune cell tolerance.

Efficacy

The first Phase III randomized clinical trial for GA was conducted amongst 251 MS patients who were randomized to either GA (20 mg SC daily) or placebo for two years [20]. At study completion, the two-year relapse rate was 1.19 +/- 0.13 in GA treated patients *vs.* 1.68 +/- 0.13 in the placebo group (29% relative reduction, p=0.0007). More patients in the treatment group experienced an improvement by > 1 point on the EDSS scale as compared to placebo.

The effect of GA in clinically isolated syndromes (CIS) was evaluated in the PRECISE trial [21]. In this study, 481 individuals with MS were randomized to either GA (20 mg SC daily) or placebo. GA significantly reduced time to MS diagnosis (hazard ratio 0.55, 95% CI 0.40-0.77; p=0.0005).

A dose-comparison study by Comi *et al.*, showed that there was no additional benefit of a higher GA dose (40 mg compared to 20 mg daily) on either relapse rate or MRI metrics of disease activity [22]. Head-to head comparison trials have also have also shown the GA has similar efficacy to both interferon beta 1a and interferon beta 1b on both relapse rate and lesion accrual on MRI [23-25]. While the side effect profile differs, tolerability also appears to be similar between GA and interferon [25].

While GA appears to have a moderate clinical effect in patients with MS, its effect on disability progression is less well established. A Cochrane review found that amongst individuals with relapsing-remitting MS there was an observed reduction in mean EDSS score on therapy at two years (-0.33, p<0.009) and 35 months (-0.45, p<0.006) as compared to placebo without an effect on proportion of individuals experiencing sustained disability progression at those time points [26].

Safety

The most common side effect of GA is a local injection-site reaction that is often characterized by erythema, swelling, and pruritis. Approximately 15% of participants treated with GA will experience an immediate post-injection reaction characterized by chest tightness, shortness of breath, palpitations, and flushing lasting 15-30 minutes [20]. These episodes are not associated with cardiovascular compromise nor are they life-threatening. However, they may be frightening to patients. GA is not associated with leukopenia, thyroid abnormalities, or alterations of liver transaminases and no blood work monitoring is required on treatment.

Natalizumab

Natalizumab (Tysabri) is a monoclonal antibody that is approved for the treatment of RRMS in the U.S., Canada, and Europe.

Mechanism of Action

Natalizumab targets the α4 subunit of the α4β1 and α4β7 leukocyte integrins inhibiting their binding to adhesion molecules on endothelial cells including VCAM-1 [27]. This has the consequence of inhibiting leukocyte trafficking across the blood brain barrier thus reducing leukocyte-mediated inflammation within the central nervous system.

Efficacy

The AFFIRM trial was a Phase III randomized placebo-controlled trial of once monthly natalizumab in individuals with RRMS [28]. After one year, annualized relapse rate in the natalizumab group was 0.26 as compared to 0.81 in the placebo group (p<0.001) translating into a relative relapse rate reduction of 68%. This difference in relapse rates was sustained at two years. Significantly more individuals in the treatment group remained relapse-free as compared to the placebo group (67% *vs.* 41% at two years, p<0.001). Disability progression was also less likely in the treatment group as compared to the placebo group. At two years the mean number of new or enhancing lesions on MRI was 11.0 +/- 15.7 in the placebo group as compared to 1.9 +/- 9.2 in the natalizumab-treated group translating into a 83% relative reduction in T2 lesion number (p<0.001).

Additional efficacy data came from SENTINEL, a trial of natalizumab plus interferon beta 1a *versus* interferon beta 1a alone in individuals with RRMS [29]. In this trial combination therapy resulted in a reduction of annualized relapse rate (0.34 in combination group *vs.* 0.75 in interferon monotherapy group, p<0.001) and with fewer new or enlarged T2 lesions on brain MRI (mean of 0.9 in combination group *vs.* 5.5 p<0.001).

Safety

In AFFIRM, adverse events occurring more commonly in natalizumab treated patients were fatigue and allergic reaction [28]. Allergic hypersensitivity reactions occurred in 9% of individuals treated with natalizumab as compared to 4% of individuals who received placebo (p=0.012) and most commonly occurred during the second infusion. Two patients died in the natalizumab group, one from malignant melanoma and one from alcohol intoxication. There was no increased risk of infection associated with natalizumab treatment.

In SENTINEL, adverse events more frequently observed in patients treated with combination therapy were anxiety, pharyngitis, sinus congestion, and peripheral edema [29]. Of note there were two cases of progressive multifocal leukoencepha-lopathy (PML) in the combination therapy group, one of which was fatal.

Significant hepatic toxicity resulting in elevated transaminases and bilirubin levels has been reported with natalizumab use [30]. Rare hematological adverse events including immune-mediated hemolytic anemia and thrombotic thrombocytopenic purpura have been described [31].

Anti-natalizumab antibodies occurred transiently in 3% and persistently remained positive in 6% in individuals treated in the AFFIRM and SENTINEL trials. Those individuals with persistent antibodies demonstrated increased relapse rate, more rapid disability accrual and more MRI T2 lesions than those treated patients who were antibody-negative [32]. In the AFFIRM trial individuals who experienced hypersensitivity reactions frequently carried anti-natalizumab antibodies [28].

PML is an often fatal infection of the central nervous system due to the JC virus. There have now been several reports of this infection with natalzumab use, and the drug was removed from the market in 2005 due to safety concerns. The drug was reintroduced to the market in 2006 along with a surveillance program (TOUCH) that requires close clinical and MRI surveillance of all individuals who are prescribed natalizumab.

Several attempts have now been made to stratify the risk of PML for any particular individual. This has been aided by the recent commercial availability JC virus antibody testing. In a recent review, the highest risk of PML was observed in those individuals who were JC virus positive, were exposed to natalizumab for greater than 24 months and had previously been treated with immunosuppressants (10.6/1000 C.I. 7.7-14.2) and the lowest risk was in those who had none of these risk factors (<0.1/1000 C.I. 0-0.56). Therefore the ideal candidate for natalizumab would be one who is JC virus negative and has not previously received immunosuppresive therapy [33].

Fingolimod

Fingolimod (FTY720) has recently been approved for use in the U.S., Canada, and several European countries for the treatment of RRMS. As the first oral agent for RRMS there is been considerable interest in this agent by both patients and practitioners.

Mechanism of Action

Fingolimod is an oral pre-drug that requires phosphorylation to be converted into its active form [34, 35]. The active metabolite then binds to the five subtypes of the sphingosine-1-phosphate receptor (S1P) [36, 37]. The S1P1-3 receptors are found on multiple cell types while S1P4 is found almost exclusively in lymphoid tissue and S1P5 is found within the central nervous system (CNS) [38-40]. Activated fingolimod binding to the S1P receptors leads to receptor modulation and ultimately receptor internalization and down-regulation of function [41, 42]. The S1P pathways are essential for lymphocyte trafficking and therefore fingolimod-mediated receptor downregulation leads to lymphocyte sequestration within lymphoid tissue and a consequent reduction of circulating lymphocytes in blood and cerebrospinal fluid (CSF) [37, 43]. Naïve and memory T cells are selectively retained in lymphoid tissue whereas effector T cells that are less dependent on S1P signaling pathways are able to enter the bloodstream [44].

Efficacy

In the 24 month Phase III randomized double-blind placebo-controlled study (FREEDOMS), patients received either fingolimod (0.5 or 1.25 mg daily) or placebo for 24 months [45]. Annualized relapse rate was 0.4 in the placebo group, 0.16 with 1.25 mg fingolimod, and 0.18 with 0.5 mg fingolimod (p<0.001 for either treatment dose *vs.* placebo) translating into an annualized relapse rate reduction of 54% in the 0.5 mg per day group and 60% in the 1.25 mg per day group. Fingolimod also reduced the risk of disability progression with a hazard ratio of 0.70 in the 0.5 mg group and 0.68 in the 1.25 mg (P=0.02 *vs.* placebo for both comparisons). The mean number of new or enlarged T2 weighted MRI lesions was significantly lower in treated patients as compared to placebo after 24 months (2.5 +/- 5.5 in the 1.25 mg group, 2.5 +/- 7.2 in the 0.5 mg group and 9.8 +/- 13.2 in the placebo group, p <0.001 for both comparisons).

TRANSFORMS was a randomized study of fingolimod (0.5 or 1.25 mg daily) *vs.* interferon beta 1a for 12 months in individuals with RRMS [46]. After 12 months the annualized relapse rate was 0.2 in the 1.25 mg group and 0.16 in the 0.5 mg group as compared to 0.33 in the interferon group (p<0.001 for each fingolimod dose as compared to interferon). In TRANSFORMS, there were fewer mean number of new or enlarged T2 lesions in the fingolimod groups as compared to interferon (1.5 +/- 2.7 in the 1.25 mg group, 1.7 +/- 3.9 in the 0.5 mg group, and 2.6 +/- 5.8 in the interferon beta 1a group; p<0.001 for 1.35 mg fingolimod and p <0.004 for 0.5 mg fingolimod as compared to interferon beta 1a respectively). There was no difference in rate of disability progression for either fingolimod dose as compared to interferon beta 1a.

Safety

In the FREEDOMS study, adverse events leading to treatment discontinuation occurred in 7.5% of participants in the 0.5 mg per day group and 14.2% of participants in the 1.25 mg group as compared to 7.5% in the placebo group [45]. Bradycardia was observed in both the TRANSFORMS and FREEDOMS trials. Bradycardia most commonly occurred in the monitoring period immediately following first dose administration with a nadir of heart rate observed at 4-5 hours (mean heart rate reduction of 8-12 beats per minute). In TRANSFORMS, second-degree atrioventricular block was observed in 0.7% of patients in the 1.25 mg fingolimod group and 0.2% of patients in the 0.5 mg fingolimod group in all cases on the first day of treatment.

Macular edema was reported in both trials was observed in 1% of patients in the 1.25 mg per day fingolimod group and 0.5% of patients in the 0.5 mg per day group. In TRANSFORMS, ten localized skin cancers were reported in participants taking fingolimod and all were successfully treated with localized excision.

In FREEDOMS, the overall infection rate was similar in the treatment and placebo groups, but lower respiratory tract infections occurred more frequently in those receiving fingolimod than in those receiving placebo (9.6% of patients taking 0.5 mg fingolimod per day, 11.4% of patients taking 1.25 mg fingolimod per day, and 6.0% of individuals taking placebo). In TRANSFORMS, two fatal infections occurred in the group that received the 1.25 mg fingolimod dose-one due to disseminated primary varicella zoster and one as a result of herpes simplex encephalitis.

The safety concerns raised by these studies have led to mandated surveillance programs in many countries as well as rigorous safety monitoring protocols including baseline electrocardiograms, first-dose observation with blood pressure and cardiac monitoring, serial assessments of blood lymphocyte counts and liver function tests, and serial ophthalmological exams. Serious cardiac adverse events have been reported, including several (11 at time of writing) deaths in treated patients internationally, which has prompted a safety review of fingolimod in some countries.

Recommendations regarding monitoring in patients who are at higher risk for cardiac complications or who may not tolerate bradycardia have been put in place in the U.S. and Canada, and include extended first dose monitoring for patients requiring intervention for first dose monitoring. Contraindications for treatment are outlined in the manufacturer's product monograph, and include individuals with a known history of cardiac conduction defects, ischemic heart disease, as well as individuals on therapy that may lower heart rate (*e.g.,* beta blockers, calcium channel blockers) or class 1a or III antiarrhythmic drugs.

Mitoxantrone

Mitoxantrone is a synthetic anthracenedione derivative that is both an antineoplastic and immunomodulatory agent.

Mechanism of Action

As a cytotoxic agent, the main presumed mechanism of action in MS is suppression of B cell, T cell, and macrophage proliferation [47-49]. In addition, mitoxantrone may impair antigen presentation by inducing apoptosis in antigen presenting cells and also reduces the production of pro-inflammatory cytokines such as interferon gamma, tumor necrosis factor, and IL-2 [50].

Efficacy

The pivotal trial for mitoxantrone as an MS therapy was conducted by the mitoxantrone in MS (MIMS) study group in which 194 patients with either worsening RRMS or SPMS were randomized to mitoxantrone (5 or 12 mg/m^2 every three months) or placebo for 24 months [51]. At study completion, patients treated with mitoxantrone were significantly improved as assessed by a composite outcome measure that included change in EDSS, change in ambulation index, adjusted total number of treated relapses time to first treated relapse and change in standardized neurological status (difference 0·30 [95% CI 0·17–0·44]; p<0·0001). Treated patients were also significantly improved in each of these measures were assessed separately. MRI analysis was performed in a subgroup of 110 patients and fewer individuals in the treatment group had enhancing lesions at 24 months as compared to the placebo group (0 *vs.* 16%, p=0.02).

Safety

In the MIMS trial, adverse events that occurred significantly more frequently with mitoxantrone than with placebo included nausea, alopecia, menstrual disorders, urinary tract infections, amenorrhea, leukopenia and elevated gamma-glutamyltranspeptidase levels. These events were usually not considered serious adverse events and generally resolved upon treatment discontinuation.

There have been multiple cases of acute myelogenous leukemia (AML) reported in MS patients treated with mitoxantrone [52-54]. The overall incidence of AML in MS patients treated with mitoxantrone is estimated to be 0.81% [55].

Cardiotoxicity has also been a concern with mitoxantrone's use. In the MIMS trial, two patients in the low-dose group and two patients in the higher-dose group developed a left-ventricular ejection fraction <50% (without overt heart failure) as compared to zero individuals in the placebo group. In the oncological literature it has been estimated that 2.6-6% of treated patients develop drug-related congestive heart failure and risk factors include total cumulative dose, use of other cardiotoxic medications and pre-existing cardiac disease [56]. In MS the rate of decreased LVEF is estimated at 12% of treated patients and the risk of heart failure is 0.4% (number needed to harm for any type of cardiotoxicity=8) [55]. Due to potential carditoxicity it is recommended that all patients have a screening cardiac evaluation and that this therapy not be offered those with a baseline LVEF <50%.

This therapy should be used with caution given the significant associated neoplastic and cardiotoxic risks.

Cyclophosphamide

Cyclophosphamide is an alkylating chemotherapeutic agent used as a second-line therapy in MS. While not actually approved in the US and Canada for use in MS, cyclophosphamide is sometimes used in the treatment of individuals who continue to have relapses despite treatment with standard first-line therapies.

Mechanism of Action

Cyclophosphamide interferes with DNA replication and mitosis in rapidly dividing cells including B cells and T cells at the level of the bone marrow [57]. In addition to more general immunosuppressive effects, cyclophosphamide appears to lead to the secretion of fewer proinflammatory cytokines (ex: IL-12 and interferon gamma) and increase the production of anti-inflammatory cytokines (ex: IL-10 and IL-4) [58]. This may be related to the alteration of T lymphocytes to a less inflammatory phenotype following cyclophosphamide exposure [59, 60].

Efficacy

Following encouraging results from small open-label trials, two large randomized trials provided conflicting results about the cyclophosphamide's efficacy in the treatment of MS. In a trial conducted by the Northeast Co-operative Multiple Sclerosis Study Group, 256 patients with progressive MS (defined as worsened EDSS in the prior 23 months) were randomized to one of: 1. IV cyclophosphamide induction and ACTH (600 mg/m^2 daily for 5 days), 2. Induction with IV boosters (700mg/m^2) every other month, 3. IV cyclophosphamide and ACTH given according to a modified regimen (600 mg/m^2 on days 1, 2, 4, 6 and 8), with boosters (700 mg/m^2) every other month or 4. Regimen 3 without IV boosters [61]. The two groups that received monthly pulse boosters had a significantly delayed time to worsening EDSS score as compared to the other two groups without boosters.

The positive results were not replicated in the Canadian Cooperative Multiple Sclerosis Group's study in which 168 MS patients were randomized to one of: (1) IV cyclophosphamide with oral prednisone; (2) oral cyclophosphamide and oral prednisone on alternate days with weekly plasma exchange (PLEX); or (3) oral placebo with sham PLEX. In this study, there was no difference in time to worsened EDSS in either of the two active treatment groups as compared to placebo.

Since the initial large randomized trials, there have been many open-label trials, the majority of which have supported the efficacy of cyclophosphamide in multiple sclerosis patients on both relapse rates and MRI measures of disease activity [62-65]. These trials used varying dosing regimens making it difficult to broadly generalize results.

Safety

The most common side effects associated with cyclophosphamide are mild alopecia and nausea or vomiting. These side effects are usually transient and reversible with therapy discontinuation. Hemorrhagic cystitis can be seen in approximately 4.5% of patients treated with cyclophosphamide [66]. Increasing fluid intake and concomitantly administering Mesna can reduce the risk of this side effect. Cyclophosphamide has also been associated with an estimated 0.3% risk of bladder carcinoma [67]. The risk of malignancy is directly related to total cumulative cyclophosphamide dose, especially when greater than 80-100g. Amenorrhea occurs in approximately 1/3 of females and azospermia may occur in males [66]. Infections, most often urinary tract or respiratory infections, also commonly occur in treated patients. Laboratory abnormalities may include leukopenia and increased liver transaminases. Laboratory monitoring protocols for individuals treated with cyclophosphamide are reviewed elsewhere [68].

EMERGING MULTIPLE SCLEROSIS THERAPIES

BG-12

BG-12 is a fumaric acid ester that was initially used in the treatment of psoriasis and is now showing promise as another oral therapy for individuals with RRMS.

Mechanism of Action

The exact mechanism of action of BG-12 remains to be full elucidated. One possible mechanism is *via* activation of nuclear factor (erythroid-derived 2)-related factor 2 (Nrf2) dependent antioxidant pathways, which increases glutathione levels and protects neurons from oxidative stress [69]. BG-12 appears to downregulate Nfκb-dependent gene expression including a number of proinflammatory cytokines and chemokines [70]. BG-12 acts to induce apoptosis of potentially pathogenic T-cells and increases Th2-related cytokines including IL-4 and IL-5 [71, 72].

Efficacy

In a Phase IIb trial, 257 patients, aged 18-55 years, with relapsing-remitting multiple sclerosis were randomly assigned to receive BG-12 at doses of 120 mg once daily, 120 mg three times daily, or 240 mg three times daily, or placebo for 24 weeks [73]. Treatment with the highest dose (240 mg three times daily) reduced the mean total number of new gadolinium-enhancing lesions by 69% as compared to placebo (1.4 *vs.* 4.5, p<0.0001). While the study was not powered for clinical outcomes, BG-12 at the highest dose reduced annualized relapse rate by 32% (0.44 in treatment group *vs.* 0.65 for placebo, p=0.272).

The results of the first Phase III clinical trial (DEFINE) have recently been presented [74]. In this study, 1237 MS patients were randomized to BG-12 at a dose of 240 mg twice daily, BG-12 240 mg three times daily, or placebo. BG-12 reduced the risk of relapse by 49 percent in the twice-daily group (HR 0.51, p<0.0001) and by 50 percent in the three times a day dosing group (HR 0.50, p<0.0001). BG-12 twice daily reduced ARR by 53%, while BG-12 TID reduced the ARR by 48% (p<0.0001 for both comparisons). BG-12 twice daily reduced the risk of disability progression by 38 percent (HR 0.62, p=0.0050), while BG-12 three times daily reduced this risk by 34 percent (HR 0.66, p=0.0128).

In a second Phase III trial (CONFIRM), 1430 individuals with RRMS were randomized to BG-12 240 mg twice daily BG-12 240 mg three times daily, glatiramer acetate (20 mg SC daily), or placebo [75]. BG-12 reduced ARR by 44% for twice daily (p<0.0001) and by 51% for three times daily dosing (p<0.0001) *vs.* placebo at two years in comparison to glatiramer acetate, which reduced the ARR by 29% (p<0.02) compared with placebo at two years. On MRI, BG-12 reduced the number of new or newly enlarging T2-hyperintense lesions by 71% for twice daily (p<0.0001) and by 73% for three times a day dosing (p<0.0001), while glatiramer acetate was associated with a 54% (p<0.0001) reduction.

Safety

In DEFINE, flushing (35% BG-12 *vs.* 5% placebo), gastrointestinal symptoms such as diarrhea (17% BG-12 *vs.* 13% placebo), nausea (13% BG-12 *vs.* 9% placebo), and abdominal pain (10% BG-12 *vs.* 5% placebo) were higher in the BG-12-combined group than the placebo group, with the highest incidence in the first month of treatment. There were no serious adverse events reported and no increased risk of infection or malignancy.

Laquinimod

Laquinimod is a derivative of linomide, a compound that had previously been investigated for the treatment of RRMS, but was associated with severe cardiopulmonary adverse events.

Mechanism of Action

Laquinimod appears to alter the balance between Th1 and Th2 towards a more anti-inflammatory profile associated with a decrease in the pro-inflammatory cytokines IFN-gamma and TNF-alpha and an increase in the anti-inflammatory cytokine IL-4 [76]. Laquinimod may also inhibit VLA-4-mediated leukocyte adhesion and central nervous system entry [77].

Efficacy

In the first Phase II study, 209 individuals with MS were randomized to laquinimod 0.1 mg daily, laquinimod 0.3 mg daily, or placebo for 24 weeks [78].

There was a significant difference in the primary outcome of mean number of active lesions (defined as a new T2 or gadolinium-enhancing lesion) between the 0.3 mg group and placebo (5.24 *vs.* 9.44 or 44% relative reduction). A Phase IIb study evaluated the effect of laquinimod 0.3 mg daily and 0.6 mg daily as compared to placebo on MRI measures of disease activity [79]. In this study, laquinimod 0.6 mg per day showed a 40.4% reduction in mean cumulative number of gadolinium-enhancing lesions per scan on the last four scans (4.2 *vs.* 2.6, p=0.0048) as compared to placebo. There was no difference between the laquinimod 0.3 mg per day group and placebo.

In a Phase III trial (ALLEGRO), 1106 patients with RRMS were randomized to either laquinimod 0.6 mg daily or placebo for 24 months [80]. Laquinimod treatment reduced ARR as compared to placebo (0.30 *vs.* 0.39, P=0.002) and reduced disability progression as defined by a sustained worsening of EDSS (11.1% *vs.* 15.7%; hazard ratio, 0.64; P=0.01). The mean number of gadolinium-enhancing lesions and new or enlarging T2-weighted lesions were lower for patients receiving laquinimod than for those receiving placebo (1.33 *vs.* 2.12 and 5.03 *vs.* 7.14, respectively; P<0.001 for both comparisons).

The results of a second Phase II clinical trial (BRAVO) were recently presented [81]. In BRAVO, 1331 MS patients were randomized to either 0.6 mg laquinimod daily, interferon beta1a 30 mcg weekly or placebo. While the initial pre-specified primary endpoint of annualized relapse rate was not statistically different between the two groups, it was felt that baseline MRI characteristics may have been different in the various treatment groups (although all participants were randomized). After adjustment, there was a difference in calculated ARR (0.37 for placebo and 0.29 for laquinimod, RR=0.787; p=0.026). There were also differences in secondary endpoints between laquinimod and placebo including risk of disability progression as measured by EDSS (33.5% decreased relative risk for laquinimod, p=0.044), and in brain volume loss (27.5% decreased relative risk of laquinimod, p<0.0001).

Safety

In clinical trials laquinimod has generally been well tolerated. Transient elevations in liver transaminases were reported in the clinical trials and exceeded three times the upper limit of normal in approximately 5% of patients [80]. There were no serious infections or malignancies. The parent compound linomide has been associated with serious cardiopulmonary toxicity (pleural effusion, myocardial infarction, and possible pulmonary embolism) [82] raising this as a potential concern for laquinimod.

Teriflunamide

Teriflunamide is an oral medication that is the active metabolite of lenflunomide. Lenflunomide has been used as a therapy for rheumatoid arthritis and other than teratogenicity has generally had a favorable side effect profile.

Mechanism of Action

Teriflunamide inhibits the enzyme dihydroorate dehydrogenase, an enzyme that is necessary for pyrimidine synthesis leading to inhibition of DNA synthesis in rapidly dividing cells including B and T lymphocytes [83, 84]. Other postulated immunomodulatory mechanisms of action may include promotion of Th2 cell proliferation, reduction of pro-inflammatory cytokines such as interferon gamma, and disrupted interaction between T cells and antigen-presenting cells [85-87].

Efficacy

In a Phase II trial, 179 patients with either RRMS or secondary progressive MS (SPMS) were randomized to teriflunamide 7 mg/day, teriflunamide 14 mg/day, or placebo for 36 weeks [88]. There was a significant effect of teriflunamide on the primary outcome of number of unique active lesions (defined as new T2 or gadolinium-enhancing lesions) at both doses with a median number of unique active lesions per scan of 0.5, 0.2, and 0.3 in the placebo, teriflunomide 7 mg/day, teriflunomide 14 mg/day, and placebo groups respectively (p<0.03 and p<0.01 respectively).There was no significant difference in relapse rates between the treatment groups and placebo.

In a phase III randomized trial (TEMSO)1088 patients with RRMS were randomized to teriflunamide 7 mg daily, teriflunamide 14 mg daily or placebo for 108 weeks. At study completion annualized relapse rates were reduced in both teriflunamide treatment groups as compared to placebo with no difference between the two teriflunamide doses (ARR 0.54 for placebo *vs.* 0.37 for both teriflunamide doses, p<0.001 for both comparisons *vs.* placebo). The number of unique active lesions per scan was also reduced in the treatment groups as compared to placebo (2.46 for placebo *vs.* 1.29 for teriflunomide 7 mg and 0.75 for teriflumomide 14 mg, p<0.001 for both comparisons *vs.* placebo). The percentage of patients with sustained disability progression was also reduced in the 14 mg daily group (but not the 7 mg daily group) as compared to placebo (20.2% *vs.* 27.3%, p=0.03).

Safety

In TEMSO, gastrointestinal symptoms (diarrhea and nausea), and hair thinning were more commonly associated with teriflunomide as compared to placebo. Elevated alanine aminotransferase was observed in treated patients (54% in 7 mg group and 57% in 14 mg group as compared to 35.9% in placebo group) and rarely exceeding three times the upper limit of normal. There was no increased risk of serious infections or malignancy. Lenflunomide has known teratogenicity and teriflunamide may also have similar fetal effects.

Alemtuzumab

Alemtuzumab is a humanized monoclonal antibody directed against CD52, which is found on the surface many cell types including lymphocytes, eosinophils, and thymocytes.

Mechanism of Action

Binding of alemtuzumab to CD52, results in depletion of B and T lymphocytes in peripheral blood and lymphoid tissue [89]. Alemtuzumab administration also increases circulating levels of cytokines including TNF-α, IL-6 and interferon-γ [89].

Efficacy

In a Phase II trial, 334 patients with RRMS were randomized to subcutaneous interferon beta1a (44 mcg) three times per week or annual intravenous alemtuzumab (either 12 mg or 24 mg per day) for 36 months [90]. Alemtuzumab at both doses reduced the risk of disability by 71% as compared to interferon beta1a (HR 0.29, p<0.001). Mean EDSS score improved by 0.39 points in both alemtuzumab groups at 36 months and increased by 0.38 points amongst those receiving interferon beta1a. As compared to interferon beta1a, alemtuzumab treatment at either dose reduced relapse rate by 74% (HR 0.26, p<0.001). Annualized relapse rate at 36 months was 0.36 for interferon beta1a and 0.10 overall for alemtuzumab (0.11 for the 12 mg dose and 0.08 for the 24 mg dose). T2 lesion load was also reduced in the alemtuzumab treated patients as compared to interferon beta1a (median reduction from baseline of 13% in interferon beta-1a group, 18.2% in 12 mg dose group, and 13.5% in 24 mg dose group; p=0.01 for 12 mg *vs.* interferon and p=0.03 for 24 mg *vs.* interferon).

There have been two recently reported Phase III trials for alemtuzumab, neither of which has been published, but results have been presented at academic meetings. In CARE-MS I, 581 MS patients were randomized to either alemtuzumab (12 mg/day by IV administration for 5-days, with a second 3-day IV administration one year later) or interferon beta1a (44 mcg three times per week). Alemtuzumab treatment resulted in a 55% reduction in relapse rate compared to interferon beta-1a over two years (p<0.0001) [91]. There was no significant difference in disability between the two groups. A significant benefit was observed for alemtuzumab as compared to interferon beta-1a in the percentage of patients with new and enlarging T2-hyperintense lesions (49% *vs.* 58%, p=0.035), new gadolinium-enhancing lesions (15% *vs.* 27%, p=0.0006), and new T1-hypointense lesions (24% *vs.* 31%, p=0.05).

In CARE-MS II, alemtuzumab (12 mg given daily IV administration for 5 days, and then again for 3 days one year later) was compared to interferon beta 1a (44 mcg three times per week) for two years [92]. A 49% decrease in relapse rate was observed in alemtuzumab-treated patients as compared to those treated with interferon-beta 1a (p<0.0001). There was also a 42% risk reduction for sustained worsening disability as measured by a sustained increase in EDSS (p=0.0084).

Safety

In the Phase II trial, alemtuzumab therapy was suspended because autoimmune thrombocytopenic purpura developed in three patients, one of whom died from an intracranial hemorrhage [90]. Other autoimmune disorders including autoimmune thyroid disease were also more common in the alemtuzumab treated group. Infections (graded mild-moderate) especially respiratory tract infections were more common in those receiving alemtuzumab as compared to those treated with interferon beta-1a. Oral herpes simplex infections were seen in three patients immediately after alemtuzumab infusion. Given the potential gravity of the described side effects, caution should be used when considering this therapy.

Daclizumab

Daclizumab is a humanized monoclonal antibody directed against CD25 (the alpha subunit of the IL-2 receptor) that has previously been used in the treatment of renal allograft rejections.

Mechanism of Action

IL-2 is secreted primarily by activated CD4+ T-cells and IL-2 binding to the IL-2 receptor is the major pathway for T cell expansion [93]. Therefore the main presumed mechanism of action for daclizumab has been inhibition of activated T cell proliferation. Daclizumab treatment also increases numbers of CD56^{bright} NK cells that inhibit activated T cell survival [94]. Daclizumab may also alter antigen presentation by dendritic cells [95].

Efficacy

The results of a Phase IIb trial (SELECT) were recently presented [96]. In SELECT, 600 individuals with RRMS were randomized to either 150 mg daclizumab every four weeks, 300 mg daclizumab every four weeks or placebo. At one year, daclizumab reduced annualized relapse rate by 54% in the 150 mg dose group (ARR=0.21, p<0.0001) and by 50% in the 300 mg dose group (ARR=0.23,p=0.0002) as compared to placebo (ARR=0.46). Daclizumab at both doses also reduced the number of new or newly enlarging T2 lesions (mean 2.4 and 1.7 in 150 mg and 300 mg groups respectively *vs.* 8.1 in placebo group, p<0.001).

In a second Phase II trial (CHOICE), 230 individuals with RRMS were randomized to receive as an add-on treatment subcutaneous daclizumab 2 mg/kg every 2 weeks, daclizumab 1 mg/kg every 4 weeks or placebo for 24 weeks [97]. The adjusted mean number of new or enlarged contrast-enhancing lesions was 4·75 in the interferon beta plus placebo group compared with 1.32 in the interferon beta plus higher-dose daclizumab group (72% relative reduction, p=0·004) and 3.58 in the interferon beta and lower-dose daclizumab group (25% relative reduction, p=0·51).

Phase III trials for daclizumab are currently in progress.

Safety

In CHOICE, there was no significant difference in the occurrence of adverse events between any of the treatment groups. In SELECT, infections, rash, and elevated liver transaminases occurred more frequently in the daclizumab treatment groups as compared to placebo. There was one death due to a psoas abscess in a patient treated with daclizumab and a contributory role for daclizumab could not be excluded.

Rituximab

Rituximab is a humanized monoclonal antibody directed against CD20 that was initially approved for the treatment of non-Hodgkin's lymphoma and rheumatoid arthritis.

Mechanism of Action

Binding of rituximab to CD20 leads to profound circulating B cell depletion through complement-dependent and antibody-mediated cellular lysis [98].

Efficacy

In a Phase II trial, 104 patients with RRMS received either rituximab (1000 mg) or placebo intravenously on Day 1 and 15. As compared to placebo, rituximab reduced total number of gadolinium-enhancing lesions and number of new enhancing lesions at week 12, 16, 20, and 24 (p<0.001 for both comparisons). At week 48 the proportion of patients who had experienced a relapse was lower in the rituximab group as compared to placebo (40% *vs.* 20%, p=0.04). Annualized relapse rate was also lower in the rituximab group as compared to placebo (0.37 *vs.* 0.72, p=0.08).

Safety

In the Phase II trial, rituximab-treated patients were more likely than those receiving placebo to experience infusion related reactions (78% *vs.* 40%) that were generally mild to moderate in severity. While the overall rate of infections was similar in the treatment group and the placebo group, urinary tract and sinus infections were commonly observed in individuals treated with rituximab. Progressive multifocal leukoencephalopathy (PML) has been reported in individuals treated with rituximab, both in combination with other immunosuppressants and when used as a monotherapy [99].

Ocrelizumab

Ocrelizumab is another humanized monoclonal antibody directed towards CD20.

Mechanism of Action

Ocrelizumab targets cell-surface CD20 located on B cells. This results in profound B cell depletion in treated patients secondary to antibody-mediated cytotoxicity [100, 101]. It has been proposed that this antibody-mediated toxicity (as opposed to complement-driven cytotoxicity) accounts for fewer infusion-related reactions seen with ocrelizumab as compared to rituximab.

Efficacy

In a Phase II trial, 220 patients with RRMS were randomized to 600 mg ocrelizumab, 2000 mg ocrelizumab (given on days 1 and 15), intramuscular interferon beta-1a (30 μg) once weekly or placebo for 24 weeks [102]. Ocrelizumab at both doses was superior to interferon beta-1a, reducing the number of gadolinium-enhancing lesions by 89% in the 600 mg dose group and by 96% in the 2000 mg dose group. Annualized relapse rate was also significantly lower in both ocrelizumab dose groups as compared to placebo (ARR 0.64 for placebo, 0.13 for 600 mg ocrelizumab, and 0.17 for 2000 mg ocrelizumab; p=0.0005 and p=0.0014 respectively).

Safety

In the Phase II trial, first dose infusion reactions were more commonly observed in the ocrelizumab groups as compared to placebo (34.5% and 43.6% *vs.* 9.3%). There were no differences in the ocrelizumab treatment groups as compared to the placebo group in number of serious infections. There was one death from acute microangiopathy in the higher-dose ocrelizumab group. In the rheumatologic literature there have been reports of opportunistic infections including Pneumocystis jirovecii pneumonia in two patients treated with ocrelizumab monotherapy [103]. In patients treated with a combination of methotrexate and ocrelizumab there have been cases of histoplasmosis infection, systemic candida, and varicella pneumonia [104].

CONCLUSIONS

There are a number of established MS therapies with varying degrees of clinical efficacy and established side effect profiles. A number of emerging MS therapies now exist, many of which carry with them the promise of improved efficacy and tolerability. The long-term side effect profiles of these medications have yet to be established. Therefore, patients and clinicians will face the challenge of making individualized treatment choices in the face of this ever-expanding armamentarium of therapeutic options.

ACKNOWLEDGEMENTS

Declared none.

CONFLICT OF INTEREST

The authors have no conflicts of interest to disclose.

DISCLOSURE

Declared none.

REFERENCES

[1]   Liu Z, Pelfrey CM, Cotleur A, Lee JC, Rudick RA. Immunomodulatory effects of interferon beta-1a in multiple sclerosis. Journal of neuroimmunology 2001;112:153-162.

[2]   Kozovska ME, Hong J, Zang YC, *et al.* Interferon beta induces T-helper 2 immune deviation in MS. Neurology 1999;53:1692-1697.

[3]   Chen M, Chen G, Nie H, *et al.* Regulatory effects of IFN-beta on production of osteopontin and IL-17 by CD4+ T Cells in MS. Eur J Immunol 2009;39:2525-2536.

[4]   Defazio G, Livrea P, Giorelli M, *et al.* Interferon beta-1a downregulates TNFalpha-induced intercellular adhesion molecule 1 expression on brain microvascular endothelial cells through a tyrosine kinase-dependent pathway. Brain Res 2000;881:227-230.

[5]   Muraro PA, Leist T, Bielekova B, McFarland HF. VLA-4/CD49d downregulated on primed T lymphocytes during interferon-beta therapy in multiple sclerosis. Journal of neuroimmunology 2000;111:186-194.

[6]   Chabot S, Williams G, Yong VW. Microglial production of TNF-alpha is induced by activated T lymphocytes. Involvement of VLA-4 and inhibition by interferonbeta-1b. The Journal of clinical investigation 1997;100:604-612.

[7]   Defazio G, Gelati M, Corsini E, *et al. In vitro* modulation of adhesion molecules, adhesion phenomena, and fluid phase endocytosis on human umbilical vein endothelial cells and brain-derived microvascular endothelium by IFN-beta 1a. J Interferon Cytokine Res 2001;21:267-272.

[8]   Jin S, Kawanokuchi J, Mizuno T, *et al.* Interferon-beta is neuroprotective against the toxicity induced by activated microglia. Brain Res 2007;1179:140-146.

[9]   Biernacki K, Antel JP, Blain M, Narayanan S, Arnold DL, Prat A. Interferon beta promotes nerve growth factor secretion early in the course of multiple sclerosis. Archives of neurology 2005;62:563-568.

[10]   IFNB Multiple Sclerosis Study Group T, University of British Columbia MS/MRI Analysis Group T. Interferon beta-1b in the treatment of multiple sclerosis: final outcome of the randomized controlled trial. Neurology 1995;45:1277-1285.

[11]   Kappos L, Polman CH, Freedman MS, *et al.* Treatment with interferon beta-1b delays conversion to clinically definite and McDonald MS in patients with clinically isolated syndromes. Neurology 2006;67:1242-1249.

[12]   Jacobs LD, Cookfair DL, Rudick RA, *et al.* Intramuscular interferon beta-1a for disease progression in relapsing multiple sclerosis. The Multiple Sclerosis Collaborative Research Group (MSCRG). Annals of neurology 1996;39:285-294.

[13]   Randomised double-blind placebo-controlled study of interferon beta-1a in relapsing/remitting multiple sclerosis. PRISMS (Prevention of Relapses and Disability by Interferon beta-1a Subcutaneously in Multiple Sclerosis) Study Group. Lancet 1998;352:1498-1504.

[14]   Jacobs LD, Beck RW, Simon JH, *et al.* Intramuscular interferon beta-1a therapy initiated during a first demyelinating event in multiple sclerosis. CHAMPS Study Group. The New England journal of medicine 2000;343:898-904.

[15]   Comi G, Filippi M, Barkhof F, *et al.* Effect of early interferon treatment on conversion to definite multiple sclerosis: a randomised study. Lancet 2001;357:1576-1582.

[16]   Sominanda A, Rot U, Suoniemi M, Deisenhammer F, Hillert J, Fogdell-Hahn A. Interferon beta preparations for the treatment of multiple sclerosis patients differ in neutralizing antibody seroprevalence and immunogenicity. Multiple sclerosis 2007;13:208-214.

[17]   Polman CH, Bertolotto A, Deisenhammer F, *et al.* Recommendations for clinical use of data on neutralising antibodies to interferon-beta therapy in multiple sclerosis. Lancet neurology 2010;9:740-750.

[18]   Teitelbaum D, Meshorer A, Hirshfeld T, Arnon R, Sela M. Suppression of experimental allergic encephalomyelitis by a synthetic polypeptide. Eur J Immunol 1971;1:242-248.

[19]   Boster A, Bartoszek MP, O'Connell C, Pitt D, Racke M. Efficacy, safety, and cost-effectiveness of glatiramer acetate in the treatment of relapsing-remitting multiple sclerosis. Ther Adv Neurol Disord 2011;4:319-332.

[20]   Johnson KP, Brooks BR, Cohen JA, *et al.* Copolymer 1 reduces relapse rate and improves disability in relapsing-remitting multiple sclerosis: results of a phase III multicenter, double-blind placebo-controlled trial. The Copolymer 1 Multiple Sclerosis Study Group. Neurology 1995;45:1268-1276.

[21]   Comi G, Martinelli V, Rodegher M, *et al.* Effect of glatiramer acetate on conversion to clinically definite multiple sclerosis in patients with clinically isolated syndrome (PreCISe study): a randomised, double-blind, placebo-controlled trial. Lancet 2009;374:1503-1511.

[22]   Comi G, Cohen JA, Arnold DL, Wynn D, Filippi M. Phase III dose-comparison study of glatiramer acetate for multiple sclerosis. Annals of neurology 2011;69:75-82.

[23]   Cadavid D, Cheriyan J, Skurnick J, Lincoln JA, Wolansky LJ, Cook SD. New acute and chronic black holes in patients with multiple sclerosis randomised to interferon beta-1b or glatiramer acetate. Journal of neurology, neurosurgery, and psychiatry 2009;80:1337-1343.

[24]   Mikol DD, Barkhof F, Chang P, *et al.* Comparison of subcutaneous interferon beta-1a with glatiramer acetate in patients with relapsing multiple sclerosis (the REbif *vs.* Glatiramer Acetate in Relapsing MS Disease [REGARD] study): a multicentre, randomised, parallel, open-label trial. Lancet Neurol 2008;7:903-914.

[25]   O'Connor P, Filippi M, Arnason B, *et al.* 250 microg or 500 microg interferon beta-1b *versus* 20 mg glatiramer acetate in relapsing-remitting multiple sclerosis: a prospective, randomised, multicentre study. Lancet neurology 2009;8:889-897.

[26]   La Mantia L, Munari LM, Lovati R. Glatiramer acetate for multiple sclerosis. Cochrane database of systematic reviews 2010:CD004678.

[27]   Ransohoff RM. Natalizumab for multiple sclerosis. The New England journal of medicine 2007;356:2622-2629.

[28]   Polman CH, O'Connor PW, Havrdova E, *et al.* A randomized, placebo-controlled trial of natalizumab for relapsing multiple sclerosis. The New England journal of medicine 2006;354:899-910.

[29]   Rudick RA, Stuart WH, Calabresi PA, *et al.* Natalizumab plus interferon beta-1a for relapsing multiple sclerosis. The New England journal of medicine 2006;354:911-923.

[30]   Bezabeh S, Flowers CM, Kortepeter C, Avigan M. Clinically significant liver injury in patients treated with natalizumab. Aliment Pharmacol Ther 2010;31:1028-1035.

[31]   Midaglia L, Rodriguez Ruiz M, Munoz-Garcia D. Severe haematological complications during treatment with natalizumab. Multiple sclerosis 2012.

[32]   Calabresi PA, Giovannoni G, Confavreux C, *et al.* The incidence and significance of anti-natalizumab antibodies: results from AFFIRM and SENTINEL. Neurology 2007;69:1391-1403.

[33]   Kappos L, Bates D, Hartung HP, *et al.* Natalizumab treatment for multiple sclerosis: recommendations for patient selection and monitoring. Lancet neurology 2007;6:431-441.

[34]   Brinkmann V, Davis MD, Heise CE, *et al.* The immune modulator FTY720 targets sphingosine 1-phosphate receptors. J Biol Chem 2002;277:21453-21457.

[35]   Billich A, Bornancin F, Devay P, Mechtcheriakova D, Urtz N, Baumruker T. Phosphorylation of the immunomodulatory drug FTY720 by sphingosine kinases. J Biol Chem 2003;278:47408-47415.

[36]   Ishii I, Fukushima N, Ye X, Chun J. Lysophospholipid receptors: signaling and biology. Annu Rev Biochem 2004;73:321-354.

[37]   Mandala S, Hajdu R, Bergstrom J, *et al.* Alteration of lymphocyte trafficking by sphingosine-1-phosphate receptor agonists. Science 2002;296:346-349.

[38]   Sanchez T, Hla T. Structural and functional characteristics of S1P receptors. J Cell Biochem 2004;92:913-922.

[39]   Chun J, Contos JJ, Munroe D. A growing family of receptor genes for lysophosphatidic acid (LPA) and other lysophospholipids (LPs). Cell Biochem Biophys 1999;30:213-242.

[40]   Watterson K, Sankala H, Milstien S, Spiegel S. Pleiotropic actions of sphingosine-1-phosphate. Prog Lipid Res 2003;42:344-357.

[41]   Graler MH, Goetzl EJ. The immunosuppressant FTY720 down-regulates sphingosine 1-phosphate G-protein-coupled receptors. Faseb J 2004;18:551-553.

[42]   Oo ML, Thangada S, Wu MT, *et al.* Immunosuppressive and anti-angiogenic sphingosine 1-phosphate receptor-1 agonists induce ubiquitinylation and proteasomal degradation of the receptor. J Biol Chem 2007;282:9082-9089.

[43]   Maeda Y, Seki N, Sato N, Sugahara K, Chiba K. Sphingosine 1-phosphate receptor type 1 regulates egress of mature T cells from mouse bone marrow. Int Immunol 2010;22:515-525.

[44]   Hofmann M, Brinkmann V, Zerwes HG. FTY720 preferentially depletes naive T cells from peripheral and lymphoid organs. Int Immunopharmacol 2006;6:1902-1910.

[45]   Kappos L, Radue EW, O'Connor P, *et al.* A placebo-controlled trial of oral fingolimod in relapsing multiple sclerosis. The New England journal of medicine 2010;362:387-401.

[46]   Cohen JA, Barkhof F, Comi G, *et al.* Oral fingolimod or intramuscular interferon for relapsing multiple sclerosis. The New England journal of medicine 2010;362:402-415.

[47]   Levine S, Gherson J. Morphologic effects of mitoxantrone and a related anthracenedione on lymphoid tissues. Int J Immunopharmacol 1986;8:999-1007.

[48]   Fidler JM, DeJoy SQ, Gibbons JJ, Jr. Selective immunomodulation by the antineoplastic agent mitoxantrone. I. Suppression of B lymphocyte function. Journal of immunology 1986;137:727-732.

[49]   Fidler JM, DeJoy SQ, Smith FR, 3rd, Gibbons JJ, Jr. Selective immunomodulation by the antineoplastic agent mitoxantrone. II. Nonspecific adherent suppressor cells derived from mitoxantrone-treated mice. Journal of immunology 1986;136:2747-2754.

[50]   Fox EJ. Mechanism of action of mitoxantrone. Neurology 2004;63:S15-18.

[51]   Hartung HP, Gonsette R, Konig N, *et al.* Mitoxantrone in progressive multiple sclerosis: a placebo-controlled, double-blind, randomised, multicentre trial. Lancet 2002;360:2018-2025.

[52]   Cattaneo C, Almici C, Borlenghi E, Motta M, Rossi G. A case of acute promyelocytic leukaemia following mitoxantrone treatment of multiple sclerosis. Leukemia 2003;17:985-986.

[53]   Brassat D, Recher C, Waubant E, *et al.* Therapy-related acute myeloblastic leukemia after mitoxantrone treatment in a patient with MS. Neurology 2002;59:954-955.

[54]   Heesen C, Bruegmann M, Gbdamosi J, Koch E, Monch A, Buhmann C. Therapy-related acute myelogenous leukaemia (t-AML) in a patient with multiple sclerosis treated with mitoxantrone. Multiple sclerosis 2003;9:213-214.

[55]   Marriott JJ, Miyasaki JM, Gronseth G, O'Connor PW. Evidence Report: The efficacy and safety of mitoxantrone (Novantrone) in the treatment of multiple sclerosis: Report of the

Therapeutics and Technology Assessment Subcommittee of the American Academy of Neurology. Neurology 2010;74:1463-1470.

[56]   Mather FJ, Simon RM, Clark GM, Von Hoff DD. Cardiotoxicity in patients treated with mitoxantrone: Southwest Oncology Group phase II studies. Cancer Treat Rep 1987;71:609-613.

[57]   Kovarsky J. Clinical pharmacology and toxicology of cyclophosphamide: emphasis on use in rheumatic diseases. Semin Arthritis Rheum 1983;12:359-372.

[58]   Smith DR, Balashov KE, Hafler DA, Khoury SJ, Weiner HL. Immune deviation following pulse cyclophosphamide/methylprednisolone treatment of multiple sclerosis: increased interleukin-4 production and associated eosinophilia. Ann Neurol 1997;42:313-318.

[59]   Karni A, Balashov K, Hancock WW, *et al.* Cyclophosphamide modulates CD4+ T cells into a T helper type 2 phenotype and reverses increased IFN-gamma production of CD8+ T cells in secondary progressive multiple sclerosis. J Neuroimmunol 2004;146:189-198.

[60]   Takashima H, Smith DR, Fukaura H, Khoury SJ, Hafler DA, Weiner HL. Pulse cyclophosphamide plus methylprednisolone induces myelin-antigen-specific IL-4-secreting T cells in multiple sclerosis patients. Clin Immunol Immunopathol 1998;88:28-34.

[61]   Weiner H, Mackin G, Orav E, *et al.* Intermittent cyclophosphamide pulse therapy in progressive multiple sclerosis: final report of the Northeast Cooperative Multiple Sclerosis Treatment Group. Neurology 1993;43:910-918.

[62]   Likosky WH FB, Elmore R, Eno G, Gale K, Goode GB, Ikeda K, Laster J, Mosher C, Rozance J, *et al.* Intense immunosuppression in chronic progressive multiple sclerosis: the Kaiser study. Journal of Neurology, Neurosuregery, and Psychiatry 1991;54:1055-1160.

[63]   Hohol MJ, Olek MJ, Orav EJ, *et al.* Treatment of progressive multiple sclerosis with pulse cyclophosphamide/methylprednisolone: response to therapy is linked to the duration of progressive disease. Mult Scler 1999;5:403-409.

[64]   Gobbini MI, Smith ME, Richert ND, Frank JA, McFarland HF. Effect of open label pulse cyclophosphamide therapy on MRI measures of disease activity in five patients with refractory relapsing-remitting multiple sclerosis. J Neuroimmunol 1999;99:142-149.

[65]   Khan OA, Zvartau-Hind M, Caon C, *et al.* Effect of monthly intravenous cyclophosphamide in rapidly deteriorating multiple sclerosis patients resistant to conventional therapy. Multiple sclerosis 2001;7:185-188.

[66]   Portaccio E, Zipoli V, Siracusa G, Piacentini S, Sorbi S, Amato MP. Safety and tolerability of cyclophosphamide 'pulses' in multiple sclerosis: a prospective study in a clinical cohort. Multiple sclerosis 2003;9:446-450.

[67]   De Ridder D, van Poppel H, Demonty L, *et al.* Bladder cancer in patients with multiple sclerosis treated with cyclophosphamide. J Urol 1998;159:1881-1884.

[68]   Awad A, Stuve O. Cyclophosphamide in multiple sclerosis: scientific rationale, history and novel treatment paradigms. Ther Adv Neurol Disord 2009;2:50-61.

[69]   Linker RA, Lee DH, Ryan S, *et al.* Fumaric acid esters exert neuroprotective effects in neuroinflammation *via* activation of the Nrf2 antioxidant pathway. Brain : a journal of neurology 2011;134:678-692.

[70]   Stoof TJ, Flier J, Sampat S, Nieboer C, Tensen CP, Boorsma DM. The antipsoriatic drug dimethylfumarate strongly suppresses chemokine production in human keratinocytes and peripheral blood mononuclear cells. Br J Dermatol 2001;144:1114-1120.

[71]   Treumer F, Zhu K, Glaser R, Mrowietz U. Dimethylfumarate is a potent inducer of apoptosis in human T cells. J Invest Dermatol 2003;121:1383-1388.

[72]   de Jong R, Bezemer AC, Zomerdijk TP, van de Pouw-Kraan T, Ottenhoff TH, Nibbering PH. Selective stimulation of T helper 2 cytokine responses by the anti-psoriasis agent monomethylfumarate. Eur J Immunol 1996;26:2067-2074.

[73]   Kappos L, Gold R, Miller DH, *et al.* Efficacy and safety of oral fumarate in patients with relapsing-remitting multiple sclerosis: a multicentre, randomised, double-blind, placebo-controlled phase IIb study. Lancet 2008;372:1463-1472.

[74]   Gold R, Kappos L, Bar-Or A, *et al.* Clinical efficacy of BG-12, an oral therapy, in relapsing-remitting multiple sclerosis: data from the phase 3 DEFINE trial. Multiple Sclerosis Journal 2011;17:S9-S52.

[75]   Fox RJ, Miller D, Phillips T, *et al.* Clinical Efficacy of BG-12 in Relapsing-Remitting Multiple Sclerosis (RRMS): Data from the Phase 3 CONFIRM Study Neurology 2012;78:S01.003.

[76]   Zou LP, Abbas N, Volkmann I, *et al.* Suppression of experimental autoimmune neuritis by ABR-215062 is associated with altered Th1/Th2 balance and inhibited migration of inflammatory cells into the peripheral nerve tissue. Neuropharmacology 2002;42:731-739.

[77]   Wegner C, Stadelmann C, Pfortner R, *et al.* Laquinimod interferes with migratory capacity of T cells and reduces IL-17 levels, inflammatory demyelination and acute axonal damage in mice with experimental autoimmune encephalomyelitis. Journal of neuroimmunology 2010;227:133-143.

[78]   Polman C, Barkhof F, Sandberg-Wollheim M, Linde A, Nordle O, Nederman T. Treatment with laquinimod reduces development of active MRI lesions in relapsing MS. Neurology 2005;64:987-991.

[79]   Comi G, Pulizzi A, Rovaris M, *et al.* Effect of laquinimod on MRI-monitored disease activity in patients with relapsing-remitting multiple sclerosis: a multicentre, randomised, double-blind, placebo-controlled phase IIb study. Lancet 2008;371:2085-2092.

[80]   Comi G, Jeffery D, Kappos L, *et al.* Placebo-controlled trial of oral laquinimod for multiple sclerosis. The New England journal of medicine 2012;366:1000-1009.

[81]   Vollmer T, Sorensen P, Arnold DL, group Bs. A placebo-controlled and active comparator phase III tria (BRAVO) for relapsing-remitting multiple sclerosis. Multiple Sclerosis Journal 2011;17:S507-S524.

[82]   Noseworthy JH, Wolinsky JS, Lublin FD, *et al.* Linomide in relapsing and secondary progressive MS: part I: trial design and clinical results. North American Linomide Investigators. Neurology 2000;54:1726-1733.

[83]   Bruneau JM, Yea CM, Spinella-Jaegle S, *et al.* Purification of human dihydro-orotate dehydrogenase and its inhibition by A77 1726, the active metabolite of leflunomide. Biochem J 1998;336 (Pt 2):299-303.

[84]   Cherwinski HM, Cohn RG, Cheung P, *et al.* The immunosuppressant leflunomide inhibits lymphocyte proliferation by inhibiting pyrimidine biosynthesis. J Pharmacol Exp Ther 1995;275:1043-1049.

[85]   Korn T, Magnus T, Toyka K, Jung S. Modulation of effector cell functions in experimental autoimmune encephalomyelitis by leflunomide--mechanisms independent of pyrimidine depletion. J Leukoc Biol 2004;76:950-960.

[86]   Zeyda M, Poglitsch M, Geyeregger R, *et al.* Disruption of the interaction of T cells with antigen-presenting cells by the active leflunomide metabolite teriflunomide: involvement of impaired integrin activation and immunologic synapse formation. Arthritis and rheumatism 2005;52:2730-2739.

[87]   Dimitrova P, Skapenko A, Herrmann ML, Schleyerbach R, Kalden JR, Schulze-Koops H. Restriction of *de novo* pyrimidine biosynthesis inhibits Th1 cell activation and promotes Th2 cell differentiation. Journal of immunology 2002;169:3392-3399.

[88]   O'Connor PW, Li D, Freedman MS, *et al.* A Phase II study of the safety and efficacy of teriflunomide in multiple sclerosis with relapses. Neurology 2006;66:894-900.

[89]   Hu Y, Turner MJ, Shields J, *et al.* Investigation of the mechanism of action of alemtuzumab in a human CD52 transgenic mouse model. Immunology 2009;128:260-270.

[90]   Coles AJ, Compston DA, Selmaj KW, *et al.* Alemtuzumab *vs.* interferon beta-1a in early multiple sclerosis. The New England journal of medicine 2008;359:1786-1801.

[91]   Coles A, Brinar V, Arnold DL, *et al.* Efficacy and safety results from CARE-MS I: a phase 3 study comparing alemtuzumab and interferon-beta-1a. Multiple Sclerosis Journal 2011;17:S507-S524

[92]   Arnold DL, Brinar V, Cohen J, *et al.* Effect of Alemtuzumab *vs.* Rebif® on Brain MRI Measurements: Results of CARE-MS I, a Phase 3 Study. Neurology 2012;8:S11.006.

[93]   Martin R. Anti-CD25 (daclizumab) monoclonal antibody therapy in relapsing-remitting multiple sclerosis. Clin Immunol 2012;142:9-14.

[94]   Bielekova B, Catalfamo M, Reichert-Scrivner S, *et al.* Regulatory CD56(bright) natural killer cells mediate immunomodulatory effects of IL-2Ralpha-targeted therapy (daclizumab) in multiple sclerosis. Proceedings of the National Academy of Sciences of the United States of America 2006;103:5941-5946.

[95]   Wuest SC, Edwan JH, Martin JF, *et al.* A role for interleukin-2 *trans*-presentation in dendritic cell-mediated T cell activation in humans, as revealed by daclizumab therapy. Nature medicine 2011;17:604-609.

[96]   Giovannoni G, Gold R, Selmaj K, *et al.* A randomized, double-blind, placebo-controlled study to evaluate the safety and efficacy of daclizumab HYP monotherapy in relapsing-remitting multiple sclerosis: primary results of the SELECT trial. Multiple Sclerosis Journal 2011;17:S507-S524.

[97]   Wynn D, Kaufman M, Montalban X, *et al.* Daclizumab in active relapsing multiple sclerosis (CHOICE study): a phase 2, randomised, double-blind, placebo-controlled, add-on trial with interferon beta. Lancet neurology 2010;9:381-390.

[98]   Reff ME, Carner K, Chambers KS, *et al.* Depletion of B cells *in vivo* by a chimeric mouse human monoclonal antibody to CD20. Blood 1994;83:435-445.

[99]   Clifford DB, Ances B, Costello C, *et al.* Rituximab-associated progressive multifocal leukoencephalopathy in rheumatoid arthritis. Archives of neurology 2011;68:1156-1164.

[100] Kausar F, Mustafa K, Sweis G, *et al.* Ocrelizumab: a step forward in the evolution of B-cell therapy. Expert Opin Biol Ther 2009;9:889-895.

[101] Genovese MC, Kaine JL, Lowenstein MB, *et al.* Ocrelizumab, a humanized anti-CD20 monoclonal antibody, in the treatment of patients with rheumatoid arthritis: a phase I/II randomized, blinded, placebo-controlled, dose-ranging study. Arthritis and rheumatism 2008;58:2652-2661.

[102] Kappos L, Li D, Calabresi PA, *et al.* Ocrelizumab in relapsing-remitting multiple sclerosis: a phase 2, randomised, placebo-controlled, multicentre trial. Lancet 2011;378:1779-1787.

[103] Harigai M, Tanaka Y, Maisawa S. Safety and efficacy of various dosages of ocrelizumab in Japanese patients with rheumatoid arthritis with an inadequate response to methotrexate therapy: a placebo-controlled double-blind parallel-group study. J Rheumatol 2012;39:486-495.

[104] Stohl W, Gomez-Reino J, Olech E, *et al.* Safety and efficacy of ocrelizumab in combination with methotrexate in MTX-naive subjects with rheumatoid arthritis: the phase III FILM trial. Ann Rheum Dis 2012;71:1289-1296.

Send Orders for Reprints on reprints@benthamscience.net

Frontiers in Clinical Drug Research - CNS and Neurological Disorders, Vol. 1, 2013, 69-115 69

CHAPTER 3

Antisense Oligonucleotides Therapy: A New Potential Tool in the Treatment of Cerebral Gliomas

Gerardo Caruso[*], Maria Caffo, Maria Angela Pino and Giuseppe Raudino

Department of Neuroscience, Neurosurgical Clinic, University of Messina, Messina, Italy

Abstract: Gliomas account for about 45% of all primary central nervous system tumors and 77% of all malignant primary cerebral tumors. Recent studies in molecular biology have better depicted the mechanisms involved in the genesis of cerebral gliomas. It is now generally understood that tumor genesis occurs either by over-expression of oncogenes or inactivation of tumor suppressor genes. The two main gene groups involved in brain tumor development are proto-oncogenes and tumor suppressor genes, respectively up-regulated and downregulated during the tumor initiation and progression. It's evident that the modulation of gene expression at more levels, such as DNA, mRNA, proteins and transduction signal pathways, may be the most effective modality to downregulate or silence specific genic functions. Nowadays, an efficacious strategy in the gliomas treatment does not yet exist. In series of patients affected by malignant gliomas mortality is still close to 100% and the survival rate in glioblastoma multiforme patients is less than 1 year.

A potential and future therapeutic approach in gliomas treatment is represented by antisense therapy targeting different antigens or signal pathway of growth factors and their receptors like IGF-I, TGF-beta2 or EGF. The antisense strategy is based on use of antisense oligonucleotides (gene therapy sensu strictu) or of antisense expressing vectors (cell gene therapy). The clinical results obtained using antisense therapy are often similar than those obtained by use of certain inhibitors (*i.e.* imatinib, getifinib) including antibodies (avastin) targeting the growth factors and their downstream element of signaling pathways. The results are especially interesting if applying the combined techniques of antisense and/or inhibitors with chemotherapy. Using either the approach of inhibitors of growth factors and their receptors, especially of EGF, or the approach of antisense IGF-I and anti TGF-beta inducing anti-tumor response, the median survival of glioblastoma patients can reach currently two years if combined with chemotherapy (temozolomide). These results constitute a progress as compared to classical treatment and underline the value of molecular biology based gene therapy, especially immune-gene therapy.

*Address correspondence to Gerardo Caruso: Neurosurgical Clinic, Department of Neuroscience, A.O.U. Policlinico "G. Martino", *via* Consolare Valeria, 1, 98125 Messina, Italy; Tel: ++39-090-2217167; Fax: ++39-090-693714; E-mail: gcaruso@unime.it

Atta-ur-Rahman (Ed)

In this chapter we describe the most relevant findings of antisense oligonucleotides application in gliomas treatment pointing out the attention on effectiveness, delivery system possibilities, targeting modalities and safety of this therapeutic strategy.

Keywords: Antisense Oligonucleotides Therapy, Gene Therapy, Glioblastoma Multiforme, Gliomas.

1. INTRODUCTION

Glial tumors, within neuroepithelial-derived lesions, are the most common intra-assial neoplastic histotypes. According to WHO classification [1] low grade gliomas include diffuse astrocytomas (fibrillar, gemistiocytic, protoplasmatic, mixed), grade II oligodendrogliomas and oligoastrocytomas. Diagnosis of anaplastic glioma is based on the presence of mitotic activity, while vascular proliferation and necrosis are typical features of glioblastoma multiform (GBM). The definition of low grade glioma does not refer perhaps to I grade astrocytomas, that differ from other gliomas in relation to their biological behavior, prognosis and genetic profile [2]. Low grade gliomas, tend perhaps to dedifferentiate: a global therapeutic approach, related to gross total (> 90%) surgical removal, radiotherapy e chemotherapy with temozolamide, leads only to a slight prolongation of overall survival, calculated from 3 to 7 years [3]. This variability depends from various factors: age at diagnosis, tumor volume, tumor crossing the midline, neurological deficits before treatment and tumor's histology [4]. Anyway, mortality is near to 100% and only 35% of patients with glioblastoma multiforme have a median survival of one year or more.

In spite of advanced techniques and instruments for the surgeon, actually, gross total removal is not always achievable in relation to the location of the tumor, to preserve vital nervous or vascular structures. Sometimes, even in cases of macroscopically total resection, microscopic neoplastic foci still remain. Furthermore, chemo and radiotherapy, have limited utility. Radiotherapy, for example, seems to be related only to a prolonged progression-free survival, with better control of seizures, but with no substantial differences in overall survival. Besides, patients treated with radiotherapy, have high risk to develop some complications such as post radiation leukoencephalopathy that is characterized by dementia, gait disturbance, incontinence, deficit in attention and executive functions. About the role of chemotherapy, one of the main problems is the presence of blood-brain barrier, so macromolecules can't cross it. Action of alkylating agents, one of the most common classes of chemotherapic drugs employed, is limited by activation of methylguanine-methyltransferase. Another limit of adjuvant drugs is the poor knowledge about genetic of gliomas. Various studies concerning genetic set of gliomas were performed previously, trying to recognize one or more molecular targets to make chemotherapy selective and effective [3]. Various genes were found mutated in glioma cells but many of these mutations are typical also of other tumors. As previous demonstrated, invasive glioma cells gain the capability to interact with extracellular matrix components (ECM), that is the first step for neoplastic progression. Based on these findings, new recent therapeutic strategies for these tumors, include gene therapy, immunotherapy, and antisense therapy. The last approach uses antisense oligonucleotides, microRNA, ribozyme that binding specific genic sequences making impossible gene transcription.

In this chapter we evaluated the principal concepts of glioma biology and of antisense oligonucleotides strategy. Moreover a brief review of the pertinent literature will, also, reported.

2. GLIOMA BIOLOGY

Gliomas account for about 45% of all primary central nervous system (CNS) tumors and 77% of all malignant primary CNS tumors. Gliomas are divided into different subtypes based on cell line from which they originate. Gliomas include astrocytomas, GBMs, oligodendrogliomas, ependymomas and other histological subtypes. The most common gliomas are astrocytomas, oligodendrogliomas and ependymomas. For the grading of brain tumors over time have been suggested several systems. The 4-level grading system proposed by WHO is the most widely accepted and widespread [1]. It is based on four histologic features: nuclear atypia, presence of mitoses, endothelial proliferation. The literature uses the term "low-grade glioma" to indicate glial tumors of WHO I and II, while use the term "high-grade glioma" to indicate glial tumors of grade III and IV. Low grade gliomas include circumscribed astrocytomas, as pilocytic astrocytomas, diffuse astrocytoma (grade II), oligodendrogliomas, some times of ependymomas. These tumors are often curable with surgery alone. Although the excision is incomplete, the tumor can be treated with radiotherapy. In the rare cases in which local treatment fails, we resort to the treatment of systemic chemotherapy [2]. High grade glioma includes anaplastic astrocytoma, anaplastic ependimoma, glioblastoma. The standard treatment of this kind of tumors is complete removal. Radiation therapy is also a standard in the treatment because it prolongs the survival period. The role of chemotherapy is controversial.

2.1. Genetic Alterations

Gliomagenesis is characterized by several biological events, such as activated growth factor receptor signaling pathways, downregulation of many apoptotic mechanisms and unbalance among proangiogenic and antiangiogenic factors. Several growth factor receptors, such epidermal growth factor receptor (EGF-R), platelet-derived growth factor receptor (PDGF-R), C-Kit, vascular endhotelial growth factor receptor (VEGF-R) and others growth factors receptors, are overexpressed, amplified and/or mutated in gliomas.

Tumor suppressor genes inactivation is a key event in the glioma progression. Tumor suppressor genes participate in the cellular growth and differentiation processes. Neoplastic transformation can develop through a direct transition from normal astrocyte to GBM or through a gradual transition from low-grade glioma, to high-grade glioma, to GBM. The mechanism for the leading to the formation of a primary GBM is due to amplification of the EGF-R, deletion or homozygous mutation of the gene for the protein cyclin-dependent kinase (CDK), mutations in tumor suppressor gene PTEN, present on chromosome 10, and deletion in the INK4a gene with loss of p14 and p16 [5].

The mechanism that leads to gradual neoplastic transformation through the transition from low-grade gliomas to high-grade gliomas involves several mechanisms, including the mutation of the TP53 gene. The effects of its mutation are the loss of cell cycle regulation, through the inactivation of $p16^{INK-4a}$, the super expression of cycline dependent kinase 4 CDK4 and CDK6 and of ubiquitin ligase Mdm2 and Mdm4. A similar mechanism is created after mutation of Rb, a tumor suppressor gene. Probably, TP53 is a gene responsible for the progression from low-grade to high-grade gliomas, but its mutation is an early event. Gliomas also show constitutive activation of tyrosine kinase dependent receptor that leads to uncontrolled function of Ras and phosphatidylinositide-3-kinase (PI3). Finally, the result is loss of control in cellular growth and proliferation. Modulation of Ras and PI3, is dependent by such genes like Nfl and PTEN. In fact, patients affected by Neurofibromatosis type 1 or by Cowden disease, the last due to mutation of PTEN gene, have an high risk to develop gliomas in their life [6].

PDGF-A and PDGF-B and their receptor are overexpressed in glioma cells, probably they interact *via* paracrine or paracrine mechanisms but their role is actually unclear and, like other factors, show an altered function in other types of cancers. Trasforming growth factor alpha (TGF-alpha) is a potent mitogen, with structural and functional similarities with the EGF, in fact it binds to the same receptor, mediating, such as the EGF, transformation and cell proliferation. It is often over-expressed in high-grade glial tumors.

The tumor suppressor gene p53 is one of the most studied and analyzed tumor suppressor. The product is a protein p53 normal of 393 amino acids whose real function is not entirely clear. It acts as a transcriptional activator and it repress the transcription of other genes, such as IL-6, c-fos, c-jun and other oncogenes. The levels of p53 increase in G1 and S phases of the cell cycle. Cells transfected with the normal gene p53 block their growth in G1, suppressing the passage S-phase. The expression of the p53 gene cause apoptosis.

Many glioma cell lines expressing p53 mutations. The mutated form of p53 has a crucial role in the regulation of neovascularization. It induces increased levels of VEGF and FGFβ through an activation of the transcription of the corresponding genes. The deletion of NFKBIA (encoding nuclear factor of κ-light polypeptide gene enhancer in B-cells inhibitor-α), an inhibitor of the EGF-R-signaling pathway, promotes tumorigenesis in GBM that do not have alterations of EGF-R. Bredel *et al.* analyzed 790 human GBMs for deletions, mutations, or expression of NFKBIA and EGF-R [7]. The authors studied the tumor-suppressor activity of NFKBIA in tumor-cell culture and compared the molecular results with the outcome of GBM patients. NFKBIA is often deleted but not mutated in GBM. Deletion and low expression of NFKBIA were associated with unfavorable outcomes. Patients who had tumors with NFKBIA deletion had outcomes that were similar to those in patients with tumors harboring EGF-R amplification. A two-gene model that was based on expression of NFKBIA and O6-methylguanine DNA methyltransferase was strongly associated with the clinical course of the disease. They conclude that deletion of NFKBIA has an effect that is similar to the effect of EGF-R amplification in the pathogenesis of GBM and is associated with comparatively short survival.

Recently, the isocitrate dehydrogenase 1 (IDH1) activity, an enzyme present in cytoplasm and peroxisomes, that catalyzing the oxidative decarboxylation of isocitrate to alpha-ketoglutarate reducing NADP+ to NADPH, has been evaluated. Novel studies have revealed that mutation of IDH1 is always found before TP53 mutation or, in case of oligodendrogliomas, before their typical deletions of chromosomes 1p and 19q [8]. Yan *et al.* studied genetic alterations of both IDH1 and of its related gene, IDH2. Genomic analysis, showed mutation of amino acid 132 of IDH1 in over 70% of patients with both low or high grade gliomas, and many of the cases negative for IDH1 mutations, had mutation involving amino acid 172 of IDH2 [8]. The role of mutated IDH is probably to favore neoangiogenesis [8, 9].

2.2. Invasion and Angiogenesis

The invasion of tumor's cells consists of an active translocation of these cells through the extracellular matrix. The basic processes in the invasion of tumor cells are the adhesion of the proteins surrounding the ECM through binding to specific receptors, the secretion of proteases, which degrade the components of the extracellular matrix, and the migration of tumor cells through the space that is created in this way.

ECM is composed of proteoglycans, glycoproteins, collagens and also contain fibronectin, laminin, tenascin, hyaluronic acid, vitronectin. Integrins are a class of adhesion molecules that have a major role in the adhesion and subsequent invasion of tumor cells. The proteolytic degradation of the basement membrane (BM) is mediated by proteases, such as the matrix metallo-proteases (MMPs), secreted by tumor and stromal cells. MMPs are secreted as proenzymes and are activated by proteolytic cleavage of their amino-terminal domain [5]. Among the MMPs, the MMP-1 is the crucial enzyme in the process of initial breaking of interstitial collagen, collagen type 1, type 2 and type 3. MMP-2 and MMP-9 are two basic metallo-proteases, largely present in malignant gliomas that act in the matrix degradation.

Angiogenesis, the proliferation of new blood vessels from pre-existing vessels, is an index of malignancy of gliomas. Cerebral gliomas are characterized by extensive microvascular proliferation and a higher degree of vasculature [5]. Neovascularization of brain tumors is correlated directly with their biological aggressiveness, with the grade of malignancy and with the possibility of relapse and is inversely proportional to the survival in the post-operative course. The presence of endothelial glomeruloid-like proliferation and of positive immunoreaction at level of BM of tumor vascular channel is predictive of active tumor invasiveness [10]. The new blood vessel growth is stimulated by the secretion of growth pro-angiogenic factors. These factors bind to receptors present on endothelial cells, so as to activate them. The expression of the receptors VEGF-R1 and VEGF-R2 is regulated on the endothelial cells in gliomas. The ligands for VEGF-R3 (VEGF-C and D) are expressed by multiple cell types that surround the angiogenic vessels, suggesting the existence of a novel pro-angiogenic paracrine signaling pathways in these neoplasms [5].

The expression of VEGF is stimulated by hypoxia and acidosis and is probably related to the production of many other growth factors and their specific receptors (EGF-R, HGF-R, PDGF-R, C-Kit, IGF-Rs). Basic fibroblast growth factor (bFGF) is expressed by vascular and tumor cells. The receptors for bFGF include FGF-R1, expressed by both tumor cells and the tumor endothelial cells, FGF-R2 expressed only by the tumor cells, whereas, FGF-R4 is not detected in malignant gliomas. The binding of VEGF on endothelial cells activates the cascade of phosphatidylinositol-3-hydroxyl kinase (PI3K)/protein kinase B (Akt). The activation of endothelial cells results in increased expression of cells adhesion receptors, such as integrins $\alpha v\beta 3$ e $\alpha 5\beta 1$, and in increased cell survival, proliferation and migration responses [5].

3. ANTISENSE THERAPY

In eukaryotic organism, pre-mRNA is transcribed in the nucleus, introns are spliced out and then the mature mRNA is exported from the nucleus to cytoplasm. The small subunit of the ribosome usually starts by binding to one end of the mRNA and is joined there by other eukaryotic initiation factors, forming the initiation complex. This multi-enzymatic complex scans along the mRNA strand until it reaches a start codon, and then the large subunit of ribosome attaches to the small subunit so that the translation of a protein begins. This process, by which the information of a gene is converted into protein, is referred to "gene expression" [5].

Gene expression can be stopped using three principal efficient strategies: 1) Inhibitor approaches based on chemical compounds [11] and monoclonal antibodies [11]; 2) Anti-gene approaches, which can be classified into three categories: a) the antisense molecules [12-14] targeted to the complementary sequence in mRNA, including antisense RNA, antisense oligodeoxynucleotide and ribozymes; b) the triple helix (TH)-forming oligomers [15, 16] targeted to the double stranded DNA gene; c) the sense oligodeoxynucleotide designed to act as decoys to trap regulatory proteins [17]; 3) siRNA approach: 21-23 mer double-stranded RNA molecules, known as siRNA effectively stopping gene expression [18, 19]. The role of 21-23 mer RNA in silencing of genes is strongly similar to that of the TH DNA mechanism [15].

The discovery of antisense approach was done by Jacob and Harland groups [12-13]. This event has been suggested to physiologically occur as the regulation mechanism of gene expression in cells. Regulating activities of untranscribed DNA strand (antisense strand) has been suggested [20]. Concerning natural antisense RNA in prokaryotes, it has been shown that they could play a regulatory role in replication, transcription or translation steps of some genes [13].

In antisense anti-tumor therapy, the approach using antisense to oncogenes (*i.e.* RAS, c-myb, Bcl-2) [21-23], and especially antisense of genes encoding growth factors (VEGF, EGF, TGF-beta, IGF-I) [23-26], are the most exploited approaches in clinical trials. As far as antisense approach was considered for gene therapy of glioma, at first antisense anti IGF-I was applied [27, 28] followed by antisense anti IGF-I receptor [29] and antisense anti-TGF-beta [30]. Antisense anti IGF-I and anti TGF-beta are introduced with success in clinical trials.

Antisense oligonucleotides (AONs) are short synthetic single-stranded DNA-sequences, 13-25 nucleotides long that bind to and induce the cleavage of homologous stretches of mRNA sequences. Their binding to this mRNA by Watson-Crick base-pairing stops translation and thereby reduces synthesis of the encoded protein. A steric blockage of the process of translation, in which the presence of the bound AON prevents the binding of translation factors and modulators and impedes the movement of the ribosome along the mRNA strand, represent an initial explanation of this process. A second mechanism that comes into play is the action of ribonuclease H (RNase H). This enzyme specifically cleaves the RNA strand of RNA-DNA duplexes. This releases the AON, which can then bind to a new mRNA strand. These result in targeted destruction of mRNA and correction of genetic aberrations. The goal of an antisense molecules-based approach is to selectively suppress the expression of a protein by exploiting the genetic sequence in which it is encoded, acting at translational level [5, 31-33].

The strand complementary to the antisense sequence is called non-transcribed strand and has the same sense sequence as the mRNA transcript. These antisense molecules may be introduced into a cell to silence genic functions through inhibiting translation of a complementary mRNA. This approach has been applied in various kinds of tumors and showed a very high specificity and efficacy against cancer cells [5, 34]. Many AONs have shown sharp *in vitro* reduction in target gene expression and promising activity against a variety of human malignancies. Several experimental *in vitro* and *in vivo* studies in cell lines cultures and animal models showed inhibition of genes involved in cell proliferation, apoptosis and angiogenesis.

The first generation of antisense agents contains backbone modifications such as replacement of the oxygen atoms of the phosphate linkage by sulphur (phosphorothioates), methyl group (methylphosphonates) or amines (phosphoroamidates). The phosphorothioates have been most widely used for gene silencing because of their sufficient resistance to nucleases and ability to induce RNase H functions [5, 14, 35].

The second generation of antisense modifications was aimed at improving these properties; substitutions of position 2' of ribose with an alkoxyl group were most successful. 2'-*O*-Methyl and 2'MOE (2'-*O*-methoxyethyl) derivatives can be further combined with a phosphorothioate linkage. 2'MOE AONs show an improved resistance against nuclease-mediated metabolism as well as tissue half-life *in vivo*, which produces a longer duration of action [14, 35].

The third generation of AONs contains structural elements, such as zwitterionic oligonucleotides, locked nucleic acids/bridged (LNA), morpholino, PNA (peptic nucleic acid) and hexitol nucleic acids. All of the modifications enhanced AONs in terms of nuclease resistance, specific binding and cellular uptake with agents such as PNA and morpholino. Morpholino (PMO: phosphorodiamidate morpholino oligo) oligomers are usually 25 bases in length. Structurally the difference between the morpholino and DNA is that bases in morpholino are bound to morpholino rings instead of deoxyribose rings and linked through phosphorodiamidate groups instead of phosphates. These modifications eliminate ionization in the usual physiological pH range, so morpholino in organisms or cells are uncharged molecules [5, 36].

AONs are taken up primarily by cells *via* endocytosis. Because of the hydrophilic and macromolecular nature of the cell membrane, permeation of AONs across is relatively difficult. Different techniques, such as scrape-loading, electroporation, microinjection or the binding to particular peptides with membrane translocation properties have been developed to overcome this permeation problem [37]. More recently, nanoparticles and AONs conjugates have shown improved cellular uptake, biodistribution and targeted delivery, especially in cancer treatment [5, 38, 39].

AONs directed regulation of messenger RNA expression may be achieved through two different pathways: a) AONs directed RNA cleavage; b) AONs directed obstruction of the processes of the mRNA metabolism through occupancy only. The first pathway permits to reduce expression of the transcript and of the target protein through RNA backbone cleavage using RNases that are expressed in the relevant system. For this pathway the AONs RNA duplex must contain an appropriate recognition element for the enzyme, such as RNase H, RNase L, RNase P and double-stranded RNase. RNase L is an interferon-induced ribonuclease which, upon activation, destroys all RNA within the cell. RNase is unique because it is a ribozyme and a ribonucleic acid that acts as a catalyst in the same way as a protein based enzyme would act. RNase H in a ubiquitous nuclease that degrades the RNA strand of a DNA/RNA duplex. The second pathway has the goal to change or obstruct a process and not cleave the massage. Oligonucleotide modifications should not be substrates or recognition elements for RNases or other enzymes involved in m RNA metabolism. The occupancy only mechanism of AONs inhibits or alters gene expression at the mRNA level. Inhibition of target protein expression at the level of the translation stage of mRNA metabolism can be achieved by using modified AONs, by targeting the 5' untraslated region and by obstructing translation initiation or by interfering translation in the elongation phase [5].

Recently, the role of surviving has been evaluated. Survivin is a family member of the inhibitor of apoptosis proteins that is expressed during mitosis in a cell cycle-dependent manner and localized to different components of the mitotic apparatus. It plays an important role in both cell division and inhibition of apoptosis and is expressed in a vast majority of human cancers. Survivin expression is often correlated with poor prognosis in a wide variety of cancer patients [40]. These features make survivin an attractive target against which cancer therapeutics could be developed. The authors have identified a survivin antisense oligonucleotide that potently down-regulated survivin expression in various human cancer including brain, lung, colon, breast, skin as measured by quantitative RT-PCR and immunoblotting analysis [40]. Apoptosis, or programmed cell death, is a highly organized physiologic event that plays an essential role in controlling cell number in many normal processes, ranging from fetal development to adult tissue homeostasis. Abnormal regulation of apoptosis has been implicated in the onset of wide range of diseases including cancer. Survivin is, also a members of the inhibitor of apoptosis proteins (IAP) family that play an important role in the negative regulation of apoptosis. Varous studies report that the AON, LY2181308, potently inhibited survivin mRNA and protein expression in various tumor cell lines. Inhibition of survivin expression in tumor cell lines by this AON resulted in the induction of caspase-3 activity, cell cycle arrest, and failure of cell division. Moreover, LY2181308 significantly inhibited survivin expression in multiple human xenograft tumors with a significant inhibition of human xenograft tumor growth when administered intravenously in sensitized tumors to gemcitabine, paclitaxel, and docetaxel.

Ribozymes are small oligo-ribo-nucleotides which have a specific base sequence with natural self-splicing activity. This activity can be directed against virtually any RNA target by the inclusion of an antisense region into the ribozyme. Ribozymes hybridize specifically to complementary mRNA and block the encoded protein by cleaving its mRNA.

A new approach permitting to modulate gene expression into cancer cells is related to the presence of RNA interference (RNAi). RNAi is mainly responsible for the post-transcriptional regulation of gene expression but is involved in transcriptional regulation, acting as a process within living cells that moderates the activity of their genes. It plays a fundamental role in diverse eukaryotic functions including viral defence, chromatin remodeling, genome rearrangement, developmental timing, brain morphogenesis, and stem cell maintenance. The power of RNAi lies in the key discovery that endogenous RNAi gene silencing machinery can be hijacked to artificially regulate genes of interest. The selective effect of RNAi on gene expression makes it a valuable research tool, both in cell culture and in living organisms because synthetic double-stranded RNA (dsRNA) introduced into cells can induce suppression of specific genes of interest. This strategy, mediated by small double stranded RNA molecules, is normally an endogenous gene silencing mechanism physiologically used by eukaryotes to regulate gene expression by reducing protein production. In mammalian cells RNAi can be effected when short (~22nt), double-stranded fragments of RNA, denominated small interfering RNAs (siRNAs), are loaded into the RNA-Induced Silencing Complex (RISC), where the strands are separated, and one strand guides cleavage by Argonaute of target mRNAs in a sequence homology-dependent manner [41]. This pathway is initiated by the enzyme Dicer, an endonuclease of the RNase III family, which cleaves long dsRNA in siRNAs. Each siRNA is unwound into two single-stranded (ss) ssRNAs, namely the passenger strand and the guide strand. The passenger strand will be degraded, and the guide strand is incorporated into the RISC. Successively, they can target specific mRNAs in a microRNA (miRNAs) sequence-dependent process and interfere in the translation into proteins of the targeted mRNAs [42]. The heart of the RISC complex and principal executer of RNAi-mediated silencing is the Argonaute protein [43]. There are four Argonaute proteins in humans (AGO 1-4), and silencing by siRNAs is accomplished *via* AGO2 [43].

4. MOLECULAR TARGETS

The introduction of molecular targeted agents is one of the most significant advances in cancer therapy in recent years. Targeted therapies block activation of oncogenic pathways, either at the ligand-receptor interaction level or by inhibiting downstream signal transduction pathways, thereby inhibiting growth and progression of cancer. Because of their specificity, targeted therapies should theoretically have better efficacy and safety profiles than systemic cytotoxic chemotherapy or radiotherapy. The choice of target gene is crucial to the potential success of the therapeutic approach. Because of the substantial neovascularization seen in malignant gliomas, targeted antiangiogenic therapies have received considerable attention [44] The main rationale for using antiangiogenic therapies is to normalize the vasculature, restoring the selective permeability of the blood-brain barrier. Other target molecules of different pathways, such as cell immortalization, apoptosis, and invasion have been studied as possible novel targets of antisense strategy.

4.1. Growth Factor Pathways

Growth factor pathways that stimulate cell proliferation can be achieved through overexpression or genetic amplification of growth factor receptor genes (EGF-R, ERBB2, MET), as well as through gene mutations that lead to ligand-independent signaling as occurs for the epidermal growth factor receptor vIII (EGF-RvIII), a mutant that sends constitutive growth signals. Alternatively, intracellular signaling pathways can be activated when the positive signaling molecules are mutated and signal constitutively, or when the negative regulators of the pathway are lost through gene loss or mutation. The main tumor suppressor pathways are those of the p53 and Rb pathways, which directly monitor cell cycle entry and progression. p53 activates the transcription of p21, a molecule that blocks cell cycle progression in the G1 phase of the cell cycle by binding and inhibiting the function of the cyclin D family of proteins [45]. These are the regulatory subunits of the cyclin/cyclin- dependent kinase complexes that regulate cell cycle entry and progression by inducing the phosphorylation and inactivation of Rb. The Rb protein can prevent cell entry into S-phase by inactivating the E2F family of transcription factors that are critical for the initiation of DNA replication. Another contributor that directs cell proliferation in gliomas is c-Myc,30 a transcription factor that drives the expression of cell cycle promoters such as cyclins and cyclin-dependent kinases while blocking the transcription of cell cycle inhibitors (CKI). Excessive growth-stimulating signals emanating from growth factor receptors or overexpressed oncogenes such as Myc are sensed by the cell and will trigger a safeguard response that results in cell death through apoptosis. Apoptosis or programmed cell death is the outcome of a physiological response that leads to cell termination. It is the result of a precise signaling process that is initiated either at the cell surface (death receptor pathway) or intrinsically by intracellular signals such as extensive DNA damage. This process involves the activation of a series of proteins called caspases, which leads to the irreversible breakdown of cellular components and ultimately cell death. To overcome this limitation in their growth, tumors will typically also genetically inactivate proapoptotic pathways or activate the overexpression of genes that can promote cell survival. The inactivation of the p53 protein by mutation abrogates proapoptotic responses in the cell because this factor controls the transcription of both cell cycle arrest (p21CKI) and proapoptotic genes (Bax, Fas, *etc.*) [45]. Another mechanism through which the tumor can overcome apoptotic induction is the overexpression of factors that will prevent activation of the apoptotic cascade. One major antiapoptotic mediator that is overexpressed in gliomas is transcription factor NFκB [46]. NFκB is well known as a mediator of immune and inflammatory responses, but it also activates the transcription of proteins that inhibit apoptosis such as members of the inhibitor of apoptosis (IAP) family (c-IAP, XIAP, and survivin). IAPs interact directly with activated caspases and can block their proapoptotic function. The transforming growth factor-beta (TGF-beta) is a multifunctional cytokine expressed in malignant gliomas that increases tumor cell motility, invasion, angiogenesis and immune escape [47]. TGF-beta acts through specific cell-surface receptors that downstream signals through their kinase activity. Malignant gliomas lines secrete TGF-beta ligands and express receptors. The presence of both TGF-beta ligands and receptors indicates the presence of an autocrine loop. The autocrine loop of TGF-beta signaling controls a variety of cellular processes including proliferation. In human malignant gliomas, overexpression and secretion of the TGF-beta2 isoforms has been described as a hallmark of its pathobiology [48].

4.2. Cell Surface Growth Factor Receptors

Epidermal growth factor receptor (EGF-R) is one of the most attractive therapeutic targets in cerebral gliomas and it is amplified and overexpressed in approximately 40% of primary GBMs. EGF-R, a ubiquitously expressed transmembrane glycoprotein belonging to the ErbB/HER family of receptor tyrosine kinases (TK), is composed of an extracellular ligand-binding domain, a hydrophobic transmembrane segment, and an intracellular TK domain. Upon ligand binding to EGF-R, the latter undergoes a conformational change that promotes homo- or heterodimerization with other members of the ErbB/HER family of receptors, followed by autophosphorylation and activation of the TK domain. EGF-R signaling activates a number of downstream effectors including the phosphatidylinositol-3-kinase (PI3K)/Akt pathway. Nearly half of tumors with EGF-R amplification also have a constitutive active EGF-R mutant known as EGF-RvIII, which has a large deletion in the extracellular domain and renders the receptor ligand independent for signaling. This deletion also engenders a unique codon, which is not found in the wild-type receptor, thereby creating a tumor-specific epitope that can be exploited for therapeutic targeting. Activation of EGF-R leads to activation of intracellular signaling pathways that regulate cell proliferation, invasion, angiogenesis, and metastasis. Inhibition of receptor activation inhibits downstream signaling pathways, resulting in decreased cell proliferation and survival. Elevated expression of EGF-R is correlated with poor prognosis. Some studies found that tumors with EGF-RvIII and intact PTEN and tumors with low phosphorylated Akt levels [49, 50] are more likely to respond to EGF-R inhibitors. Current trials in patients with malignant glioma are evaluating irreversible EGF-R inhibitors such as BIBW 2992 and PF-00299804, the dual EGF-R and VEGF-R inhibitor vandetanib (ZD6474), and the humanized monoclonal antibody against EGF-R, nimotuzumab. CDX-110, a peptide vaccine against the unique epitope of EGF-RvIII, has a favorable toxicity profile and currently is being studied in combination with temozolomide in patients with newly diagnosed GBM [51]. Glioma cells produce many different proangiogenic factors, including VEGF. For these reasons, VEGF and VEGF-R have been targeted for potential treatment of gliomas. Strategies for targeting this ligand and its receptor include using VEGF-R TKIs, VEGF antibodies, and protein kinase C (PKC)-β inhibitors. PDGF and its receptors play an important role in tumor interstitial pressure, tumor growth, and angiogenesis. As with other growth factors and their receptors, over-expression of PDGF and its cognate receptors in gliomas has caused PDGF and PDGF-R to become targets for antitumor treatments. The PDGF-R subtypes α and β and PDGF ligands A and B are also overexpressed in malignant gliomas [52]. The PDGF-R inhibitor imatinib mesylate was reported to have significant antitumor activity both *in vitro* and in orthotopic glioma models [53].

Focal adhesion kinase (FAK) is a ubiquitously expressed nonreceptor protein tyrosine kinase that localizes at focal adhesions, mediates signaling induced by integrins, and plays an important role in many cellular functions [54]. Activated FAK, marked by autophosphorylation at Y397, recruits a number of SH2 and SH3 domain-containing proteins, including c-Src. The binding of c-Src to FAK is proposed to disrupt the intramolecular interaction between the c-Src SH2 domain and the negative regulatory carboxy-terminal Y529. Activated Src, in turn, phosphorylates FAK and further enhances FAK activity, thereby forming a positive feedback loop leading to the activation of downstream signaling molecules, such as extracellular signal-regulated kinase 1 and 2 (ERK1/2) and phosphatidylinositol 3-kinase (PI3-K)/AKT [55, 56]. FAK can exert control over the rate of focal adhesion turnover and, therefore, cell motility [56]. FAK is a positive regulator of normal cell migration, and its overexpression or activation has been shown to promote cancer cell metastasis of some tumor types [55, 56].

4.3. IGF Pathway

The IGF system involves three ligands (insulin, IGF-I and IGF-II), their receptors IGF-IR and IGF-IIR, and insulin-like growth factor binding proteins (IGFBPs, BP1 to BP6) [57]. Both IGF-I and IGF-II are ligands for IGF-IR [57, 58]. The action of IGF-I on cellular metabolism depends on IGFBPs, which prolong the half-life of this growth factor and modify its interaction with its receptor [59, 60]. IGFBPs have growth regulatory actions that are probably independent of their capacity to bind IGF-I [59, 60]. IGF-II is a fetal growth factor whereas IGF-I, a basic 70 amino acid polypeptide, which plays a fundamental role in fetal, normal and neoplastic tissue differentiation [57]. They reappear in neoplastic developing neuroglial-derived tissues including GBM [61]. The presence of IGF-II in neoplastic neural cells led to a consideration of stopping neoplastic glial development by targeting IGF-I. Targeting the IGF-I system has emerged as a potentially important and useful method to reduce glial malignant development [62, 63]. IGF-I is involved in neural development, neurogenesis, glial differentiation and glucose metabolism, acting locally with autocrine/paracrine fashion, with a predominant role compared with other growth factors [64]. Its overproduction is considered to be a participating factor in brain cancer development [59, 64]. The blockade of IGF-I producing immunogenic anti-tumor phenomena and apoptosis [65], was demonstrated for the first time in 1996-1998 [66-68]. IGF-I is implicated in malignant tumors of the CNS, where neurons and glial cells have lost the ability of programmed cell death related to signal transduction. Moreover, IGF-I has a role in regulating cell motility, as IGF-I treatment stimulates motility in malignant cells, which is mediated through IGF-IR, PI3K activation. Concerning IGF-IR as a tumor diagnostic marker, the neoplastic progression in some malignant tumors, including glioma, may be associated with increased expression of this receptor, although this over-expression seems to be less common than that for ERBB2 in malignant tumors [59, 69].

4.4. PI3K/Akt Pathway

The PI3K/Akt pathway is a critical regulator of tumor cell metabolism, growth, proliferation, and survival and is initially activated at the cell membrane, where the signal for activation is propagated through class IA PI3K. Activation of PI3K can occur through tyrosine kinase growth factor receptors such as EGF-R and IGF-IR, cell adhesion molecules such as integrins, G-protein-coupled receptors (GPCRSs), and oncogenes such as Ras.

Among those kinases, PI3K and PI3K-related kinases (PIKK) belong to a family of high molecular mass kinases whose catalytic domains show a strong resemblance. This family and the ribosomal protein P70S6K, mTOR, the DNA-dependent protein kinase, the ataxia telangiectasia mutated gene (ATM), the ataxia-telangiectasia related (ATR) protein and key components of the histone acetylase complex are involved in checkpoint regulation of cell cycle, DNA repair, telomere length and cell death [70]. Activated PI3K phosphorylates inositol lipids at the 30 position of the ring inositol, generating the lipid products PI3-phosphate [PI(3)P], PI3,4-biphosphonate [PI(3,4)P2] and PI3,4,5-triphosphate

[PI(3,4,5)P3]. These lipid products are involved in a number of cellular processes including cell proliferation, survival, cytoskeletal reorganisation, membrane trafficking, cell adhesion, motility, angiogenesis and insulin action [71]. Glycogen synthase kinase (GSK3) appears negatively regulated by AKT-dependent phosphorylation. Reduced GSK-3 activity leads to increased levels of the growth stimulator beta-catenin. On the contrary, AKT indirectly activates mTOR *via* TSC, which in turn phosphorylates and activates several targets involved in translation of specific mRNAs, apoptosis and/or cell cycle [71].

Downstream effectors of mTOR have an array of biological functions that promote hypoxic adaptation and protein translation [72]. In malignant gliomas, PI3K/Akt/mTOR signaling is frequently increased because of receptor tyrosine kinase over-activity (EGF-R, PDGF-R, and mesenchymal-epithelial transition factor [MET]), mutated oncogenic PI3K subunits, and/or loss of PTEN tumor suppressor activity.

The tumor suppressor phosphatase and tensin homolog deleted on chromosome ten (PTEN) antagonizes PI3kinase by dephosphorylating PI(3,4,5)P3 and (PI(3,4)P2), thereby preventing the activation of Akt and PDK-1. Amplification of Akt1 has been described in human gastric adenocarcinomas, and amplification of Akt2 has been described in ovarian, breast, and pancreatic carcinomas [73]. Moreover AKT also modifies cell cycle progression by inhibiting p27 and p21, up-regulates transcription of cyclin D and promotes nuclear entry of mdm2. AKT enhances telomerase activity and promotes cell invasiveness and angiogenesis by stimulating secretion of MMPs and by activating endothelial nitric oxide synthase [5]. Immunohistochemical studies using antibodies that recognize Akt phosphorylated at S473 have demonstrated that activated Akt is detectable in cancers including head and neck cancers [74]. Moreover, using antibodies against S473 and T308, two sites of Akt phosphorylation, Akt activation was shown to be selective for neoplastic tissue *versus* normal tissue, and phosphorylation of Akt at both sites was shown to be a better predictor of poor prognosis in brain tumor than phosphorylation at S473 alone [75].

Previous reports show an increased glycogen content in GBM, in induced tumors using rat derived-GBM cells, and in cultured cancer cells [76]. These data suggest a critical role of glycogen metabolism in meeting the metabolic demands imposed by cancer progression and malignancy. The glycogen synthase (GS), is the key enzyme of glycogenesis. *In vitro* rat C6-glioma cells, transfected with antisense glycogen synthase (C6-AS cells) exhibited a decreased expression of glycogen synthase and reduced activity of glycogen synthesis, along with attenuated invasiveness. *In vivo* tumors induced by C6-AS cells in nude mice exhibited a significant reduction in tumor growth compared with controls. More importantly, the successful inhibition of glycogen synthase through an antisense approach, and the demonstration of the resultant anti-tumorigenic effects of this inhibition in rat C6-glioma cells, provide compelling validation of a new target for antisense gene therapy for malignant gliomas.

Another frequent genetic event occurring in human cancer is loss of function of the tumor suppressor PTEN, which normally suppresses activation of the PI3K/Akt/mTOR pathway by functioning as a lipid phosphatase. Loss of PTEN function in cancer can occur through mutation, deletion, or epigenetic silencing. Mutation, deletion, or epigenetic silencing of PTEN has been shown to correlate with poor prognosis and reduced survival in patients with various types of cancer, with loss of PTEN being a common mechanism for activation of the PI3K/Akt/mTOR pathway and poor prognosis. Activation of PI3K has been described in human tumors that may result from the amplification, overexpression, or mutation of the p110 catalytic or p85 regulatory subunit. Amplification of the 3q26 chromosomal region, which contains the *PI3KCA* gene that encodes the p110α catalytic subunit of PI3K, has been observed in 40% of ovarian and 50% of cervical carcinomas [77]. Somatic mutations of this gene have been detected in several cancer types, with mutant PI3K having increased kinase activity relative to wild type [78]. Any of the alterations in individual components of the PI3 kinase/Akt pathway would result in its activation, and activation of this pathway has been reported to be among the most frequent molecular alterations in tumors [78]. Several mTOR inhibitors are undergoing evaluation in malignant gliomas, including sirolimus (rapamycin), temsirolimus (CCI-779), everolimus (RAD001), and ridaforolimus (AP23573). To date, mTOR inhibitors have demonstrated only minimal single agent activity against malignant gliomas [79]. In addition to the PI3K/Akt/mTOR pathway, signal transduction from activated tyrosine kinases such as EGF-R and PDGF-R is mediated by the Ras/Raf/mitogen-activated protein kinase pathway. Activation of Ras requires localization to the intracellular surface of the cell membrane, a critical step that depends on farnesylation. Farnesyl transferase inhibitors (FTI) interfere with this process and have demonstrated promising activity in glioma models. Histone deacetylase (HDAC) inhibitors interfere with transcriptional regulation and can induce growth arrest, terminal differentiation, and apoptosis of tumor cells. The HDAC inhibitor vorinostat (sub-eroylanilide hydroxamic acid) proved effective in pre-clinical models but only modestly prolonged the 6-month PFS (PFS6) in a phase 2 trial in patients with recurrent GBM [80]. Ongoing studies are investigating the combination of vorinostat with temozolomide and radiation.

4.5. Protein Kinase C

Protein Kinase C (PKC) is a family of 14 protein kinases that regulate cell growth, proliferation, and angiogenesis. The protein kinase C-α (PKC-α), translated from the mRNA of PKC-α gene, located on chromosome 16p11.2-q12.1, is involved in growth-factor-mediated signal transduction pathways. PKC is downstream of growth factor receptors such as EGF-R and PDGF-R and is activated *via* signaling from the extracellular signal-regulated Ras/MAPK pathway. Also, there have been some data to show that there is cross-talk between PKC and the PI3K/AKT pathway [81]. Its overexpression has been implicated in various kinds of tumors such as breast, colon, lung, ovarian cancer, melanoma, brain tumors [82]. ISIS 3521, an AON directed against PKC-α, is the most experimented, and the effects of the PKC-α down-regulation are evident in ovarian cancer, NSCLCs, kidney clear cell carcinoma and low-grade lymphoma [83].

4.6. Extracellular Matrix Protein

ECM is a complex network of different collagens, proteoglycans, hyaluronic acid, laminin, fibronectin, and many other glycoproteins, including proteolytic enzymes involved in degradation and remodeling of the ECM. Extracellular matrix exists in two forms: interstitial matrix that fills in the intercellular space and the more specialized BM, which is a thin sheet of extracellular matrix underlying the epithelium. ECM provides the microenvironment for the cells and serves as a tissue scaffold, guiding cell migration during embryonic development and wound repair. Beyond that, it also functions as the repository and modulator of growth factors and cytokines, and therefore is responsible for transmitting environmental signals to the cells. Cell responses to contact with these networks depend on the cell's repertoire of ECM receptors. Connections from the matrix through these receptors determine the organization of cytoskeletal structures and the localization and activation of signaling molecules leading to unique tissue specific cell functions. ECM modification aids the loss of contact inhibition allowing tumor cells to freely migrate and invade the surrounding tissues. Various ECM components, such us hyaluronan, vitronectin, fibronectin, tenascin-C, osteopontin may be considered crucial target of molecular therapies able to down-regulate invasion pathways during brain tumor infiltration [84]. Changes in these ECM components are felt to modulate brain tumor growth, proliferation and invasion, although specific interactions and exact mechanisms are unknown [5, 33]. Cell adhesion is the binding of the cells to each other and to the ECM through cell adhesion molecules such as integrins, selectins, cadherins, the Ig (immunoglobulin) superfamily and lymphocyte homing receptors. Cell adhesion mediates cell attachment, migration, and signaling to and from the extracellular matrix.

The integrins are a family of heterodimeric cell surface receptors that have important roles during development and in the adult organism [85]. One of the major function of integrins is to provide mechanical transmembrane linkages between ECM proteins such as fibronectin, collagen and laminin and the actin cytoskeleton. Integrins also play an important role in regulating several signal transduction pathways. A number of integrins are differentially expressed between tumors and normal tissue [86], with the integrin $\alpha v\beta 3$ being of particular interest in this context since it is highly expressed in angiogenic endothelia and in some types of tumor cells [87]. Integrins of the $\beta 1$ and αv classes are also expressed on neurons, glial cells, meningeal and endothelial cells. $\beta 2$ integrins are specifically expressed by leukocytes and they are found on microglia and on infiltrating leukocytes within the CNS, and $\alpha v\beta 3$ integrin is associated with elevated levels of its ligands and with vitronectin in human gliomas [88]. Thus integrins are relatively abundantly expressed, are efficiently internalized by endocytosis, and some members of this family are differentially expressed on tumor cells; in aggregate these characteristics suggest that targeting integrins could be an excellent approach for delivery of therapeutic oligonucleotides.

4.7. Bcl-2, Telomerase

Bcl-2 gene is a proto-oncogenes located at the breakpoints of t(14;18) chromosomal translocations in low-grade B-cell non-Hodgkin's lymphomas. Bcl-2 encodes an oncogenic mitochondrial protein with transport functions and its expression is very increased in follicular non Hodgkin lymphoma, leukemias and lung, breast, colorectal, gastric, prostate, renal cancer and neuroblastoma. High expression levels of bcl-2 protein have been found in many tumors without such chromosomal abnormality, and upregulation of bcl-2 has been shown to be a key element in tumoral malignancy [89] and drug resistance [90]. Bcl-2 overexpression has been observed in several glioma cell lines and in glioma biopsies, regardless of histological grade [91]. However, the precise role of bcl-2 in gliomas remains a matter of disagreement. Some studies have shown that increased expression of bcl-2 is associated with improved survival, but without showing the relationship between bcl-2 expression and histological grade of glioma [92]. Others have demonstrated that bcl-2 immunohistochemical positivity is inversely correlated with survival and that bcl-2 protein promotes migration and invasiveness of human glioma cells [93]. Bcl-2 protein prevents the release of cytochrome C from the mitochondria [94] leading to inhibition of apoptosis.

Telomeres are distinctive DNA-protein structures which cap the ends of linear chromosomes. Three core components of the telomerase holoenzyme are human telomerase RNA(hTR), human telomerase reverse transcriptase (hTERT) and telomerase-associated protein (TP1). HTR contains a template domain that serves as a reading frame for the extension of the G-rich 3' end of the telomeres by hTERT [95]. The role of human telomerase (hTR) is to maintain telomeric length and promote cell proliferation potential. Previous studies have demonstrated that hTR and telomerase activity may play a key role in the development and progression of malignant gliomas, and that the inhibition or down-regulation of telomerase activity may potentially be used in the clinic as a novel target of gene therapy for malignant gliomas [96, 97]. The activation of telomerase is considered a critical factor in cellular immortality.

5. ANTISENSE THERAPY AND BRAIN DELIVERY

The brain is probably one of the least accessible organs for the delivery of active pharmacological compounds. In the CNS there are two physiological barriers: the blood-brain barrier (BBB) and the blood-cerebrospinal fluid barrier (BCFB), that separate the brain from its blood supply. Their function is to maintain a constant internal environment inside the brain by strictly regulating the composition of the cerebral extra-cellular fluid and to protect the brain against potentially toxic substances [5]. In fact, a large-molecule pharmaceutics, including peptides, recombinant proteins, monoclonal antibodies, RNA interference-based drugs, generally do not cross the barrier [95, 96]. New biologic drugs such as proteins and nucleic acids require novel delivery technologies that will minimize side effects and lead to better patient compliance. As current advances in biotechnology and related areas are aiding the discovery and rational design of many new classes of drugs, it is crucial to improve specific drug-delivery methods to turn these new advances into clinical effectiveness. Several drugs are limited by their poor solubility, high toxicity, and high dosage, aggregation due to poor solubility, nonspecific delivery, *in vivo* degradation and short circulating half-lives. Innovative drug delivery systems may make it possible to use certain chemical entities or biologic that were previously impractical because of toxicities or because they were impossible to administer.

Nano-structured drug carriers will help to penetrate or overcome these barriers to drug delivery. Drug delivery carriers are macromolecular assemblies that can incorporate imaging and therapeutic compounds of distinct nature, such as small chemicals, fluorophores and biosensors, peptides and proteins, oligonucleotides and genes. Using targeted NPs to deliver chemotherapeutic agents in cancer therapy offers many advantages to improve drug delivery and to overcome many problems associated with conventional chemotherapy [39, 98]. Nanoparticles (NPs) *via* either passive targeting or active targeting have been shown to enhance the intracellular concentration of drugs in cancer cells while avoiding toxicity in normal cells [99]. They can be designed to improve the solubility of these cargo molecules and their bioavailability, and also to control their circulation, biodistribution in the body, and release rate, together enhancing their efficacy [100]. Surface properties' modifications confer advantageous properties to the particles, such as increased solubility and biocompatibility useful in the crossing of biophysical barriers [99]. Advantages of nanostructure-mediated drug delivery include the ability to deliver drug molecules directly into cells and the capacity to target tumors within healthy tissue. DNA and RNA packaged within a nanoscale delivery system can be transported into the cell to fix genetic mutations or alter gene expression profiles. The mechanisms of cellular uptake of external particulates include clathrin- and caveoli-mediated endocytosis, pinocytosis, and phagocytosis. Drug delivery carriers can be functionalized to improve control of their circulation and biodistribution in the body at the tissue, cellular, and sub-cellular level. This can be achieved by incorporating immune-evading moieties and/or affinity molecules that favor adhesion to either general or specific biological markers, depending on the degree of selectivity required. Intracellular targeting refers to the delivery of therapeutic agents to specific compartments or organelles within the cell and the delivered cargoes must gain access to intracellular compartments where their molecular targets are located. Gene therapy is dependent on the entry of the vector into the cell while protecting the contents from the harsh environment of the cytoplasm. Transport of the gene into the nucleus must then occur without imparting damage from the vehicle itself. Gene therapy is a promising new approach for treating a variety of genetic and acquired diseases. These macromolecules are unstable and show a poor cellular uptake and are rapidly degraded by nucleases. To overcome these limitations, various chemical modifications of oligonucleotides have been tried. These modifications have disadvantages such as decreased mRNA hybridization, elevated cytotoxicity, and increased nonspecific targeting. In order to overcome the disadvantages of viral carriers (high cytotoxicity, cost, small transgene size) nonviral carriers have been developed. The advantages associated with nonviral carriers include facile large scale manufacture, low immunogenic response, versatile modifications, and the capacity to carry large inserts. The major route of intracellular therapeutic uptake is through endocytosis. This strategy is ideal in the case of delivery of therapeutic agents whose action is required at said sub-cellular compartments, such as in the case of carrier-assisted delivery of enzyme replacement for lysosomal storage disorders. Carriers themselves can also be designed to overcome endosomal membranes, such as in the case of pH-sensitive poly(acid) carriers and temperature-responsive poly(electrolyte) hydrogels [101].

Other strategies have been designed to directly overcome the plasmalemma; these include electroporation and ultrasound, where a local electric or ultrasound pulse is exerted in the immediate post-administration period causing transient enhancement of the plasmalemma permeability [102], and biolistic particle delivery systems, where penetration into cells is gained by means of tungsten or gold particles that are propelled by a "gene gun" across the plasma membrane [103]. Amphiphilic and biodegradable cationic copolymers are efficient gene delivery systems which can condense nucleic acid and form controlled nanosized complexes. Recently, has been demonstrated *in vivo* experiment that a significantly higher amount of antisense oligonucleotides could be found in tumor cells and also in normal cerebral cells if the oligonucleotides were bound to nanoparticles coated with Polysorbate 80 as compared to pure oligonucleotide injections [104]. One of the most investigated approaches to gene therapy uses liposomes as sub-micron delivery vehicles. Liposomes consist of a lipid shell surrounding a core containing a therapeutic molecule or gene. They are particularly useful as gene therapy devices due to their ability to pass through lipid bi-layers and cell membranes and several groups have recently reported convincing results following local delivery. Lipid encapsulation is an attractive delivery approach because of the biocompatibility of the constituents and facile assembly of the complexes, which requires only mixing and incubation of components. In addition, these complexes can be engineered for specific delivery through conjugation of targeting moieties directly to the lipid molecules prior to liposome production. Liu *et al.* have shown that composite liposomes containing poly(cationic lipid) and cholesterol showed much higher transfection in the liver than naked DNA alone [105]. This work focused on lipoplexes consisting of poly(cationic lipid), cholesterol, and DNA injected directly into portal circulation following a partial hepatectomy. The reporter gene expression (luciferase) was much higher in these lipoplexes than in naked DNA alone [105]. Cohen-Sacks *et al.* have characterized the delivery of 300 nm-diameter nanospheres loaded with platelet-derived growth factor beta-receptor antisense for the treatment of restenosis. This group observed cell internalization, inhibition of smooth muscle cell proliferation, and roughly a 22% reduction in restenosis relative to controls [106]. Polyamidoamine (PAMAM) and poly[2-(dimethylamino) ethyl methacrylate] (PDMAEMA) are low toxic polymers which have shown great potential as carriers. Polycaprolactone (PCL) is another promising delivery system. PCL-*g*-PDMAEMA nanoparticle/DNA complexes could escape from the endosome and release their payloads effectively in cytoplasm, which may be induced by the enhanced interaction between the complexes and cell membrane due to hydrophobic modification [107].

6. ANTISENSE OLIGONUCLEOTIDES AND CEREBRAL GLIOMAS

Antisense therapy has been applied in various kinds of tumors and showed a very high specificity and efficacy against cancer cells. Many AONs have shown sharp *in vitro* reduction in target gene expression and promising activity against a variety of human malignancies. Several experimental *in vitro* and *in vivo* studies in cell lines cultures and animal models showed inhibition of genes involved in cell proliferation, apoptosis and angiogenesis. Clinical trials demonstrated a good tolerability and no increment of toxicity of AONs [5, 33].

Growth factors such as epidermal growth factor, vascular-endothelial growth factor, transforming growth factor, and insulin-like growth factor type I, are present during the development of the CNS. When they reappear in the mature brain they are overexpressed in neoplastic glia, participating in the development of malignant gliomas.

6.1. Antisense Oligonucleotides

EGF-R overexpression and immunoreactivity are common in primary tumors [50]. Recently was evaluated the suitability of folate-PAMAM dendrimer conjugates for efficient EGF-R AON delivery into glioma cells, wherein they release the AON from the FA-PAMAM to knock down EGF-R expression in C6 glioma cells, both *in vitro* and *in vivo*. Folic acid was coupled to the surface amino groups of G5-PAMAM dendrimer (G5D) through a 1-[3-(dimethylamino)propyl]-3-ethylcarbodiimide bond, and AONs corresponding to rat EGF-R were then complexed with FA-PAMAM. The AON transfection rates mediated by FA-PAMAM and PAMAM resulting in greater suppression of EGF-R expression and glioma cell growth. Stereotactic injection of EGFR-AON:FA-PAMAM complexes into established rat C6 intracranial gliomas resulted in greater suppression of tumor growth and longer survival time of tumor-bearing rats compared with PAMAM and oligofectamine-mediated EGF-R-AON therapy [108]. In a previous study, was demonstrated that antisense-EGF-R transfection inhibited the growth and transforming phenotype of U87MG cells. In this research, the cotransfection of U87MG cells with wild-type PTEN and antisense EGF-R constructs could inhibit the cellular growth by 91.7%. In addition, these cotransfected cells showed a differentiated form and expressed much lower telomerase activity than cells transfected with wild-type PTEN or antisense-EGF-R alone [109].

VEGF and VEGF-R have been targeted for potential treatment of gliomas. Strategies for targeting this ligand and its receptor include using VEGF-R TKIs, VEGF antibodies, and protein kinase C-β inhibitors. Edema in glioma tumors is considered one of the most pathological characteristics, but the mechanism of regulating vascular permeability is still unclear. In tumorigenic mice generated by subcutaneous injection of glioma cell lines, overexpression of antisense VEGF (C6-VEGF(-/-) mice) significantly suppressed tumor growth, decreased angiogenesis and reduced tumoral edema. Further studies by electron microscope revealed that tumor-induced hyperpermeability was mediated by formation of vesiculo-vacuolar organelles (VVO), specifically reducing the number of vesicle and caveolae in VVO, and this effect was partially blocked by antisense VEGF [110]. In a successive study, in C6 cells with expression vectors containing sense (C6/VEGF+) or antisense (C6/VEGF-) VEGF complementary DNA (cDNA) or an empty vector (C6/vec), VEGF expression, water content, and morphological characteristics were investigated. The observations demonstrated that VEGF can aggravate edema in tumor tissues and plays critical roles in the stickiness of tumor cells to vessel wall and in the integrity and continuity of the basal lamina of vessels [111].

Down-regulation of IGF-I-R using antisense strategy has resulted in inhibition of tumor growth and metastases in a preclinical glioma model [112]. The rationale for this was based on the observation that inhibition of IGF-I-R in the experimental glioma model results in apoptosis of tumor cells, inhibition of tumor genesis and immune anti-tumor response. Rats that were given injections of C6 glioma cells expressing antisense IGF-I-R did not develop tumors, and were protected from a subsequent challenge with wild-type C6 glioma cells for at least 3 months. IGF-1R AONs induced autologous glioma cells toward apoptosis [113].

The crucial involvement of two signal transduction pathways, activated by focal adhesion kinase (FAK) and insulin-like growth factor I receptor (IGF-IR) kinase into gliomagenesis, has been evidenced in an *in vitro* and *in vivo* study (U87 and LN229 glioma cell lines and mice bearing intracranial glioma xenografts) [114]. The authors demonstrated that TAE226, an ATP-competitive inhibitor of several tyrosine kinases, down-regulated proliferation and invasion of glioma cells and inhibited cell cycle progression at the G2-M checkpoint. TAE226 also gave rise to a concomitant reduction of the expression of p-cdc2 (Tyr15) and cyclin B1, increase in apoptosis and reduction in glioma invasion in an *in vitro* Matrigel [115]. Anti-FAK phosphotothioate AONs packaging into liposomes in U251 MG cells down-regulate the expression levels of FAK and activation of apoptosis, through increase in caspase-3 activity, the key-mediator of apoptosis in mammalian cells. In fact, FAK activates the PI3K survival pathway with the concomitant activation of nuclear factor Kappa B (NF-kB) and induction of inhibitor of apoptosis [116].

Inhibition of PKC-α expression by a synthetic AON inhibits proliferation of C6 glioma cells. In addition, inhibition of PKC-α expression, has also showed growth inhibition of transformed U-87 cells transfected with antisense anti-PKC-α oligonucleotide [116, 117]. PKC-α antisense oligonucleotide treatment in GBM and in A172 GBM cells was accompanied by reduction in PKC-α levels and the induction of wild-type p53 and insulin-like growth factor-binding protein-3 (IGFBP3) 24-72 h after treatment. Grossman *et al.* studied the therapeutic efficacy and toxicity of a phosphorothioate AON (Aprinocarsen, Eli Lilly LY9000003) directed against PKC-α in patients with recurrent malignant gliomas. In phase II study no therapeutic benefit was seen, probably because of tumor growth or the effect of Aprinocarsen on BBB depend on mechanisms unknown [118].

In a murine model, was showed the down-regulation of proliferation rate of C6 glioma cells transfected with antisense AKT2 cDNA construct. Parental C6 cells and C6 cells, transfected with antisense construct, were implanted through lipofectamine complexes. Dominant-negative (DN-AKT2) and antisense AKT2 constructs (AS-AKT2) were transfected into rat C6 glioma cells with elevated endogenous AKT2 expression. AKT2 expression was inhibited in C6 cells transfected with AS-AKT2 but did not significantly change in cells transfected with DN-AKT2. The cell migration distance was reduced in cells transfected with DN-AKT2 or AS-AKT2 compared to the control cells and gelatin zymography showed that the production of MMP2 and MMP9 was inhibited in transfected cells [119]. In a successive study, antisense inhibition of AKT2 caused a reduction of growth rate and proliferation of C6 cells, and in the induction of apoptosis [120]. A recent study was designed to investigate regulation of antisenseAKT2 (AS-AKT2) and dominant-negative AKT2 (DN-AKT2) constructs on proliferation and apoptosis of glioma cell line TJ905. AKT2 expression was significantly inhibited in TJ905 cells transfected with AS-AKT2, while no significant change observed in TJ905 cells transfected with DN-AKT2 [121]. Edwards *et al.* targeted the phosphatidylinositol 3-kinase/protein kinase B (PKB)/Akt and the Ras/MAPK pathway for their involvement in cell survival and cell proliferation. The glioblastoma cell lines U87MG, SF-188, and U251MG were transiently transfected with an antisense oligonucleotide targeting ILK (ILK-AON) alone or in combination with the Raf-1 inhibitor GW5074 or with the MEK inhibitor U0126. GBM cells transfected with ILKAON exhibited reduced levels of ILK and phosphorylated PKB/Akt on Ser473. These results confirmed that combinations targeting ILK and components of the Ras/MAPK pathway result in synergy and could potentially be more effective against GBM than monotherapy [122].

Early preclinical and clinical studies of anti- TGF-beta strategies in the treatment of malignant gliomas have demonstrated that TGF-beta might represent a potential therapeutic target. The TGF-beta2 is overexpressed in more than 90% of high-grade gliomas, and its levels are closely related to tumor progression [123]. Recently the efficacy of specific antisense therapy in down-regulation of TGF-beta2 through stimulation of antitumoral immunosurveillance has been demonstrated [124]. In HTZ-153, HTZ-209 and HTZ-243 glioblastoma and malignant astrocytomas cell lines, TGF-beta2 specific phosphorotioate AONs enhanced lymphocyte proliferation up to 2.5 folds and autologous tumor cytotoxicity up to 60% [124]. In a phase I clinical trial, in grade IV astrocytomas, was assessed the safety of a whole-cell vaccine comprising autologous tumor cells genetically modified by TGF-beta2 antisense vector. Blocking secretion of the immunosuppressive molecule TGF-beta should inhibit one of the mechanisms by which tumor cells evade immune surveillance. TGF-beta2 secretion by the tumor cells used to vaccinate patients was inhibited by 53–98%. There were indications of humoral and cellular immunity induced by the vaccine [125]. TGF-beta2 binds to TGF-beta receptors (TBR) and gives the start to a signaling cascade *via* cytoplasmic signaling mediators (SMAD) into the nucleus, inducing regulation of target gene expression [126]. The antisense compound AP 12009 (AON anti TGF-beta2 mRNA) enhanced the immune cell mediated cytotoxic antitumor response [26, 127, 128]. Trabedersen (AP-12009) is a synthetic antisense oligodeoxynucleotide designed to block the production of TGF- beta2. Preclinical studies demonstrated that trabedersen reduced the secretion of TGF-beta2 in cultured tumor cells and exhibited antitumor activity *ex vivo*. In a phase IIb trial, improved survival was observed in patients with brain tumors who were intratumorally administered trabedersen, compared with patients receiving standard chemotherapy [129]. A randomized, controlled international Phase III study compare trabedersen 10 microM *versus* conventional alkylating chemotherapy in patients with recurrent or refractory anaplastic astrocytoma after standard radio- and chemotherapy [130]. In the entire study population, the 2-year survival rate for 10 mM trabedersen was 39% *vs.* 22% for standard chemotherapy and the median survival of patients with AA exceeded that of standard chemotherapy by 17.4 months (39.1 *vs.* 21.7 months) [131]. Schneider *et al.* have examined a "double-punched" approach to overcome the escape of GBM cells to the immune surveillance, through an active specific immunization (ASI) with Newcastle-Disease-Virus infected tumor cells and blocked the TGF-beta production by delivery of TGF-beta AONs using polybutyl cyanoacrylate NPs [104]. This approach induced a significant decrease in plasma TGF-beta2 level and, at the same time, an increase in rate of high affinity IL-2 receptor (CD25) on lymphocytes and consequently of antitumoral cytotoxicity [104].

During progression of human gliomas, the expression of capillary BM laminins containing $\alpha 4$ chain switches from the predominant laminin-9 into lanimin-8. Effects of antisense inhibition of laminin-8 expression in glioma therapy through an *in vitro* model using human GBM cell lines M059K and U-87MG co-cultured with normal human brain microvascular endothelial cells (HBMVEC) have been observed [132]. Laminin-8 and its receptor, integrins $\alpha 3\beta 1$ and $\alpha 6\beta 1$, are important for the functioning of endothelial cell BMs, which play a role in the maintenance of the BBB. Laminin-8 plays an important role in glioma cell invasiveness, in combination with other proteins associated with glioma progression, such as tenascin-C, MMP-2 and MMP-9. In fact, up-regulation of laminin-8 expression in both glioma cells and glioma-adjacent capillary endothelial cells may reduce glial cell adhesion and enhance migration. The result of the study was a significant reduction of invasion of co-cultures through Matrigel and through the use of morpholino oligos against α4 and β1 chains of laminin-8 [132]. α4 chain is a component of laminin-8 and laminin-9 both of which are overexpressed in approximately 75% of GBM cases and in low grade gliomas, respectively. Laminin α4 chain appears to be an important factor in glioma migration and invasion, both *in vitro* and *in vivo*. The downregulation of laminin α4 chain using AONs inhibits the motility of human glioma cells was shown [133]. Up to now, there are only a few reports of direct conjugation of integrin ligands to oligonucleotides. Some studies [134-136] have reported on various conjugates with linkages of thiazolidine and oxime. An oligonucleotide with an aldehyde moiety was coupled to RGD bearing a cysteine residue for thiazolidine formation, and an aminooxy group for oxime formation. These effectively reacted without a protection strategy and under mild aqueous conditions. Alam *et al.* used a bivalent form of RGD to target an antisense oligonucleotide to αvβ3-expressing human melanoma cells [137]. This oligonucleotide was designed to correct splicing of a luciferase reporter cassette that contained an aberrant intron, thus up-regulating luciferase expression and providing a positive read-out for oligonucleotide delivery. The RGD-oligonucleotide produced a several-fold enhancement of effect as compared to unmodified antisense. The mechanism of uptake seemed to track the caveolar uptake pathway known for the αvβ3 integrin. The surface amino groups of dendrimers can be modified by attaching molecular ligands for selective targeting, for example, to recognize certain receptors expressed in disease sites. Furthermore, decreasing the *in vivo* expression of the integrin β1 subunit by an antisense strategy in the intracranial C6 glioma model leads to an inhibition of glioma associated angiogenesis at the invasive edge of the tumors [137]. Down-regulated *β*1 integrin protein levels *in vivo* probably affect interactions of glioma cells with ECM components, leading to reduced migration along vascular basement membranes.

The antisense modulation of Bcl-2 expression could increase the effectiveness of conventional chemotherapeutic agent [82]. Antisense human bcl-2 cDNA was transfected into human malignant glioma cells. The effects of bcl-2 protein down-regulation on glioma cell morphology, *in vitro* tumor growth, and tumorigenicity in nude mice, as well as chemosensitivity to cisplatin, were studied. Expression of antisense bcl-2 cDNA decreased bcl-2 protein by more than sixfold. Antisense bcl-2 stable transfectants (AS-bcl-2) showed profound morphological change, marked retarded cell growth *in vitro* and significantly increased cytotoxicity of cisplatin [138]. Phase I and II studies are also being done to test G3139 in combination with docetaxel in patients with advanced breast cancer, hormone-refractory prostate cancer, and other solid tumors. By using AONs against the first six codons of the human Bcl-2 gene transfected into malignant glioma cells (Jon52 and Roc GBM cell lines), has been demonstrated a decrease in cell growth and an increase in apoptotic death [139]. Anti-bcl-XL AONs (ISIS 16009, ISIS 16967) in M059K GBM cell lines has showed a valid correlation between reduction of bcl-XL protein expression, induction of intrinsic apoptotic pathway and enhancement of cytotoxic responses to paclitaxel treatment [140].

It has been demonstrated that anti-hTR antisense treatment suppressed tumor cell growth and survival by inducing apoptosis by treating malignant glioma cells (U251-MG cells) in culture with 2-5A AON against telomerase RNA [141]. *In vitro* study showed that human telomerase antisense inhibition induced two different pathways: apoptosis and differentiation in subpopulations of malignant glioma cells that escape from apoptosis. Probably, malignant glioma cells are cells in G2/M phase in contrast with the major percentage of apoptotic cells that are in S phase of the cell cycle. These results indicate that anti-hTR antisense strategy reduced cell cycling and caused an accumulation of cells in G2/M. In the same study, treatment of surviving nonapoptotic cells with AONs against p27 (a cyclin-dependent kinase inhibitors–CDKIs) induced apoptotic cell death; this suggests that p27 may have a protective role for the survival of differentiating glioma cells [141]. Antisense hTR cDNA inhibited TJ905 human glioma cells proliferation and tumor growth *in vitro* and *in vivo* [142]. The efficacy of 2-5A anti-hTR and recombinant adenovirus p53 (Ad5CMV-p53) was demonstrated in malignant glioma cell lines with mutant p53 and *in vivo* GBM murine model [143]. By using an AON against hTER (RNA component of hTR), was detected many telomerase-positive tumor cells in the vast majority of malignant gliomas [144]. Telomerase activation is a critical event in cell immortalization, and an increase in human telomerase reverse transcriptase (hTERT) expression is the key step in activating telomerase. The phosphatase and tensin homolog (PTEN) gene encodes a double-specific phosphatase that induces cell cycle arrest, inhibits cell growth, and causes apoptotic cell death. In a recent study, has been evaluated a combined PTEN and antisense hTERT gene therapy for experimental glioma *in vitro* and *in vivo*. Infection with antisense-hTERT and wild-type-PTEN adenoviruses significantly inhibited human U251 glioma cell proliferation *in vitro* and glioma growth in a xenograft mouse model. The efficacy of therapy was obviously higher in the tumor xenografts infected with both PTEN and antisense hTERT than in the gliomas infected with either agent alone at the same total viral dose [145]. Tamoxifene treatment and the down-regulation of hTR expression by antisense hTR, in human GBM cell line TJ905 and in rat glioma cell line C6, resulted in a significant suppression of cell growth and the induction of cell apoptosis through the inhibition of telomerase activity. The antitumor effect of treatment with TAM alone was found to be mediated in part by the down-regulation of telomerase activity and of PKC and IGF-I expression [146]. In a recent *in vitro* (U87-MG and U373-MG human malignant glioma cells cultures, cell cultures of human astrocytes expressing telomerase with or without oncogenic Ras and *in vivo* study, it was demonstrated that treatment with 2-5A-anti-hTR in the presence of N, N'-bis (2-chloroethyl)-N-nitrosourea (BCNU), cisplatin (CDDP) and temozolomide (TMZ), better enhanced the cell-killing effect if done sequentially [147]. Treatment with 2-5A-anti-hTR in the presence of paclitaxel (PTX) and γ irradiation always enhanced this effect. The enhanced cell-killing resulted by up-regulation of apoptosis. In an animal xenograft study model, TMZ enhanced the anti-tumor effect of 2-5A anti-hTR *via* induction of apoptosis [147]. In a murine model, created by implantation of U-87 MG malignant glioma cells in mouse, it has been demonstrated the suppression of the ability of glioma cells to form tumors in mice [148]. This result was obtained after transfection of antisense VEGF cDNA in an antisense orientation through the recombinant adenoviral vector Ad5CMV-αVEGF. Infection of U-87 MG malignant glioma cells caused the reduction of the endogenous VEGF mRNA level and of the targeted VEGF secretory form production [148].

c-Met, a receptor tyrosine kinase, and its ligand, hepatocyte growth factor, are critical in cellular proliferation, motility, and invasion and are overexpressed in gliomas. It was demonstrated *in vitro* and *in vivo* that AONs against c-Met (FAM-labeled c-Met nonsense AONs-LIPOFECTAMINE PLUS^{TM}) markedly suppressed the expression of c-Met mRNA in human glioma cells and cell growth and enhanced significantly the cytotoxic effect of radiation on human U251 glioma cells in culture [149]. Recently has also been demonstrated that c-Met antisense oligodeoxy nucleotides increase the sensitivity of human glioma cells to paclitaxel. A combination of paclitaxel with antisense c-Met oligodeoxynucleotides inhibited cell growth, induced apoptosis and induced c-Met protein expression in U251 and SHG44 human glioma cells more significantly than either paclitaxel or the oligodeoxynucleotides on their own [150].

6.2. Antisense RNA

Studies in human GBM cells suggest that EGF-RvIII, a mutated EGF-R isoform, promotes tumor growth and progression *via* constitutive activation of the PI3-K/Akt pathway; it also induces up-regulation of cell proliferation *via* the MAPK/ERK1/2 signal transduction pathway. The use of antisense RNA into murine models of human glioma facilitates tumor targeting and induces reduction of HER1/EGFR expression increasing survival [151].

In C6 rat glioma cell line was comparatively transfected *in vitro* with IGF-I antisense (pMT-Anti-IGF-I) or IGF-I receptor antisense (pMT-Anti-IGF-I-R) expression vectors. The transfected antisens cells showed total inhibition of IGF-I. The both transfected cultures of IGF-I and of IGF-IR type were positively stained for MHC-I. Moreover antisense IGF-IR cells as compared to IGF-I-antisense cells showed slightly higher expression of MHC-I. The transfected cells showed also the feature of apoptosis in 60% of cells [152]. In another experiment, the IGF-I antisense or IGF-I TH-transfected rat glioma and mouse embryonic carcinoma cells were hybridized with activated dendritic cells (cell hybridomas). The analysis of the cell hybridomas demonstrated an absence of IGF-I and a presence of MHC-I, MHC-II and B7. The hybridomas composed of IGF-I TH cells fused with activated dendritic cells showed higher levels of MHC-I and B7 than the TH-transfected non-fused cells. The injection of these IGF-I TH hybridomas into

Lewis rats bearing glioma suppressed the tumors in five of six of cases [153]. The expression of insulin-like growth factor I receptor (IGF-IR) antisense mRNA inhibits the growth of C6 rat glioblastoma cells both *in vitro* and *in vivo* and the injection of C6 cells expressing an antisense mRNA to the IGF-IR into syngeneic rats prevents subsequent wild-type tumorigenesis and induces regression of established tumors. IGF-IR antisense cells were weakly tumorigenic, exhibiting a six- to tenfold increase in tumor latency. Injection of IGF-IR antisense C6 cells mildly delayed the development of wild-type tumors, and did not induce the regression of established wild-type C6 tumors in athymic nude mice [154].

Recently, Piwecka *et al.* have implemented the experimental therapy of patients suffering from malignant brain tumors based on application of double-stranded RNA (dsRNA) specific for tenascin-C (TN-C) mRNA. That therapeutic agent, called ATN-RNA, induces RNAi pathway to inhibit the synthesis of TN-C, the extracellular matrix protein which is highly overexpressed in brain tumor tissue [155]. The expression of fibronectin and other extracellular matrix components by glioma is directly linked to the expression of integrins on cell surface. Binding of integrins with its ligand triggers downstream signal transduction cascades, of which activation of Src kinase has been deemed to be important for tumor development [156, 157]. Fibronectin is a ligand for the integrin β1 receptors and owns a pivotal role in cell signaling pertinent to tumor progression and in the proliferation of different types of cancers, including glioblastoma [158, 159].

7. CONCLUSIONS

The treatment of brain cancer is one of the most difficult challenges in neurosurgery and oncology. Malignant gliomas involve, in their progression, multiple aberrant signaling pathways and the BBB restricts the delivery of many chemotherapeutic agents. Multimodal therapeutic approaches and molecular-based targeted therapies have successfully been applied in cancers, but their efficacy remains low in cerebral gliomas [33]. The studies reported are few and limited to an experimental phase. Evidently, exist an objective lack of information about the real efficacy and the long term safe oh these novel therapeutic approaches. There are several factors underlying the disappointing results in brain cancer therapeutics including limited tumor cell drug uptake, intracellular drug metabolism, inherent tumor sensitivity to chemotherapy, and cellular mechanisms of resistance. Understanding the genetic bases of gliomas and of the invasive behavior, in particular the differentiated gene expression in distinct areas of the same tumor during glioma progression, may suggest new molecular targets to overcome the mechanisms of multi-drug-resistance of the actual therapeutic approaches to gliomas and to attack at the same time different crucial biological events of gliomagenesis. An optimal realization of a system that overcomes the problems associated with novel strategics in brain tumor treatments requires the identification of specific neoplastic markers, the development of technology for the biomarker-targeted delivery of therapeutic agents, and the simultaneous capability of avoiding biological and biophysical barriers. Considering the multitude of molecular entities and signaling pathways regulating the proliferation and cellular survival/cell death [5, 33], the inhibition of a singular target gene, in general of growth factor and its receptor, tyrosine kinase, principally IGF-I-R and EGF-R could be sufficient to a suppress neoplastic progression [29, 50, 57, 58]. It is clear that the cross-talk between signal transduction pathways of growth factor receptor play the potential role of targeting a single receptor like IGF-I-R [59, 72, 160]. Concerning IGF-I-R, the neoplastic progression in malignant tumors, including glioma, is associated with increased expression of this receptor, although this over-expression seems to be less common than that for ERBB2 in malignant tumors. Even if IGF-I-R's level of expression is low, this receptor seems to have a principal role in regulating proliferation and differentiation of malignant tumors [59, 60, 69].

ECM proteases, cell adhesion molecules and their related signaling pathways show an important role in glioma cell migration and invasion and could be selectively attack to inhibit the glioma invasive rim. In this complex field it seems to be very important to improve specific selective drugs delivery systems to lead to diffusion of drugs, antisense oligonucleotides, RNA interference, engineered monoclonal antibodies and other therapeutic molecules into CNS overcoming BBB. Particularly interesting is the role of antisense RNA and DNA oligonucleotides in cell-specific gene expression regulation and phenotype determination. Recently long more than 21 nucleotides, RNA regulator molecules named long non-coding RNA, have been discovered. All these nucleotidic sequences could easily move into the cell and among different cells to regulate gene expression and cellular phenotype.

Many antisense oligonucleotides have shown sharp *in vitro* reduction in target gene expression and promising activity against a variety of human malignancies. Several experimental *in vitro* and *in vivo* studies in cell lines cultures and animal models showed inhibition of genes involved in cell proliferation, apoptosis and angiogenesis. Clinical trials demonstrated a good tolerability and no increment of toxicity of antisense oligonucleotides [5]. Ther are several future interesting molecular target for specific antisense glioma therapy. The inhibitor of apoptosis family of protein constitute a group of apoptosis suppressors (XIAP, c-IAP1, c-IAP2, NAIP, livin) that act as direct inhibitors of certain caspases. Other possible targets, tested in preclinical study, may be clusterin, eIF-4E (eukaryotic initiation factor-4E), metalloproteinases, interleukin and other key molecules involved in invasion and angiogenesis. Heat shock proteins (HSP) are among the most conserved proteins known and include a number of different families. HSP expression in certain cancer types correlates with poor prognosis amd resistance to treatment.

Nanotechnology provides a unique opportunity to combat cancer on the molecular scale through careful engineering of nanomedicines to specifically interact with cancer cells and inhibit cancer cell function. Beyond use of NPs as nanomedicines, more focus should be placed on use of NPs as tools to learn more about cancer biology and failure of treatments. It is also possible to take into neoplastic tissue novel selective contrast enhancement molecules to visualize brain tumor and to study *in vivo* all of its characteristics, such as cellular proliferation, angiogenesis, necrosis, tumor-safe tissue interface, edema. It is known that in malignant brain tumors BBB is normally breakdown. Nanoparticles-based delivery systems could increase the overcoming of BBB by the drug with a targeted-cell specificity modality. This approach permits the use of a lower dose of drug, a selective drug delivery to target tumor cells, both into the central core of tumor and into the distal foci of tumor cells within areas often characterized from integrity of BBB. Obviously more detailed researches about knowledge of cellular biology, and cancer biology will permit important progress in the implications of nanotechnology in cancer and naturally in malignant gliomas treatment [38, 39, 99].

Antisense targeted therapy is very fascinating and probably will be one of the most important strategies to treat high-grade glioma, into a multimodal approach for brain tumor treatment. Moreover, clinical results of this molecular approach are currently very promising despite the relatively short time of application of this kind of strategy in brain tumors in last five years, using the single molecular target chosen for targeted therapy - TGF-beta and especially IGF-I (median survival between 18-24 months); the antisense anti TGF-beta and anti IGF-I approaches induce immune anti-tumor response mediated generally by activated CD8+lymphocytes which bypass the anatomic and physiological barriers existing in CNS [26, 62, 63, 123, 128, 130]. Another important aspect to valuate is represented by the similarity between the overall survival and progression free survival in glioma patients treated according the standard protocol with surgery, chemotherapy and radiotherapy *versus* new experimental molecular multimodal protocol. Moreover, a careful assessment of future controlled studies is needed to confirm antisense-mediated down-regulation of the target protein in a large number of patients. The potential value of antisense therapy depend, principally, on our increasing knowledge of genes and their functions. On the base of these preliminary results and the limits of actual standard therapeutic protocol, we think that antisense therapy may be an interesting approach to modify the biological development of gliomas, probably trying to modulate crucial pathways of gliomagenesis during precocious steps of tumor progression and possibly two or more molecular targets of the same pathway or of two different pathways.

ACKNOWLEDGEMENT

Declared none.

CONFLICT OF INTEREST

The author(s) confirm that this chapter content has no conflict of interest.

DISCLOSURE

Declared none.

REFERENCES

[1]   Louis DN, Oghaki H, Wiestler OD, *et al.* The 2007 WHO classification of tumours of the central nervous system. Acta Neuropathol 2007; 114(2); 97-109.

[2]   Pouratian N, Schiff D. Management of low-grade glioma. Curr Neurol Neurosci Rep 2010; 10(3); 224-31.

[3]   Weller M. Novel diagnostic and therapeutic approaches to malignant glioma. Swiss Med Wkly 2011; 141.

[4]   Pignatti F., van den Bent M, Curran D, *et al.* Prognostic factors for survival in adult patients with cerebral low-grade glioma. J Clin Oncol 2002; 20(8); 2076-84.

[5]   Caruso G, Caffo M, Raudino G, Alafaci C, Salpietro FM, Tomasello F. Antisense oligonucleotides as innovative therapeutic strategy in the treatment of high-grade gliomas. Recent Pat CNS Drug Discov 2010, 5(1); 53-69.

[6]   Lim SK, Llaguno SR, McKay RM, Parada LF. Glioblastoma multiforme: a perspective on recent findings in human cancer and mouse models. BMB Rep 2011; 44(3); 158-64.

[7]   Bredel M, Scholtens DM, Yadav AK *et al.* NFKBIA deletion in glioblastomas. N Engl J Med 2011; 364(7); 627-7.

[8]   Yan H, Parsons DW, Jin G, *et al.* IDH1 and IDH2 mutations in gliomas. N Engl J Med 2009; 360(8); 765-73.

[9]   Watanabe T, Nobusawa S, Kleihues P, Ohgaki H. IDH1 mutations are early events in the development of astrocytomas and oligodendrogliomas. Am J Pathol 2009; 174(4); 1149-53.

[10]   Caffo M, Germanò A, Caruso G, *et al.* An immunohistochemical study of extracellular matrix proteins laminin, fibronectin and type IV collagen in paediatric glioblastoma multiforme. Acta Neurochir (Wien) 2004; 146(10); 1113-8.

[11]   Butowski N, Chang SM. Small molecule and monoclonal antibody therapies in neurooncology. Cancer Control 2005; 12(2); 116-24.

[12]   Rubenstein JL, Nicolas JF, Jacob F. Nonsense RNA: a tool for specifically inhibiting the expression of a gene *in vivo*. CR Acad Sci III 1984; 299(8); 271-4.

[13]   Izant JG, Weintraub H. Science 1985; 229(4711); 345-52.

[14]   Dias N, Stein CA. Antisense oligonucleotides: basic concepts and mechanisms. Mol Cancer Ther 2002; 1(5); 347-55.

[15]   Helene C. Control of oncogenes expression by antisense nucleic acids. Eur J Cancer 1994; 30A(11); 1721-6.

[16]   Shevelev A, Burfeind P, Schulze E, *et al.* Potential triple helix-mediated inhibition of IGF-I gene expression significantly reduces tumorigenicity of glioblastoma in an animal model. Cancer Gene Ther 1997; 4(2); 105-12.

[17]   Morishita R, Higaki J, Tomita N, Ogihara T. Application of transcription factor "decoy" strategy as means of gene therapy and study of gene expression in cardiovascular disease. Circ Res 1998; 82(10); 1023-8.

[18]   Sharp PA. RNA interference. Genes Dev 2001; 15(5); 485-90.

[19]   Boado RJ. RNA interference and nonviral targeted gene therapy of experimental brain cancer. NeuroRx 2005; 2(1); 139-50.

[20]   Ring BZ, Roberts JW. Function of a nontranscribed DNA strand site in transcription elongation. Cell 1994;78(2); 317-24.

[21]   Abaza MS, Al-Attiyah RJ, Al-Saffar AM, Al-Sawan SM, Moussa NM. Antisense oligodeoxynucleotide directed against c-myb has anticancer activity and potentiates the

antiproliferative effect of conventional anticancer drugs acting by different mechanisms in human colorectal cells. Tumour Biol 2003; 24(5); 241-57.

[22] Nahta R, Esteva FJ. Bcl-2 antisense oligonucleotides: a potential novel strategy for the treatment of breast cancer. Semin Oncol 2003; 30(5 suppl 16); 143-9.

[23] Morioka CY, Saito S, Machado MC, *et al.* Antisense therapy specific to mutated K-ras gene in hamster pancreatic cancer model. Can it inhibit the growth of 5-FU and MMC-resistant metastatic and remetastatic cell lines? *In Vivo* 2004; 18(2); 113-7.

[24] Ly A, Duc HT, Kalamarides M, *et al.* Human glioma cells transformed by IGF-I triple helix technology show immune and apoptotic characteristics determining cell selection for gene therapy of glioblastoma. Mol Pathol 2001; 54(4); 230-9.

[25] Riedel F. Expression of VEGF and inhibition of tumor angiogenesis by abrogation of VEGF in head and neck cancer. Laryngorhinootologie 2003; 82(6); 436-7.

[26] Schlingensiepen KH, Schlingensiepen R, Steinbrecher A, *et al.* Targeted tumor therapy with the TGF-beta 2 antisense compound AP 12009. Cytokine Growth factor Rev 2006; 17(1-2); 129-39.

[27] Trojan J, Blossey BK, Johnson TR, *et al.* Loss of tumorigenicity of rat glioblastoma directed by episome-based antisense cDNA transcription of insulin-like growth factor I. Proc Natl Acad Sci U S A 1992; 89(11); 4874-8.

[28] Trojan J, Johnson TR, Rudin SD, Ilan J, Tykocinski ML, Ilan J. Treatment and prevention of rat glioblastoma by immunogenic C6 cells expressing antisense insulin-like growth factor I RNA. Science 1993; 259(5091); 94-7.

[29] Resnicoff M, Sell C, Rubini M, *et al.* Rat glioblastoma cells expressing an antisense RNA to the insulin-like growth factor-1 (IGF-1) receptor are nontumorigenic and induce regression of wild type tumors. Cancer Res 1994; 54(8); 2218-22.

[30] Fakhrai H, Dorigo O, Shawler DL, *et al.* Eradication of established intracranial rat gliomas by transforming growth factor beta antisense gene therapy. Proc Natl Acad Sci U S A 1996; 93(7); 2909-14.

[31] Tamm I. Antisense therapy in malignant disease: status quo and quo vadis? Clin Sci 2006; 110(4); 427-42.

[32] Wacheck V, Zangemeister-Wittke U. Antisense molecules for targeted cancer therapy. Crit Rev Oncol Hematol 2006; 59(1); 65-73.

[33] Caruso G, Caffo M, Raudino G, Alafaci C, Tomasello F. New therapeutic strategies in gliomas treatment. In: Abujamra AL , Ed. Brain tumors - current and emerging therapeutic strategies. 1st ed. Rijeka: INTECH 2011; pp. 281-306.

[34] Pawlak W, Zolnierek J, Sarosiek T, Szczylik C. Antisense therapy in cancer. Cancer Treat Rev 2000; 26(5); 333-50.

[35] Dean NM, Bennett FC. Antisense oligonucleotide-based therapeutics for cancer. Oncogene 2003; 22(56); 9087-96.

[36] Summerton J. Morpholino antisense oligomers: the case for an RNase H-independent structural type. Biochim Biophys Acta 1999; 1489(1); 141-58.

[37] Partridge M, Vincent A, Matthews P, Puma J, Stein D, Summerton J. A simple method for delivering morpholino antisense oligos into the cytoplasm of cells. Antisense Nucleic Acid Drug Dev 1996; 6(3); 169-75.

[38] Caruso G, Raudino G, Caffo M, *et al.* Nanotechnology platforms in diagnosis and treatment of primary brain tumors. Recent Pat Nanotechnol 2010; 4(2); 119-24.

[39]   Caruso G, Caffo M, Raudino G, Tomasello C, Alafaci C, Tomasello F. In: Souto EG, ed. Patenting Nanomedicines. Berlin Heidelberg: Springer-Verlag; 2012; pp. 167-204.

[40]   Carrasco RA, Stamm NB, Marcusson E, Sandusky G, Iversen P, Patel BK. Antisense inhibition of survivin expression as a cancer therapeutic. Mol Cancer Ther 2011, 10(2); 221-32.

[41]   Zamore PD, Tuschl T, Sharp PA, Bartel DP. RNAi: double-stranded RNA directs the ATP-dependent cleavage of mRNA at 21 to 23 nucleotide intervals. Cell 2000; 101(1); 25-33.

[42]   Lages E, Ipas H, Guttin A, *et al.* MicroRNAs: molecular features and role in cancer. Front Biosci 2012; 17; 2508-40.

[43]   Hutvagner G, Simar MJ. Argonaute proteins: key players in RNA silencing. Nat Rev Mol Cell Biol 2008; 9(1); 22-32.

[44]   Chi AS, Sorensen AG, Jain RK, *et al.* Angiogenesis as a therapeutic target in malignant gliomas. Oncologist 2009; 14(6); 621-36.

[45]   Yin S, Van Meir EG. p53 pathway alterations in brain tumors. In: Van Meir EG, ed. CNS cancer: models, markers, prognostic factors, targets and therapeutic approaches. New York: Humana Press (Springer); 2009; pp. 283-314.

[46]   Laver T, Nozell S, Benveniste EN. The NF-kB signaling pathway in GBMs: implications for apoptotic and inflammatory responses and exploitation for therapy. In: Van Meir EG, ed. CNS cancer: models, markers, prognostic factors, targets and therapeutic approaches. New York: Humana Press (Springer); 2009; pp. 1011-36.

[47]   Rich JN. The role of transforming growth factor-β in primary brain tumors. Front Biosci 2003; 8; E245-60.

[48]   Platten M, Wick W, Weller M. Malignant glioma biology: role for TGF-beta in growth, motility, angiogenesis, and immune escape. Microsc Res Tech 2001; 52(4), 401-10.

[49]   Haas-Kogan DA, Prados MD, Tihan T, *et al.* Epidermal growth factor receptor, protein kinase B/Akt, and glioma response to erlotinib. J Natl Cancer Inst 2005; 97(12); 880-7.

[50]   Sampson JH, Archer GE, Mitchell DA, *et al.* An epidermal growth factor receptor variant III-targeted vaccine is safe and immunogenic in patients with glioblastoma multiforme. Mol Cancer Ther 2009; 8(10); 2773-9.

[51]   Walker PR, Prins RM, Dietrich P-Y, Liau LM. Harnessing T-cell immunity to target brain tumors. In: Van Meir, EG., editor. CNS Cancer: Models, Markers, Prognostic Factors, Targets and Therapeutic Approaches. 1. New York: Humana Press (Springer); 2009. p. 1165-1217.

[52]   Verhaak RGW, Hoadley KA, Purdom E, *et al.* An integrated genomic analysis identifies clinically relevant subtypes of glioblastoma characterized by abnormalities in PDGFRA, IDH1, EGFR and NF1. Cancer Cell 2010;17(1); 98-110.

[53]   Kilic T, Alberta JA, Zdunek PR, *et al.* Intracranial inhibition of platelet-derived growth factormediated glioblastoma cell growth by an orally active kinase inhibitor of the 2-phenylaminopyrimidine class. Cancer Res 2000; 60(18); 5143-50.

[54]   McLean GW, Carragher NO, Avizienyte E, Evans J, Brunton VG, Frame MC. The role of focal-adhesion kinase in cancer. A new therapeutic opportunity. Nat Rev Cancer 2005; 5(7); 505-15.

[55]   Hanks SK, Ryzhova L, Shin NY, Brabek J. Focal adhesion kinase signaling activities and their implications in the control of cell survival and motility. Front Biosci 2003; 8; d982-96.

[56]   Parsons JT. Focal adhesion kinase: the first ten years. J Cell Sci 2003; 116(8); 1409-16.

[57] Baserga R, Sell C, Porcu P, Rubini M. The role of the IGF-I receptor in the growth and transformation of mammalian cells. Cell Prolif 1994; 27(2); 63-71.

[58] Sun X, Tu X, Baserga R. A mechanism for cell size regulation by the insuline and insulin-like growth factor-I receptors. Cancer Res 2006; 66(23); 11106-9.

[59] Pollak MN, Schernhammer ES, Hankinson SE. Insulin-like growth factors and neoplasia. Nat Rev Cancer 2004; 4(7); 505-18.

[60] Baserga R. The insulin-like growth factor-I receptors as a target for cancer therapy. Expert Opin Ther Targets 2005; 9(4); 753-68.

[61] Wrensch M, Rice T, Miike R, *et al.* Diagnostic, treatment, and demographic factors influencing survival in a population-based study of adult glioma patients in the San Francisco Bay Area. Neuro Oncol 2006; 8(1); 12-26.

[62] Trojan J, Cloix JF, Ardourel MY, Chatel M, Anthony DD. Insulin-like growth factor type I biology and targeting in malignant gliomas. Neuroscience 2007; 145(3); 795-811.

[63] Trojan J, Pan YX, Wei MX, *et al.* Methodology for anti-gene anti-IGF-I therapy of malignant tumours. Chenother Res Pract 2012; 2012; 721873.

[64] Adhami VM, Afaq F, Mukhtar H. Insulin-like growth factor-I axis as a pathway for cancer chemoprevention. Clin Cancer Res 2006; 12(19); 5611-4.

[65] Kooijman R. Regulation of apoptosis by insulin-like growth factor (IGF)-I. Cytokine Growth Factor Rev 2006; 17(4), 305-23.

[66] Trojan J, Johnson TR, Rudin SD, *et al.* Gene therapy of murine teratocarcinoma: separate functions for insulin-like growth factors I and II in immunogenicity and differentiation. Proc Natl Acad Sci U S A 1994; 91(13); 6088-92.

[67] Trojan J, Duc HT, Lafarge-Frassynet C, *et al.* Immunotherapy of tumors expressing IGF-I. C R Seances Soc Biol Fil 1996; 190(1); 165-9.

[68] Lafarge-Frassynet C, Duc HT, Frassynet C, *et al.* Antisense insulin-like growth factor I transferred into a rat hepatoma cell line inhibits tumorigenesis by modulating major histocompatibility complex I cell surface expression. J Hepatol 1998; 29(5); 807-18.

[69] Jiang R, Mircean C, Shmulevich I, *et al.* (2006) Pathway alterations during glioma progression revealed by reverse phase protein lysate arrays. Proteomics 2006; 6(10); 2964-71.

[70] Krasilnikov MA. Phosphatidylinositol-3 kinase dependent pathways: the role in control of cell growth, survival, and malignant transformation. Biochemistry 2000; 65(1); 59-67.

[71] Beckner ME, Gobbel GT, Abounader R, *et al.* Glycolytic glioma cells with active glycogen synthase are sensitive to PTEN and inhibitors of PI3K and gluconeogenesis. Lab Invest 2005; 85(12); 1457-70.

[72] Vignot S, Faivre S, Aguirre D, Raymond E. mTor-targeted therapy of cancer with rapamycin derivatives. Ann Oncol 2005; 16(4); 525-37.

[73] Cheng JQ, Ruggeri B, Klein WM, *et al.* Amplification of AKT2 in human pancreatic cancer cells and inhibition of AKT2 expression and tumorigenicity by antisense RNA. Proc Natl Acad Sci U S A 1996; 93(8); 3636-41.

[74] Gupta AK, McKenna WG, Weber CN, *et al.* Local recurrence in head and neck cancer: relationship to radiation resistance and signal transduction. Clin Cancer Res 2002; 8(1); 885-92.

[75] Tsurutani J, Fukuoka J, Tsurutani H, *et al.* Evaluation of two phosphorylation sites improves the prognostic significance of Akt activation in non-small-cell lung cancer tumors. J Clin Oncol 2006; 24(2); 306-14.

[76]   Ardourel M, Blin M, Moret J, *et al.* A new putative target for antisense gene therapy of glioma. Cancer Biol Ther 2007; 6(5); 719-23.

[77]   Ma YY, Wei SJ, Lin YC, *et al.* PIK3CA as an oncogenes in cervical cancer. Oncogene 2000; 19(23); 2739-44.

[78]   Samuels Y, Ericson K. Oncogenic PI3K and its role in cancer. Curr Opin Oncol 2006; 18(1); 77-82.

[79]   Galanis E, Buckner JC, Maurer MJ, *et al.* Phase II trial of temsirolimus (CCI-779) in recurrent glioblastoma multiforme: a North Central Cancer Treatment Group Study. J Clin Oncol 2005; 23(23); 5294-304.

[80]   Galanis E, Jaeckle KA, Maurer MJ, *et al.* Phase II trial of vorinostat in recurrent glioblastoma multiforme: a north central cancer treatment group study. J Clin Oncol 2009; 27(19); 2052-8.

[81]   Balendran A, Hare GR, Kieloch A, *et al.* Further evidence that 3-phosphoinositide-dependent protein kinase-1 (PDK1) is required for the stability and phosphorylation of protein kinase C (PKC) isoforms. FEBS Lett 2000; 484(3); 217-23.

[82]   Pirollo KF, Rait A, Sleer SL, *et al.* Antisense therapeutics: from theory to clinical practice. Pharmacol Ther 2003; 99(1); 55-77.

[83]   Nemunaitis J, Holmlund JT, Kraynak M, *et al.* Phase I evolution of ISIS 3521, an antisense oligodeoxynucleotide to protein kinase-c-α, in patients with advanced cancer. J Clin Oncol 1999; 17(11); 3586-95.

[84]   Caruso G, Caffo M, Raudino G, Raudino F, Venza M, Tomasello F. In: Antisense Oligonucleotides in the Treatment of Malignant Gliomas. Erdmann VA, Barciszewski (eds.) From Nucleic Acid Sequences to Molecular Medicine, RNA Technologies. Springer-Verlag Berlin Heidelberg 2012, pp. 215-46.

[85]   Hynes RO. Integrins: bidirectional, allosteric signaling machines. Cell 2002;110(6); 673-87.

[86]   Moschos SJ, Drogowski LM, Reppert SL, Kirkwood JM. Integrins and cancer. Oncology (Williston Park) 2007;21(9); 13-20.

[87]   Stupack DG, Cheresh DA. Integrins and angiogenesis. Curr Top Dev Biol 2004; 64; 207-38.

[88]   Gladson CL. Expression of integrin alpha v beta 3 in small blood vessels of glioblastoma tumors. J Neuropathol Exp Neurol 1996; 55(11); 1143-9.

[89]   Oliver L, Tremblais K, Guriec N, *et al.* Influence of bcl-2-related proteins on matrix metalloproteinase expression in a rat glioma cell line. Biochem Biophys Res Commun 2000, 273(2); 411-6.

[90]   Jansen B, Schlagbauer-Wadl H, Brown BD, *et al.* bcl-2 antisense therapy chemosensitizes human melanoma in SCID mice. Nat Med 1998; 4(2), 232-4.

[91]   Krajewski S, Krajewska M, Ehrmann J, *et al.* Immunohistochemical analysis of Bcl-2, Bcl-X, Mcl-1, and Bax in tumors of central and peripheral nervous system origin. Am J Pathol 1997; 150(3); 805-14.

[92]   McDonald FE, Ironside JW, Gregor A, *et al.* The prognostic influence of bcl-2 in malignant glioma. Br J Cancer 2002; 86(12), 1899-904.

[93]   Fels C, Schafer C, Huppe B, *et al.* Bcl-2 expression in higher-grade human glioma: a clinical and experimental study. J Neurooncol 2000; 48(3); 207-16.

[94]   Gross A, McDonnell JM, Korsmeyer SJ. BCL-2 family members and the mitochondria in apoptosis. Genes Dev 1999; 13(15); 1899-911.

[95] Lingner J, Hughes TR, Shevchenko A, Mann M, Lundblad V, Cech TR. Reverse transcriptase motifs in the catalytic subunit of telomerase. Science 1997; 276(5132); 561-7.

[96] George J, Banik NL, Ray SK. Combination of hTERT knockdown and IFN-gamma treatment inhibited angiogenesis and tumor progression in glioblastoma. Clin Cancer Res 2009; 15(23); 7186-95.

[97] You Y, Geng X, Zhao P, *et al.* Evaluation of combination gene therapy with PTENand antisense hTERT for malignant glioma *in vitro* and xenografts. Cell Mol Life Sci 2007; 64(5); 621-31.

[98] Heidel JD, Davis ME. Clinical developments in nanotechnology for cancer therapy. Pharm Res 2011; 28(2); 187-99.

[99] Caruso G, Caffo M, Alafaci C, *et al.* Could nanoparticle systems have a role in the treatment of cerebral gliomas? Nanomedicine 2011; 7; 744-52.

[100] Panyam J, Labhasetwar V. Biodegradable nanoparticles for drug and gene delivery to cells and tissue. Adv Drug Deliv Rev 2003; 55(3); 329-47.

[101] Oishi M, Kataoka K, Nagasaki Y. pH-responsive three-layered PEGylated polyplex micelle based on a lactosylated ABC triblock copolymer as a targetable and endosome-disruptive nonviral gene vector. Bioconjug Chem 2006; 17(3); 677-88.

[102] Trollet C, Scherman D, Bigey P. Delivery of DNA into muscle for treating systemic diseases: advantages and challenges. Methods Mol Biol 2008; 423; 199-214.

[103] O'Brien JA, Holt M, Whiteside G, Lummis SC, Hastings MH. Modifications to the hand-held Gene Gun: improvements for *in vitro* biolistic transfection of organotypic neuronal tissue. J Neurosci Methods 2001; 112(1); 57-64.

[104] Schneider T, Becker A, Ringe K, Reinhold A, Firsching R, Sabel BA. Brain tumor therapy by combined vaccination and antisense oligonucleotide delivery with nanoparticles. J Neuroimmunol 2008; 195(1-2); 21-7.

[105] Liu L, Zern M, Lizarzaburu M, Nantz MH, Wu J. Poly(cationic lipid)-mediated *in vivo* gene delivery to mouse liver. Gene Ther 2003; 10(2); 180-7.

[106] Cohen-Sacks H, Najajreh Y, Tchaikovski V, *et al.* Novel PDGFbetaR antisense encapsulated in polymeric nanospheres for the treatment of restenosis. Gene Ther 2002; 9(23); 1607-16.

[107] Guo S, Huang Y, Zhang W, *et al.* Ternary complexes of amphiphilic polycaprolactone-graft-poly(N,N-dimethylaminoethylmethacrylate), DNA and polyglutamic acid-graft-poly(ethylene glycol) for gene delivery. Biomaterials 2011; 32(18); 4283-92.

[108] Kang C, Yuan X, Li F, *et al.* Evaluation of folate-PAMAM for the delivery of antisense oligonucleotides to rat C6 glioma cells *in vitro* and *in vivo*. J Biomed Mater Res A 2010; 93(2); 585-94.

[109] Tian XX, Zhang YG, Du J, *et al.* Effects of cotransfection of antisense-EGFR and wild-type PTEN cDNA on human glioblastoma cells. Neuropathology 2006, 26(3); 178-87.

[110] Lin ZX, Yang LJ, Huang Q, *et al.* Inhibition of tumor-induced edema by antisense VEGF is mediated by suppressive vesiculo-vacuolar organelles (VVO) formation. J Cancer Sci 2008; 99(12); 2540-6.

[111] Yang L, Lin Z, Huang Q, *et al.* Effect of vascular endothelial growth factor on remodeling of C6 glioma tissue *in vivo*. J Neurooncol 2011; 103(1); 33-41.

[112] Resnicoff M, Tjuvajev J, Rotman HL, *et al.* Regression of C6 rat brain tumors by cells expressing an antisense insulin-like growth factor I receptor. J Exp Ther Oncol 1996; 1(6); 385-9.

[113] Andrews DW, Resnicoff M, Flanders AE, *et al.* Results of a pilot study involving the use of an antisense oligodeoxynucleotide directed against the insuline-like growth factor type I receptor in malignant astrocytomas. J Clin Oncol 2001; 19(8); 2189-200.

[114] Liu TJ, LaFortune T, Honda T, *et al.* Inhibition of both adhesion kinase and insulin-like growth factor-I receptor kinase suppresses glioma proliferation *in vitro* and *in vivo*. Mol Cancer Ther 2007; 6(4); 1357-67.

[115] Wu ZM, Yuan XH, Jiang PC, *et al.* Antisense oligonucleodes targeting the focal adhesion kinase inhibit proliferation, induce apoptosis and cooperate with cytotoxic drugs in human glioma cells. J Neurooncol 2006; 77(2); 117-23.

[116] Baltuch GH, Dooley NP, Rostworowski KM. Protein kinase C isoform alpha overexpression in C6 glioma cells and its role in cell proliferation. J Neurooncol 1995; 24(3); 241-50.

[117] Pollack IF, Kawecki S, Lazo JS. Blocking of glioma proliferation *in vitro* and *in vivo* and potentiating the effects of BCNU and cisplatin: UCN-01, a selective protein kinase C inhibitor. J Neurosurg 1996; 84(6), 1024-32.

[118] Grossman SA, Alavi JB, Supko JG, *et al.* Efficacy and toxicity of the antisense oligonucleotide aprinocarsen directed against protein kinase C-α delivered as a 21-day continuous intravenous infusion in patients with recurrent high-grade astrocytomas. Neuro Oncol 2005; 7(1); 32-40.

[119] Pu P, Kang C, Li J, Jiang H. Antisense and dominant-negative AKT2 cDNA inhibits glioma cell invasion. Tumour Biol 2004; 25(4); 172-8.

[120] Pu P, Kang C, Li J, Jiang H, Cheng J. The effects of antisense AKT2 RNA on the inhibition of malignant glioma cell growth *in vitro* and *in vivo*. J Neurooncol 2006; 76(1); 1-11.

[121] Kang CS, Pu PY, Li J, Wang GX. Inhibitory effect of antisense and dominant-negative AKT2 constructs on proliferation of glioma cell line TJ905. Ai Zheng 2004; 23(11); 1267-72.

[122] Edwards LA, Verreault M, Thiessen B, *et al.* Combined inhibition of the phosphatidylinositol 3-kinase/Akt and Ras/mitogen-activated protein kinase pathways results in synergistic effects in glioblastoma cells. Mol Cancer Ther 2006; 5(3); 645-54.

[123] Hau P, Jachimczak P, Schlaier J, Bogdahn U. TGF-β2 signaling in high-grade gliomas. Curr Pharm Biotechnol 2011; 12(12); 2150-7.

[124] Jachimczak P, Hessdorfer B, Fabel-Schulte K, *et al.* Transforming growth factor-beta mediated autocrine growth regulation of gliomas with phosphorothioate antisense oligonucleotides. Int J Cancer 1996; 65(3); 332-7.

[125] Fakhrai H, Mantil JC, Liu L, *et al.* Phase I clinical trial of a TGF-b antisense-modified tumor cell vaccine in patients with advanced glioma. Cancer Gene Ther 2006; 13(12); 1052-60.

[126] Nickl-Jockschat T, Arslan F, Doerfelt A, Bogdahn U, Bosserhoff A, Hau P. An imbalance between Smad and MAPK pathways is responsible for TGF-beta tumor promoting effects in high-grade gliomas. Int J Oncol 2007; 30(2); 499-507.

[127] Hau P, Jachimczak P, Schlingensiepen R, *et al.* Inhibition of TGF-β2 with AP 12009 in recurrent malignant gliomas: From preclinical to phase I/II studies. Oligonucleotides 2007; 17(2); 201-12.

[128] Schlingensiepen KH, Fischer-Blass B, Schmaus S, Ludwig S. Antisense therapeutics for tumor treatment: the TGF-beta2 inhibitor AP 12009 in clinical development against malignant tumors. Recent Results Cancer Res 2008; 177; 137-50.

[129] Vallieres L. Trabedersen, a TGF beta 2-specific antisense oligonucleotides for the treatment of malignant gliomas and other tumors overexpressing TGF beta 2. I Drugs 2009; 12(7); 445-53.

[130] Hau P, Jachimczack P, Bogdahn U. Treatment of malignant gliomas with TGF-beta2 antisense oligonucleotides. Expert Rev Anticancer Ther 2009; 9(11); 1663-74.

[131] Boghdan U, Hau P, et al. Targeted therapy for high-grade glioma with the TGF-b2 inhibitor trabedersen: results of a randomized and controlled phase IIb study. Neuro-Oncology 2011; 13(1); 132-142.

[132] Khazenzon NM, Ljubimov AV, Lakhter AJ, et al. Antisense inhibition of laminin-8 expression reduces invasion of human gliomas in vitro. Mol Cancer Ther 2003; 2(10); 985-94.

[133] Nagato S, Nakagawa K, Harada H, et al. Downregulation of laminin α4 chain expression inhibits glioma invasion in vitro and in vivo. Int J Cancer 2005; 117(1); 41-50.

[134] Edupuganti OP, Singh Y, Defrancq E, Dumy P. New strategy for the synthesis of 3',5'-bifunctionalized oligonucleotides conjugates through sequential formation of chemoselective oxime bonds. Chemistry 2004; 10(23); 5988-95.

[135] Forget D, Boturyn D, Defrancq E, Lhomme J, Dumy P. Highly efficient synthesis of peptide-oligonucleotide conjugates: chemoselective oxime and thiazolidine formation. Chemistry 2001; 7(18); 3976-84.

[136] Spinelli N, Edupuganti OP, Defrancq E, Dumy P. New solid support for the synthesis of 3'-oligonucleotide conjugates through glyoxylic oxime bond formation. Org Lett 2007; 9(2); 219-22.

[137] Alam MR, Dixit V, Kang H, et al. Intracellular delivery of an anionic antisense oligonucleotides via receptor-mediated endocytosis. Nucleic Acids Res 2008; 36(8); 2764-76.

[138] Zhu C, Li YB, Wong MC. Expression of antisense bcl-2 cDNA abolishes tumorigenicity and enhances chemosensitivity of human malignant glioma cells. J Neurosci Res 2003; 74(1); 60-6.

[139] Julien T, Frankel B, Longo S, et al. Antisense-mediated inhibition of the bcl-2 gene induces apoptosis in human malignant glioma. Surg Neurol 2000; 53(4); 360-9.

[140] Guensberg P, Wacheck V, Lucas T, et al. Bcl-xL antisense oligonucleotides chemosensitize human glioblastoma cells. Chemotherapy 2002; 48(4); 189-95.

[141] Kondo S, Tanaka Y, Kondo Y, et al. Antisense telomerase treatment: induction of two distinct pathways, apoptosis and differentiation. FASEB J 1998; 12(10); 801-11.

[142] You Y, Pu P, Huang Q, et al. Antisense telomerase RNA inhibits the growth of human glioma cells in vitro and in vivo. Int J Oncol 2006; 28(5);1225-32.

[143] Komata T, Kondo Y, Koga S, et al. Combination therapy of malignant glioma cells with 2-5A-antisense telomerase RNA and recombinant adenovirus p53. Gene Ther 2000; 7(24); 2071-9.

[144] Mukai S, Kondo Y, Koga S, et al. (2000) 2-5A antisense telomerase RNA therapy for intracranial malignant gliomas. Cancer Res 2000; 60(16); 4461-7.

[145] You Y, Geng X, Zhao P, et al. Evaluation of combination gene therapy with PTEN and antisense hTERT for malignant glioma in vitro and xenografts. Cell Mol Life Sci 2007; 64(5); 621-31.

[146] Wang Y, Zhou Z, Luo H, et al. Combination of tamoxifen and antisense human telomerase RNA inhibits glioma cell proliferations and anti-apoptosis via suppression of telomerase activity. Mol Med Report 2010; 3(6); 935-40.

[147] Iwado E, Daido S, Kondo Y, *et al.* Combined effect of 2-5A-linked antisense against telomerase RNA and conventional therapies on human malignant glioma cells *in vitro* and *in vivo.* Int J Oncol 2007; 31(5); 1087-95.

[148] Im SA, Gomez-Manzano C, Fueyo J, *et al.* Antiangiogenesis treatment for gliomas: Transfer of antisense-vascular endothelial growth factor inhibits tumor growth *in vivo.* Cancer Res 1999; 59(4); 895-900.

[149] Chu SH, Yuan X, Li Z, Jiang P, Zhang J. C-Met antisense oligodeoxynucleotide inhibits growth of glioma cells. Surg Neurol 2006; 65(6); 533-8.

[150] Chu SH, Ma YB, Feng DF, Zhang J, Qiu JH, Zhu Za. c-Met antisense oligodeoxynucleotides increase sensitivity of human glioma cells to paclitaxel. Oncol Rep 2010; 24(1); 189-94.

[151] Halatsch ME, Schmidt U, Behnke-Mursch J, Unterberg A, Rainer Wirtz C. Epidermal growth factor receptor inhibition for the treatment of glioblastoma multiforme and other malignant brain tumors. Cancer Treat Rev 2006; 32(2); 74-89.

[152] Szpechcinski A. Trzos R, Jarocki P, *et al.* Presence of MHC-I in rat glioma cells expressing antisense IGF-I-receptor RNA. Rocz Akad Med Bialymst 2004; 49(1); 98-104.

[153] Trojan LA, Kopinski P, Wei MX, *et al.* IGF-I triple helix gene therapy of rat and human gliomas. Rocz Akad Med Bialmyst 2003; 48; 18-27.

[154] Sheng Z, Li L, Zhu L, *et al.* A genome-wide RNA interference screen reveals an essential CREB3L2/ATF5/MCL1 survival pathway in malignant glioma with therapeutic implications. Nat Med 2010; 16(6); 671-7.

[155] Piwecka M, Rolle K, Wyszko E, *et al.* Nucleic acid-based technologies in therapy of malignant gliomas. Curr Pharm biotechnol 2011, 12(11); 1805-22.

[156] Desgrosellier JS, Barnes LA, Shields DJ, *et al.* An integrin α(v)β(3)-c-Src oncogenic unit promotes anchorage-independence and tumor progression. Nat Med 2009; 15(10); 1163-9.

[157] Putnam AJ, Schulz VV, Freiter EM, Bill HM, Miranti CK. Src, PKCα, and PKCδ are required for αvβ3 integrin-mediated metastatic melanoma invasion. Cell Commun Signal 2009; 7; 10.

[158] Rooprai HK, Vanmeter T, Panou C, *et al.* The role of integrin receptors in aspects of glioma invasion *in vitro.* Int J Dev Neurosci 1999; 17(5-6); 613-23.

[159] Han S, Khuri FR, Roman J. Fibronectin stimulates non-small cell lung carcinoma cell growth through activation of Akt/mammalian target of rapamycin/S6 kinase and inactivation of LKB1/AMP-activated protein kinase signal pathways. Cancer Res 2006; 66(1); 315-23.

[160] Premkumar DR, Arnold B, Jane EP, Pollack IF. Synergistic interaction between 17-AAG and phosphatidylinositol 3-kinase inhibition in human malignant glioma cells. Mol carcinog 2006; 45(1); 47-59.

Send Orders for Reprints on reprints@benthamscience.net

CHAPTER 4

A Novel Pharmacological Approach for the Treatment of Posttraumatic Stress Disorder Using Cognitive Enhancers

Shigeto Yamamoto[*] and Shigeru Morinobu

Department of Psychiatry and Neurosciences, Applied Life Sciences Institute of Biomedical & Health Sciences, Hiroshima University, Japan

Abstract: Posttraumatic stress disorder (PTSD) is a complex disorder associated with an intricate biological and psychological symptom profile and various common comorbidities. Despite the availability of various treatment strategies, PTSD is often difficult to treat. Several researchers have proposed that impaired fear extinction is involved in the pathophysiology of PTSD. Therefore, enhancing fear extinction using cognitive enhancers could be a new modality for the treatment of PTSD. To date, various cognitive enhancers that alter GABAergic, glutamatergic, dopaminergic, noradrenergic, and cannnabinoid pathways have been investigated in animal models of fear extinction. The present review focuses on D-cycloserine (DCS), a partial agonist of N-methyl-D-aspartate (NMDA) receptors, and on histone deacetylase (HDAC) inhibitors. Herein, we provide an overview of the effects of these agents, the clinical implications and drawbacks associated with their use, and directions for future research. Many preclinical and clinical studies of DCS have demonstrated a facilitating effect on fear extinction. Thus, among various cognitive enhancers, DCS seems to be the most promising agent for the treatment of PTSD. However, recent clinical studies of DCS in PTSD have not demonstrated sufficient efficacy. Several preclinical studies have also revealed that administration of HDAC inhibitors with exposure therapy can enhance fear extinction. However, no clinical studies have been conducted to assess the efficacy of HDAC inhibitors in PTSD; therefore, it is recommended that clinical trials of HDAC inhibitors be conducted in the future.

Keywords: Cognitive enhancers, D-cycloserine, fear extinction, histone, HDAC inhibitors, posttraumatic stress disorder, vorinostat, NMDA receptor, NR2B, hippocampus.

INTRODUCTION

Posttraumatic stress disorder (PTSD) is a complex disorder associated with an intricate

Address correspondance to Shigeto Yamamoto: Department of Psychiatry and Neurosciences, Applied Life Sciences Institute of Biomedical & Health Sciences, Hiroshima University 1-2-3 Kasumi, Minami-ku, 734-8551, Hiroshima, Japan; Tel: +81-82-257-5208; Fax: +81-82-257-5209; Email: shigeto1888@yahoo.co.jp

biological and psychological symptom profile and various common comorbidities. Although the precise mechanism of PTSD remains to be uncovered, several effective therapeutic strategies have been recently developed. Selective serotonin re-uptake inhibitors (SSRIs) are recommended as first-line treatment for the pharmacotherapy of PTSD, and several randomized-controlled studies have also demonstrated the efficacy of cognitive behavioral therapy (CBT) [1]. However, even with SSRI and CBT combination therapy, the therapeutic response is inadequate in a significant percentage of patients with PTSD [2].

Patients with PTSD often exhibit long-lasting reexperience of traumatic events and avoidance of the trauma-related stimuli, even though they recognize that the traumatic events are no longer occurring. Although the exact mechanism of reexperience of trauma remains unknown, it has been recently suggested that the impairment of fear extinction plays an important role in the development of clinical symptoms, such as reexperiencing of trauma, in PTSD [3-5]. In support of this possibility, neuroimaging studies have demonstrated that PTSD patients show diminished activity in the medial prefrontal cortex and increased activity in the amygdala [6, 7]. This dysfunctional circuitry is involved in the impairment of fear extinction [8].

Given that CBT relies on an extinction-based mechanism, enhancing fear extinction through cognitive enhancers may be a new strategy for the treatment of PTSD. Several cognitive enhancers that alter GABAergic, glutamatergic, dopaminergic, noradrenergic, and cannnabinoid pathways have so far been investigated using animal models of fear extinction [9]. In particular, N-methyl-D-aspartate (NMDA) receptors play a critical role in fear extinction, and D-cycloserine (DCS) is the most studied cognitive enhancer [10, 11]. Systemic administration of NMDA-receptor antagonists before or after extinction training induces impairment of within-session extinction or extinction retention, whereas the NMDA-receptor partial agonist DCS facilitates fear extinction, suggesting that NMDA receptors are involved in consolidation or encoding of extinction memory [10].

Furthermore, epigenetic mechanisms that are involved in fear extinction have recently been examined using histone deacetylase (HDAC) inhibitors, which alter chromatin structure and gene expression [12-15]. HDAC inhibitors block the activity of histone deacetylases and thus increase histone acetylation. Consequently, the increased acetylation leads to transcription of some genes that are critical for memory formation. Through this process, HDAC inhibitors can enhance fear extinction when administered in combination with extinction training, which leads to the reevaluation of the stimulus as safe [13]. Although HDAC inhibitors have been primarily studied in the treatment of cancer, they have recently been investigated in preclinical studies for the development of a novel regimen to treat memory disturbances. For example, chronic or acute injections of vorinostat, an HDAC inhibitor, into specific brain regions have been shown to increase synaptic plasticity and facilitate memory formation in mice [16, 17].

Based on these findings, it is hypothesized that a drug which enhances fear extinction may be appropriate as a novel treatment for PTSD patients exhibiting reexperience of trauma. In this context, we have focused herein on DCS and HDAC inhibitors, and provided an overview of these agents' effects, clinical implications, drawbacks, and potential for future development.

COGNITIVE ENHANCERS

D-Cycloserine (DCS)

A large body of literature suggests that glutamate acting at the NMDA receptor is critically involved in learning and memory [18]. Several studies have demonstrated that administration of NMDA-receptor antagonists before training blocks the extinction of the fear-potentiated startle response, contextual fear conditioning, inhibitory avoidance, and eye blink conditioning [19-23]. Since NMDA antagonists block fear extinction, it is reasonable to speculate that enhancing the function of this receptor would facilitate fear extinction. To test this possibility, several studies have examined the effects of DCS on fear extinction [24, 25]. For example, Walker and coworkers administered DCS either systemically or directly into the amygdala of rats before extinction training and then tested retention of extinction on the next day [25]. When DCS was given in combination with repeated exposure to the fear stimulus without a shock, extinction retention was enhanced. This did not occur in control rats that received the drug alone without extinction training. On the basis of these results, it was argued that the positive effects of DCS were associated with extinction and did not result from a general dampening of fear expression.

Ledgerwood and colleagues subsequently conducted a series of experiments showing various effects of DCS [24, 26, 27]. DCS given immediately after training was shown to dose-dependently enhance extinction of conditioned freezing in rats tested 24 h after extinction training. In addition, increasing the delay between the end of extinction training and administration of DCS led to a linear decrease in the effectiveness of DCS [24]. This finding indicates that DCS causes a time-dependent enhancement of extinction consolidation. Further, Ledgerwood and associates assessed the effect of DCS on the generalization of fear extinction [27]. Rats were first exposed to light-noise and tone-noise pairings, and light-alone trials extinguished the fear response to light. The results showed less fear conditioning to the tone in the DCS group compared with the control group. Despite these findings showing the facilitating effect of DCS on fear extinction, there have to date been no studies using an animal model of PTSD. However, we recently confirmed the effect of DCS on fear extinction in an animal model of PTSD in which rats were subjected to a single prolonged stress (SPS) [28].

To establish the therapeutic value of DCS, it is important to determine whether DCS can prevent fear extinction relapse. Several studies have examined the effects of DCS on the return of fear manipulation. DCS was found to reduce spontaneous recovery and reinstatement, but not renewal nor rapid reacquisition of conditioned fear [26, 27, 29]. For example, Ledgerwood and coworkers tested the effects of DCS on extinction when an unsignalled shock was given prior to the extinction retention test [26]. As expected, saline-treated rats showed a return of conditioned fear, whereas DCS-treated rats exhibited much less return of fear, suggesting that DCS blocked the reinstatement effect. Woods and Bouton conducted an experiment to determine whether DCS reduces the renewal effect [29]. Rats were exposed to fear conditioning in one context, and then extinguished with extinction training in a second context. When tested in the extinction training context, DCS-treated rats exhibited superior extinction compared with saline-treated rats. However, when re-exposed in the original conditioning context, DCS-treated rats showed a substantial renewal of fear, similarly to saline-treated rats. These findings indicate that although DCS may enhance fear extinction and prevent reinstatement, it does not alter context-specific learning and may not protect against relapse.

Some caveats with respect to the clinical application of DCS have emerged from preclinical studies. Langton and Richardson demonstrated that DCS facilitates extinction but not re-extinction of a conditioned fear response [30]. In that study, rats were first trained with a conditioned stimulus (CS) (light) and this fear was then extinguished. Subsequently, rats were retrained to fear the same CS (light) or a new CS (white noise). When given a second extinction session, DCS facilitated extinction of the new CS but not the original CS. This result suggests that DCS may be less effective in patients who have relapsed after treatment. A study by Parnas and colleagues suggested that chronic exposure to DCS decreases its effectiveness [31]. DCS- and saline-treated rats received 0, 1, or 5 injections prior to fear conditioning. The results showed that DCS facilitated extinction of conditioned freezing to a light CS when no drug pre-exposure had occurred, but pre-exposure to DCS just prior to conditioning disrupted the facilitation of the extinction effect. When a 28-day period was interposed between pre-exposure and conditioning, the enhancing effect of DCS on extinction was restored, suggesting that DCS should be used acutely rather than chronically. A study by Lee and associates suggested that DCS can enhance not only fear extinction but also fear reconsolidation, depending on the number of extinction trials [20]. In that study, rats were exposed to either 1 or 10 extinction trials, and DCS was administered 30 min before the extinction trials. Twenty-four hours after the extinction trials, DCS-treated rats showed a lower fear response as compared with controls when they had received 10 extinction trials, whereas DCS-treated rats showed a heightened fear response compared with controls when they had received 1 extinction trial. This result indicates a potential problem with the clinical use of DCS. Namely, if the exposure session is not long enough, DCS could enhance reconsolidation of the fear memory.

Ressler and colleagues were the first to demonstrate the clinical effect of DCS in acrophobia [32]. The study was a double-blinded randomized controlled clinical trial conducted in 27 patients with acrophobia. Patients were assigned randomly to groups receiving placebo (n = 10) or DCS (n = 17) in conjunction with virtual reality exposure (VRE). Patients underwent a suboptimal amount of exposure therapy for acrophobia (2 VRE sessions) and were instructed to take a single dose of study medication 2 to 4 h before each session. Assessments at 1 to 2 weeks and at 3 months following the exposure showed that patients in the DCS group exhibited significantly greater improvement than did patients in the placebo group. In addition, at 3 months post-treatment, patients in the DCS group reported greater real-life exposure to heights and higher self-ratings of improvement relative to patients in the placebo group. Since then, Hofmann and associates reported that DCS enhances exposure therapy for social anxiety disorder (public speaking anxiety) [33]. Twenty-seven participants were randomly assigned to exposure therapy plus DCS (n = 12) or exposure therapy plus placebo (n = 15). Participants received 5 therapy sessions; the first session provided an introduction to the treatment model and was followed by 4 sessions emphasizing exposure to increasingly challenging public speech situations. One hour prior to each session, participants received DCS or placebo. Patients who took DCS prior to 4 social situation sessions showed significantly greater reduction in general social anxiety symptoms, both immediately post-treatment and at 1-month follow-up, compared with the placebo group. Controlled effect sizes were in the medium to large range. Similarly positive results were obtained in a randomized controlled trial of 56 participants with social anxiety disorder [34]. Other clinical studies have demonstrated a positive effect of DCS with exposure therapy in patients with panic disorder [35] and obsessive-compulsive disorder [36].

Norberg and colleagues recently performed a meta-analysis of preclinical and clinical studies on the ability of DCS to facilitate fear extinction/exposure therapy [37]. In this study, post-treatment results were compared between animal and human studies at post-treatment, human studies were compared with animal studies. This comparison showed a significant difference, with greater effects seen in the animal studies. Both animal and human studies nevertheless were associated with significant effect sizes; the effects were large and robust in the animal studies (d = 1.19) and small in the human studies (d =0.42). The human clinical studies (excluding nonclinical samples) showed a moderate effect (d =0.60). Examining all studies together, DCS was associated with a significant overall effect size *versus* placebo when added to extinction/exposure (d =0.90). Taken together, the results of this meta-analysis suggest that DCS facilitates the effects of fear extinction/exposure therapy in both animals and humans.

Clinical studies evaluating the effect of DCS with exposure therapy for the treatment of PTSD have been long awaited, and 2 such clinical studies were recently published in succession [38, 39]. One of the studies, by de Kleine and colleagues, showed that DCS did not enhance overall treatment effects [38]. The treatment effect protocol was completed by 45 participants, 24 receiving exposure with DCS and 21 receiving exposure with placebo. The majority of patients (80.6%) were female, and the mean age of the cohort was 38.3 years. The traumatic events underlying PTSD were mixed and comprised sexual abuse, violent nonsexual assault, a road traffic or other accident, war-zone experiences, and miscellaneous. Comorbid Axis I disorders were depressive disorder (53.7%) and anxiety disorders (41.8%). The primary outcome measure was assessed with the Clinician-Administered PTSD Scale (CAPS) and the Posttraumatic Stress Symptom Scale-Self Report (PSS-SR). DCS or placebo was given 1 h prior to the start of an exposure session. All participants who completed treatment before the eighth session were considered as early completers and those who received the eighth to tenth session as regular completers. The enhancement effects of DCS were found only in patients with more severe PTSD needing longer treatment. Specifically, DCS enhanced outcomes only in a subgroup of patients with severe symptoms who had completed regular sessions. The early completers did not show the enhancing effect of DCS. This may be explained by a ceiling effect; *i.e.*, for the early completers, there would be no need to enhance the efficacy with pharmacological therapy. However, other possibilities should be considered in the interpretation of these results. Interestingly, a previous study reported the effects of DCS specifically in patients with severe symptoms [40], and another study showed no effect of DCS on subclinical fear [41]. These negative outcomes indicate that DCS may be useful only in patients with clinically significant, maladaptive fear. Alternatively, there may be a different optimal dose of DCS for patients with less severe symptoms. This issue needs to be addressed in the future in order to clarify the mechanism of action of DCS.

The other recent clinical trial on DCS with exposure therapy in PTSD patients, by Litz and associates, was a randomized, double-blind, placebo-controlled trial examining whether DCS could augment exposure therapy for PTSD in veterans returning from Iraq and Afghanistan [39]. Twenty-six veterans of the Iraq and

Afghanistan wars who had a primary diagnosis of PTSD participated in the trial. Patients were randomly assigned to exposure therapy plus DCS (n = 13) or exposure therapy plus placebo (n = 13). The most common comorbid disorders were major depressive disorder (27%), alcohol abuse (19%), and social anxiety (8%). The primary outcome measure was assessed with CAPS. Participants attended a total 6 sessions of 60 to 90 min duration. Results indicated that PTSD patients undergoing exposure therapy with DCS treatment experienced significantly less symptom reduction than did patients who received exposure therapy plus placebo over the course of treatment. Taken together with the results of the 2 trials for PTSD, DCS did not show sufficient efficacy. It is difficult to exactly explain why DCS did not enhance the treatment effect in PTSD, as it did in the treatment of other anxiety disorders. It was unclear whether comorbid diseases (*e.g.*, depression) affected the treatment outcome of DCS in these trials. One possible explanation is that DCS facilitated fear reconsolidation, but not fear extinction, in line with the work by Lee and colleagues. Alternatively, it should be noted that DCS might act preferentially on low- rather than high-order learning [42]. Notwithstanding the effectiveness of DCS in rodents, some studies of laboratory-based fear conditioning in humans have shown no effect of DCS on fear extinction [43, 44]. Evidence indicates that conditioning in rodents is, in general, a low-level automatic process [45], whereas laboratory-based conditioning in humans relies essentially on high-level cognitive learning [46]. The pathophysiology of PTSD could be more complicated than that of other anxiety disorders (*e.g.* phobias), and accordingly, CBT for PTSD entails exposure and cognitive restructuring, but not exposure alone. On the basis of these findings, DCS may be less effective for the treatment of PTSD because DCS has poor efficacy in high-level cognitive learning.

Histone Deacetylase (HDAC) Inhibitors

Fear memory extinction requires memory consolidation mediated *via* gene transcription [47]. Transcription is regulated by the concerted action of transcription factors and cofactors that modify and remodel the structure of chromatin [48]. Because epigenetic manipulations cause remodeling of chromatin structure, thereby leading to long-term cellular and molecular changes, a clear understanding of the epigenetic mechanisms of fear extinction and associated alterations in gene expression would provide opportunities for the development of novel therapies. In particular, histone acetylation, which alters the compact chromatin structure and changes the accessibility of DNA to regulatory proteins, is emerging as a fundamental mechanism for regulating gene expression [49].

Several recent studies have demonstrated that histone acetylation plays a role in learning behavior and synaptic plasticity [50-52]. For example, Levenson and Sweatt demonstrated that increased acetylation of histone H3 was correlated with long-term memory formation, and HDAC inhibitors, at least in part, contributed to the induction of long-term potentiation and long-term memory [50]. Similarly, Korzus and associates reported that a decrease in the activity of HAT disturbed the formation of long-term memory, and that the administration of HDAC inhibitors ameliorated this disturbance [53]. Furthermore, HDAC inhibitors have recently been shown to enhance not only contextual fear conditioning but also fear extinction [12, 15, 54, 55]. For example, Lattal and colleagues showed that administration of HDAC inhibitors, such as sodium butyrate (NaB) systemically or trichostatin A intrahippocampally, prior to a contextual extinction session produced greater effects than vehicle treatment in contextual fear extinction [12]. Bredy and associates reported that valproic acid enhanced fear extinction through histone H4 acetylation around the promoter region of the brain-derived neurotrophic factor (BDNF) gene and also increased BDNF mRNA expression in the prefrontal cortex [15].

However, little is known about whether fear memory consolidation or extinction is preferentially enhanced under the condition of increased histone acetylation by HDAC inhibitors. To this end, Stafford and associates examined the enhancing effect of NaB on fear extinction and initial conditioning of fear in mice using intra-hippocampal or systemic administration [56]. Across a variety of conditions, the effects of NaB on extinction were larger and more persistent compared to its effects on initial fear memory. Curiously, they found a smaller NaB-induced enhancement of initial fear memory, inconsistent with other studies [15, 54]. More recently, we have examined the behavioral and molecular effects of systemic injection of vorinostat, an HDAC inhibitor, on fear extinction in the rat hippocampus, and found that vorinostat enhanced both the original conditioned fear and extinction of the conditioned fear equally, depending on the timing of administration [54]. Our results are consistent with those of Bredy and colleagues regarding enhancement of both fear extinction and the original conditioned fear in response to administration of HDAC inhibitors. In their study, valproic acid either strengthened reconsolidation of the original fear memory or enhanced long-term memory for extinction, depending on the conditions of memory retrieval [15]. The difference in these results seems to be due to methodological differences. However, as Stafford and colleagues noted, the reason is that the rate of extinction may be slower than the rate of initial acquisition. A slower rate of learning during extinction would theoretically leave more room for enhancements than the relatively fast rate of learning associated with initial acquisition. Thus, the learning that occurs during extinction may be more susceptible to pharmacological manipulations compared with initial conditioning. At any rate, we should take into consideration the fact that HDAC inhibitors could enhance fear reconsolidation under certain conditions in clinical use, similarly to DCS

In addition to studies involving hippocampus-dependent contextual fear conditioning or fear extinction tasks, enhancement of hippocampus-dependent memory induced by treatment with an HDAC inhibitor was recently observed for object-recognition memory [57] and object-location memory [58]. Given these findings, it is conceivable that HDAC inhibitors may enhance new memory formation (CS with no unconditioned stimulus representation) with respect to the mechanisms through which fear extinction is enhanced. That is, the effect of HDAC inhibitors on fear extinction would possibly lie in the enhancement of memory consolidation. Previous reports indicate that this process may occur by increasing histone acetylation at certain gene promoters and thereby altering gene expression [13, 54]. However, it is important to note that in addition to the acetylation of histone proteins, the acetylation of non-histone proteins has also been implicated in the mechanism of memory enhancement. It has been reported that many HDACs deacetylate non-histone proteins such as α-tubulin, p53, NF-κB, and various transcription factors [59]. In particular, the NF-κB pathway plays an important role in memory enhancement in response to HDAC inhibitors [60].

Although evidence indicates that the interconnections among the amygdala, hippocampus and medial prefrontal cortex (mPFC) play an important role in extinction of conditioned fear [61], the exact brain region where chromatin remodeling through the enhancement of histone acetylation is undertaken, remains unknown. Previous studies reported that the PFC or hippocampus can be critical for the facilitation of fear extinction by administration of HDAC inhibitors [12, 14]. The hippocampus has strong reciprocal connections with the mPFC, and pharmacological lesions in the infralimbic cortex (IL) impair the retrieval of extinction [61]. In addition, evidence indicates that prelimbic cortex (PL) activity is necessary for fear expression but not extinction memory, whereas IL activity is necessary for extinction memory but not fear expression [62]. Both the PL and IL, which are involved in extinction, exert bidirectional regulation of fear expression. In this context, Stafford and associates demonstrated that the network between the hippocampus and IL, but not the PL, is critical for fear extinction from the aspect of epigenetic machinery [56]. Specifically, they investigated the effects of intra-hippocampal NaB after extinction on histone acetylation (H3) and c-Fos expression in the mPFC to understand how modulating the hippocampus affects transcriptional events in brain regions important for extinction. In addition, they infused NaB into the IL or PL to further examine the specificity. They observed that NaB infusion into the hippocampus resulted in increases in histone acetylation and c-Fos expression in the IL but not the PL. The involvement of the IL was confirmed by the finding that infusion of NaB into the IL but not the PL induced persistent extinction enhancement.

In a series of experiments, Stafford and colleagues also observed changes in c-Fos expression in the hippocampus or IL, whereas c-Fos is only one of the transcription factors that regulate gene transcription through the activator protein-1 binding site. Thus, it is necessary to identify downstream target genes with increased expression that are critical for fear extinction. In our recent study, we observed that vorinostat increased the levels of NR2B mRNA and protein as well as the levels of acetylated histone H3 and H4 at the promoter of the NR2B gene in the hippocampus, along with enhancement of fear extinction [54]. This finding suggests that the NR2B gene in the hippocampus is a candidate gene involved in fear extinction. However, the possibility cannot be ruled out that alterations in the expression of NR2B were coincident with the enhancement of fear extinction in response to vorinostat. That is, it is possible that the NR2B gene is only one of the genes whose expression is affected by vorinostat, but that it is not the key target. In fact, fear extinction appears to involve not only NMDA receptors, but also neurotransmitter-mediated signals including GABA, AMPA receptors, and metabotropic glutamate receptors [47]. Further studies are warranted to identify the target genes involved in fear extinction in the future.

Although recent pharmacological interventions have been successful in the enhancement of fear extinction, spontaneous recovery and related phenomena (*e.g.*, renewal, reinstatement), which sometimes follow extinction treatments, are a major challenge [13]. To date, little is known about the persistence of the effects of HDAC inhibitors on fear extinction. Stafford and associates demonstrated the persistent effect of intra-hippocampal NaB, which inhibits spontaneous recovery [56]. As for spontaneous recovery, Bouton and colleagues noted that the passage of time represents a gradually changing context [63]. In that model, extinction may be specific to the environmental context in which it occurs, as well as the temporal context. Bouton *et al*. likened spontaneous recovery to the renewal effect, with the changed context being a different time rather than a different environment. Based on this idea, it is possible that the hippocampus would be associated with the underlying mechanisms of spontaneous recovery as well as fear renewal.

To our knowledge, the exposure-enhancing effects of HDAC inhibitors have not been investigated in animal models of PTSD or in a PTSD population. We have more recently investigated the effect of vorinostat on fear extinction using single prolonged stress paradigm in rats as an animal model of PTSD. Animals were randomly assigned to 2 groups (placebo or vorinostat). Twenty-four hours after the footshock, extinction training was conducted for 2 consecutive days. Extinction training was defined as the repetitive exposure to the contextual cue (the apparatus) for 20 min in the absence of footshock. Vorinostat or placebo was administered systemically immediately after the second extinction training session. On day 3, the test was conducted for 5 min (Fig. **1A**). The result showed that the freezing time in SPS rats with vorinostat was significantly shorter than that with placebo (Fig. **1B**) (unpublished). Similar to our recent study [54], vorinostat was also capable of enhancing fear extinction in SPS rats. This is the first study demonstrating the effect of vorinostat on extinction using an animal model of PTSD, and the results indicate that vorinostat might be useful in the treatment of PTSD. Several preclinical studies have suggested that administration of HDAC inhibitors in conjunction with exposure therapy can be a promising tool for the treatment of anxiety disorders including PTSD. However, there have been no clinical studies examining the effects of HDAC inhibitors on fear extinction. Therefore, it is recommended that clinical trials of HDAC inhibitors be conducted in the future.

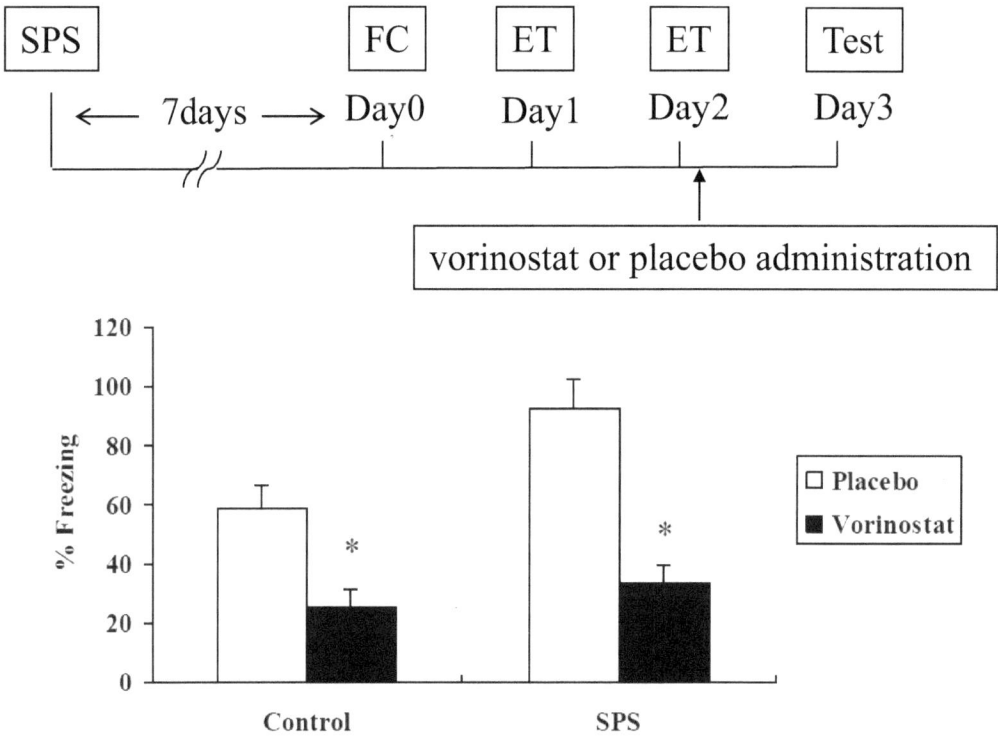

Figure 1: (A) Experimental design. Twenty-four hours after footshock, extinction training was conducted for 2 consecutive days. Vorinostat or placebo was administered systemically immediately after the second extinction training session. On day 3, the test was conducted for 5 min. SPS = single prolonged stress, FC = fear conditioning, ET = extinction training **(B)** The effect of vorinostat on fear extinction in SPS rats. Data are expressed as mean ± SEM of 8 rats per group. In the SPS group, vorinostat-treated rats showed the enhancement of fear extinction relative to placebo-treated rats, similar to the control group. * $p<0.05$; unpaired Student's t-test.

CONCLUSIONS

Recent research in the treatment of PTSD has brought about the development of a novel therapeutic approach involving cognitive enhancers. Many preclinical and clinical studies of DCS have demonstrated a facilitating effect on extinction. DCS seems to be the most promising cognitive enhancer for the treatment of PTSD.

However, there are some concerns about the clinical use of DCS. More research is needed to determine the long-term effects of DCS on fear extinction and to understand the optimal timing, number, and size of doses. Further, clinical studies in PTSD populations have just begun; therefore, there is clearly a need for further study of DCS.

Although several preclinical studies have shown the efficacy of HDAC inhibitors, the research is still in the preliminary stage. Among the HDAC inhibitors, vorinostat is the first drug to be approved for clinical use by the U.S. Food and Drug Administration for the treatment of refractory cutaneous T-cell lymphoma. Our previous study suggested that administration of vorinostat in conjunction with exposure therapy can provide a promising new avenue for the treatment of PTSD [54]. However, there are problems with the clinical application of vorinostat because of its adverse effects. The most common clinical adverse events of any grade are diarrhea (52%), fatigue (52%), nausea (41%), and anorexia (24%) [64]. Therefore, in the treatment of PTSD it may be problematic to use vorinostat chronically but not acutely, in the same manner as it is used for cancer therapy. However, our study indicates that acute administration of vorinostat with exposure therapy could produce a positive effect on the outcome with acceptable safety profiles.

To date, there have been some negative findings regarding the use of cognitive enhancers for the treatment of PTSD. It can be speculated that cognitive enhancers were ineffective in some studies because the psychopathology of PTSD is more complex than that of other anxiety disorders (*e.g.*, specific phobias). One possible explanation is that confronting fear-evoking trauma-related stimuli during exposure may not be sufficient, because avoidance of reminders of the trauma is inherent in PTSD and is one of the cardinal symptoms of the disorder. It is important to consider the impact of distraction during exposure, which prevents full engagement and habituation to the feared stimuli. Another explanation is that in the context of PTSD, stimulus generalization can occur with relative ease, by which conditioned fear from danger cues transfers to safety cues with overlapping features. To overcome this issue, it is recommended to utilize a variety of different feared stimuli during exposure. Also, conducting exposure therapy in a variety of different environments and contexts would be a recommended method.

Clinicians should keep in mind these considerations before evaluating the effectiveness of cognitive enhancers. Incorporating several diverse techniques noted above into CBT for PTSD would be beneficial in the treatment of PTSD. Moreover, such efforts in combination with cognitive enhancers could aid in the development of a new therapeutic strategy leading to a better treatment outcome in the clinical management of PTSD.

ACKNOWLEDGEMENTS

This work was supported by a grant-in-aid for general scientific research from the Ministry of Education, Science, and Culture of Japan, a Health Science Research Grant for Research on Brain Science from the Ministry of Health and Welfare of Japan, and a grant from Core Research for Evolutional Science and Technology (CREST) of Japan Science and Technology (JST).

CONFLICT OF INTEREST

The author(s) confirm that this chapter content has no conflict of interest.

DISCLOSURE

Declared none.

REFERENCES

[1]   Powers MB, Halpern JM, Ferenschak MP, Gillihan SJ, Foa EB. A meta-analytic review of prolonged exposure for posttraumatic stress disorder. Clin Psychol Rev 2010; 30 (6): 635-641.

[2]   Schneier FR, Neria Y, Pavlicova M, Hembree E, Suh EJ, Amsel L, Marshall RD. Combined prolonged exposure therapy and paroxetine for PTSD related to the World Trade Center attack: a randomized controlled trial. Am J Psychiatry 2012; 169 (1): 80-88.

[3]   Rothbaum BO, Davis M. Applying learning principles to the treatment of post-trauma reactions. Ann N Y Acad Sci 2003; 1008: 112-121.

[4]   Rauch SL, Shin LM, Phelps EA. Neurocircuitry models of posttraumatic stress disorder and extinction: human neuroimaging research--past, present, and future. Biol Psychiatry 2006; 60 (4): 376-382.

[5]   Milad MR, Rauch SL, Pitman RK, Quirk GJ. Fear extinction in rats: implications for human brain imaging and anxiety disorders. Biol Psychol 2006; 73 (1): 61-71.

[6]   Milad MR, Pitman RK, Ellis CB, Gold AL, Shin LM, Lasko NB, Zeidan MA, Handwerger K, Orr SP, Rauch SL. Neurobiological basis of failure to recall extinction memory in posttraumatic stress disorder. Biol Psychiatry 2009; 66 (12): 1075-1082.

[7]   Shin LM, Wright CI, Cannistraro PA, Wedig MM, McMullin K, Martis B, Macklin ML, Lasko NB, Cavanagh SR, Krangel TS, Orr SP, Pitman RK, Whalen PJ, Rauch SL. A functional magnetic resonance imaging study of amygdala and medial prefrontal cortex responses to overtly presented fearful faces in posttraumatic stress disorder. Arch Gen Psychiatry 2005; 62 (3): 273-281.

[8]   Kim MJ, Loucks RA, Palmer AL, Brown AC, Solomon KM, Marchante AN, Whalen PJ. The structural and functional connectivity of the amygdala: from normal emotion to pathological anxiety. Behav Brain Res 2011; 223 (2): 403-410.

[9]   Kaplan GB, Moore KA. The use of cognitive enhancers in animal models of fear extinction. Pharmacol Biochem Behav 2011; 99 (2): 217-228.

[10]   Myers KM, Carlezon WA, Jr., Davis M. Glutamate receptors in extinction and extinction-based therapies for psychiatric illness. Neuropsychopharmacology 2011; 36 (1): 274-293.

[11]   Davis M, Ressler K, Rothbaum BO, Richardson R. Effects of D-cycloserine on extinction: translation from preclinical to clinical work. Biol Psychiatry 2006; 60 (4): 369-375.

[12]   Lattal KM, Barrett RM, Wood MA. Systemic or intrahippocampal delivery of histone deacetylase inhibitors facilitates fear extinction. Behav Neurosci 2007; 121 (5): 1125-1131.

[13]   Stafford JM, Lattal KM. Is an epigenetic switch the key to persistent extinction? Neurobiol Learn Mem 2011; 96 (1): 35-40.

[14]   Bredy TW, Wu H, Crego C, Zellhoefer J, Sun YE, Barad M. Histone modifications around individual BDNF gene promoters in prefrontal cortex are associated with extinction of conditioned fear. Learn Mem 2007; 14 (4): 268-276.

[15]   Bredy TW, Barad M. The histone deacetylase inhibitor valproic acid enhances acquisition, extinction, and reconsolidation of conditioned fear. Learn Mem 2008; 15 (1): 39-45.

[16]   Guan JS, Haggarty SJ, Giacometti E, Dannenberg JH, Joseph N, Gao J, Nieland TJ, Zhou Y, Wang X, Mazitschek R, Bradner JE, DePinho RA, Jaenisch R, Tsai LH. HDAC2 negatively regulates memory formation and synaptic plasticity. Nature 2009; 459 (7243): 55-60.

[17]   Peleg S, Sananbenesi F, Zovoilis A, Burkhardt S, Bahari-Javan S, Agis-Balboa RC, Cota P, Wittnam JL, Gogol-Doering A, Opitz L, Salinas-Riester G, Dettenhofer M, Kang H, Farinelli L, Chen W, Fischer A. Altered histone acetylation is associated with age-dependent memory impairment in mice. Science 2010; 328 (5979): 753-756.

[18]   Newcomer JW, Krystal JH. NMDA receptor regulation of memory and behavior in humans. Hippocampus 2001; 11 (5): 529-542.

[19]   Szapiro G, Vianna MR, McGaugh JL, Medina JH, Izquierdo I. The role of NMDA glutamate receptors, PKA, MAPK, and CAMKII in the hippocampus in extinction of conditioned fear. Hippocampus 2003; 13 (1): 53-58.

[20]   Lee JL, Milton AL, Everitt BJ. Reconsolidation and extinction of conditioned fear: inhibition and potentiation. J Neurosci 2006; 26 (39): 10051-10056.

[21]   Lee HJ, Choi JS, Brown TH, Kim JJ. Amygdalar nmda receptors are critical for the expression of multiple conditioned fear responses. J Neurosci 2001; 21 (11): 4116-4124.

[22]   Baker JD, Azorlosa JL. The NMDA antagonist MK-801 blocks the extinction of Pavlovian fear conditioning. Behav Neurosci 1996; 110 (3): 618-620.

[23]   Falls WA, Miserendino MJ, Davis M. Extinction of fear-potentiated startle: blockade by infusion of an NMDA antagonist into the amygdala. J Neurosci 1992; 12 (3): 854-863.

[24]   Ledgerwood L, Richardson R, Cranney J. Effects of D-cycloserine on extinction of conditioned freezing. Behav Neurosci 2003; 117 (2): 341-349.

[25]   Walker DL, Ressler KJ, Lu KT, Davis M. Facilitation of conditioned fear extinction by systemic administration or intra-amygdala infusions of D-cycloserine as assessed with fear-potentiated startle in rats. J Neurosci 2002; 22 (6): 2343-2351.

[26]   Ledgerwood L, Richardson R, Cranney J. D-cycloserine and the facilitation of extinction of conditioned fear: consequences for reinstatement. Behav Neurosci 2004; 118 (3): 505-513.

[27]   Ledgerwood L, Richardson R, Cranney J. D-cycloserine facilitates extinction of learned fear: effects on reacquisition and generalized extinction. Biol Psychiatry 2005; 57 (8): 841-847.

[28]   Yamamoto S, Morinobu S, Fuchikami M, Kurata A, Kozuru T, Yamawaki S. Effects of single prolonged stress and D-cycloserine on contextual fear extinction and hippocampal NMDA receptor expression in a rat model of PTSD. Neuropsychopharmacology 2008; 33 (9): 2108-2116.

[29]   Woods AM, Bouton ME. D-cycloserine facilitates extinction but does not eliminate renewal of the conditioned emotional response. Behav Neurosci 2006; 120 (5): 1159-1162.

[30]   Langton JM, Richardson R. D-cycloserine facilitates extinction the first time but not the second time: an examination of the role of NMDA across the course of repeated extinction sessions. Neuropsychopharmacology 2008; 33 (13): 3096-3102.

[31]   Parnas AS, Weber M, Richardson R. Effects of multiple exposures to D-cycloserine on extinction of conditioned fear in rats. Neurobiol Learn Mem 2005; 83 (3): 224-231.

[32]   Ressler KJ, Rothbaum BO, Tannenbaum L, Anderson P, Graap K, Zimand E, Hodges L, Davis M. Cognitive enhancers as adjuncts to psychotherapy: use of D-cycloserine in phobic individuals to facilitate extinction of fear. Arch Gen Psychiatry 2004; 61 (11): 1136-1144.

[33]   Hofmann SG, Meuret AE, Smits JA, Simon NM, Pollack MH, Eisenmenger K, Shiekh M, Otto MW. Augmentation of exposure therapy with D-cycloserine for social anxiety disorder. Arch Gen Psychiatry 2006; 63 (3): 298-304.

[34]   Guastella AJ, Richardson R, Lovibond PF, Rapee RM, Gaston JE, Mitchell P, Dadds MR. A randomized controlled trial of D-cycloserine enhancement of exposure therapy for social anxiety disorder. Biol Psychiatry 2008; 63 (6): 544-549.

[35]   Otto MW, Tolin DF, Simon NM, Pearlson GD, Basden S, Meunier SA, Hofmann SG, Eisenmenger K, Krystal JH, Pollack MH. Efficacy of d-cycloserine for enhancing response to cognitive-behavior therapy for panic disorder. Biol Psychiatry 2010; 67 (4): 365-370.

[36]   Kushner MG, Kim SW, Donahue C, Thuras P, Adson D, Kotlyar M, McCabe J, Peterson J, Foa EB. D-cycloserine augmented exposure therapy for obsessive-compulsive disorder. Biol Psychiatry 2007; 62 (8): 835-838.

[37]   Norberg MM, Krystal JH, Tolin DF. A meta-analysis of D-cycloserine and the facilitation of fear extinction and exposure therapy. Biol Psychiatry 2008; 63 (12): 1118-1126.

[38]   de Kleine RA, Hendriks GJ, Kusters WJ, Broekman TG, van Minnen A. A Randomized Placebo-Controlled Trial of d-Cycloserine to Enhance Exposure Therapy for Posttraumatic Stress Disorder. Biol Psychiatry 2012; 71 (11): 962-968.

[39]   Litz BT, Salters-Pedneault K, Steenkamp MM, Hermos JA, Bryant RA, Otto MW, Hofmann SG. A randomized placebo-controlled trial of d-cycloserine and exposure therapy for posttraumatic stress disorder. J Psychiatr Res 2012; 46 (9): 1184-1190.

[40]   Siegmund A, Golfels F, Finck C, Halisch A, Rath D, Plag J, Strohle A. D-cycloserine does not improve but might slightly speed up the outcome of *in vivo* exposure therapy in patients with severe agoraphobia and panic disorder in a randomized double blind clinical trial. J Psychiatr Res 2011; 45 (8): 1042-1047.

[41]   Guastella AJ, Dadds MR, Lovibond PF, Mitchell P, Richardson R. A randomized controlled trial of the effect of D-cycloserine on exposure therapy for spider fear. J Psychiatr Res 2007; 41 (6): 466-471.

[42]   Grillon C. D-cycloserine facilitation of fear extinction and exposure-based therapy might rely on lower-level, automatic mechanisms. Biol Psychiatry 2009; 66 (7): 636-641.

[43]   Kalisch R, Holt B, Petrovic P, De Martino B, Kloppel S, Buchel C, Dolan RJ. The NMDA agonist D-cycloserine facilitates fear memory consolidation in humans. Cereb Cortex 2009; 19 (1): 187-196.

[44]   Guastella AJ, Lovibond PF, Dadds MR, Mitchell P, Richardson R. A randomized controlled trial of the effect of D-cycloserine on extinction and fear conditioning in humans. Behav Res Ther 2007; 45 (4): 663-672.

[45]   Squire LR, Zola SM. Structure and function of declarative and nondeclarative memory systems. Proc Natl Acad Sci U S A 1996; 93 (24): 13515-13522.

[46]   Lovibond PF, Shanks DR. The role of awareness in Pavlovian conditioning: empirical evidence and theoretical implications. J Exp Psychol Anim Behav Process 2002; 28 (1): 3-26.

[47]   Myers KM, Davis M. Mechanisms of fear extinction. Mol Psychiatry 2007; 12 (2): 120-150.

[48]   Holliday R. Epigenetics: a historical overview. Epigenetics 2006; 1 (2): 76-80.

[49]   Goldberg AD, Allis CD, Bernstein E. Epigenetics: a landscape takes shape. Cell 2007; 128 (4): 635-638.

[50]   Levenson JM, Sweatt JD. Epigenetic mechanisms in memory formation. Nat Rev Neurosci 2005; 6 (2): 108-118.

[51]   Haettig J, Stefanko DP, Multani ML, Figueroa DX, McQuown SC, Wood MA. HDAC inhibition modulates hippocampus-dependent long-term memory for object location in a CBP-dependent manner. Learn Mem 2011; 18 (2): 71-79.

[52]   Fischer A, Sananbenesi F, Mungenast A, Tsai LH. Targeting the correct HDAC(s) to treat cognitive disorders. Trends Pharmacol Sci 2010; 31 (12): 605-617.

[53]   Korzus E, Rosenfeld MG, Mayford M. CBP histone acetyltransferase activity is a critical component of memory consolidation. Neuron 2004; 42 (6): 961-972.

[54]   Fujita Y, Morinobu S, Takei S, Fuchikami M, Matsumoto T, Yamamoto S, Yamawaki S. Vorinostat, a histone deacetylase inhibitor, facilitates fear extinction and enhances expression of the hippocampal NR2B-containing NMDA receptor gene. J Psychiatr Res 2012; 46 (5): 635-643.

[55]   Vecsey CG, Hawk JD, Lattal KM, Stein JM, Fabian SA, Attner MA, Cabrera SM, McDonough CB, Brindle PK, Abel T, Wood MA. Histone deacetylase inhibitors enhance memory and synaptic plasticity *via* CREB:CBP-dependent transcriptional activation. J Neurosci 2007; 27 (23): 6128-6140.

[56]   Stafford JM, Raybuck JD, Ryabinin AE, Lattal KM. Increasing histone acetylation in the hippocampus-infralimbic network enhances fear extinction. Biol Psychiatry 2012; 72 (1): 25-33.

[57]   Stefanko DP, Barrett RM, Ly AR, Reolon GK, Wood MA. Modulation of long-term memory for object recognition *via* HDAC inhibition. Proc Natl Acad Sci U S A 2009; 106 (23): 9447-9452.

[58]   Hawk JD, Florian C, Abel T. Post-training intrahippocampal inhibition of class I histone deacetylases enhances long-term object-location memory. Learn Mem 2011; 18 (6): 367-370.

[59]   Buchwald M, Kramer OH, Heinzel T. HDACi--targets beyond chromatin. Cancer Lett 2009; 280 (2): 160-167.

[60]   Yeh SH, Lin CH, Gean PW. Acetylation of nuclear factor-kappaB in rat amygdala improves long-term but not short-term retention of fear memory. Mol Pharmacol 2004; 65 (5): 1286-1292.

[61]   Quirk GJ, Mueller D. Neural mechanisms of extinction learning and retrieval. Neuropsychopharmacology 2008; 33 (1): 56-72.

[62]   Sierra-Mercado D, Padilla-Coreano N, Quirk GJ. Dissociable roles of prelimbic and infralimbic cortices, ventral hippocampus, and basolateral amygdala in the expression and extinction of conditioned fear. Neuropsychopharmacology 2011; 36 (2): 529-538.

[63]   Bouton ME. Context and behavioral processes in extinction. Learn Mem 2004; 11 (5): 485-494.

[64]   Mann BS, Johnson JR, Cohen MH, Justice R, Pazdur R. FDA approval summary: vorinostat for treatment of advanced primary cutaneous T-cell lymphoma. Oncologist 2007; 12 (10): 1247-1252.

Send Orders for Reprints on reprints@benthamscience.net

Frontiers in Clinical Drug Research - CNS and Neurological Disorders, Vol. 1, 2013, 135-148 **135**

CHAPTER 5

Lactate Metabolism as a New Target for the Therapeutics of Schizophrenia

Takashi Uehara[*] and Tomiki Sumiyoshi

Department of Neuropsychiatry, University of Toyama Graduate School of Medicine and Pharmaceutical Sciences, 2630 Sugitani, 930-0194 Toyama, Japan

Abstract: Accumulated evidence indicates that lactate plays a significant role in energy production in the brain, especially during acute neural activation. The astrocyte-neuron lactate shuttle hypothesis has been proposed to explain how lactate is supplied to neurons. Furthermore, the ability of lactate to protect neurons against excite toxicity and oxidative stress has been reported. There is a growing interest in the role for serotonin (5-HT)$_{1A}$ receptors in the treatment of negative symptoms and cognitive disturbances of schizophrenia. This is based on clinical evidence that 5-HT$_{1A}$ agonists, such as tandospirone, perospirone, and lurasidone, improve cognitive function in patients with the illness. Data from rodent studies also support the hypothesis that stimulation of 5-HT$_{1A}$ receptors is advantageous in treating cognitive deficits of schizophrenia.

Rats experiencing transient N-methyl-D-aspartate receptor (NMDA-R) blockade during the neonatal stage show reduction of the stress-induced increase in extracellular lactate concentrations in the medial prefrontal cortex around the pubertal period. Treatment with 5-HT$_{1A}$ receptor agonists, such as tandospirone, reversed the decreased lactate production in these model animals. These observations indicate that lactate metabolism provides a novel probe for the development of psychotropic compounds to treat core symptoms of schizophrenia.

Keywords: Lactate, energy metabolism, cognition, neuroprotection, 5-HT1A receptor, schizophrenia.

INTRODUCTION

Disturbances of cognitive function, evaluated by psychological and neurophysiological methods, have been shown to predict outcome in patients with schizophrenia [1,2]. Currently, the disease is treated mainly with antipsychotic drugs through actions on various neurotransmitter receptors. The second-generation

*Address correspondence to Takashi Uehara:** Department of Neuropsychiatry, University of Toyama Graduate School of Medicine and Pharmaceutical Sciences, 2630 Sugitani, 930-0194 Toyama, Japan; Tel: 81-76-434-7323; Fax: 81-76-434-5030; E-mail: uehara@med.u-toyama.ac.jp

antipsychotic drugs, or atypical antipsychotic drugs (AAPDs), have been shown to be effective for partially alleviating cognitive impairment, which may be related to their relatively high affinity for serotonin (5-HT)$_{2A}$ receptors compared with dopamine (DA)-D$_2$ receptors [2,3]. Several lines of recent evidence suggest that 5-HT$_{1A}$ partial agonists also improve cognitive deficits of schizophrenia, a concept supported by data from rodents treated with N-methyl-D-aspartate (NMDA) receptor antagonists [4]. The cognitive benefits of 5-HT$_{1A}$ agonists are likely mediated by glutamate and γ-aminobutyric acid (GABA) neurons [5-7].

Schizophrenia is considered as "neurodegenerative" [8,9] and "neurodevelopmental" [10, 11] disorder. While these two concepts may appear contradictory, they are, in fact, neural processes to explain the pathophysiology of the illness, if coupled with the temporal factor. Each process may predominate at different stages of schizophrenia, which may contribute to the variety of symptoms [12]. Specifically, negative symptoms and cognitive impairment may be related to the morphological and functional changes of the brain in response to neurotoxic stress intrinsic to the disease [12,13].

Since the proposal of the astrocyte-neuron lactate shuttle (ANLS) hypothesis [14], lactate has been found to play a crucial role in energy metabolism in the brain [15,16]. Lactate also has been shown to protect neurons against glutamatergic excitotoxicity and oxidative stress [17,18]. In this article, we propose the concept that lactate metabolism provides a novel strategy for the development of therapeutics of schizophrenia. Preclinical and clinical data are presented to suggest a putative role for lactate production in the pro-cognitive effects of serotonin (5-HT)-5-HT$_{1A}$ receptor agonists.

Lactate Metabolism as a New Energy Source in the Brain

Astrocyte-Neuron Lactate Shuttle (ANLS) Hypothesis

Traditionally, lactate was considered a waste product with no substantial function in the metabolic process when eukaryotic cells have sufficient oxygen. However, mounting evidence suggests that lactate is an efficient energy substrate for neurons, and is used preferentially by neurons to maintain synaptic transmission, particularly during intense activity [15]. The role for glucose transporter (GLUT) and monocarboxylate transporter (MCT) is important in moving glucose and lactate between neurons and astrocytes [15,19]. Thus, exposure of astrocytes to glutamate has been shown to directly enhance glucose transport *via* GLUT1, accompanied with increased glucose utilization, while glucose transport is reduced in neurons on exposure to glutamate [19]. Lactate formed in astrocytes is then released in the extracellular space *via* the high capacity of MCT1 and MCT4. Moreover, ascorbic acid increases the activity of MCT2 on neurons and enhances lactate uptake by them.

Neural Activity and Lactate Metabolism

Since the proposal of the ANLS hypothesis, the role for lactate in energy production in brain cells has been a subject of intense debate [20-23]. The ANLS hypothesis predicts that this mechanism solely accounts for the energy metabolism during neural activation [24,25]. The hypothesis also supports the traditional view that glucose plays a role in providing neurons with energy [25].

It should be noted that lactate behaves differently depending on the phase of neural activation (early *vs.* late), its intensity, and length of stimulation period [25,26]. For example, "compartmentalization" of energy metabolism in neurons and glia cells using a mathematical modeling reveals that the ANLS could be valid during the first tens of seconds of sustained neural activation and during the post-stimulus period [27]. In fact, when glucose and lactate are present in physiological concentrations in the extracellular fluid, the proportion of energy source for glucose and lactate is 81 % and 19 %, respectively, as demonstrated by primary cultures of cortical neurons and astrocytes from the rat brain [28].

Nonetheless, lactate is a primary substrate for energy metabolism in the brain of humans [29] or rodents [30] under working conditions if both lactate and glucose are sufficiently available, as demonstrated by *in vivo* studies [30]. Accordingly, a positron emission tomography study [29] showed that the rate of glucose uptake in the whole brain was significantly reduced during intravenous infusion of lactate in humans.

5-HT$_{1A}$ Agonists in the Treatment of Cognitive Disturbances of Schizophrenia

5-HT$_{1A}$ Receptors and Cognition; Clinical Evidence

Disturbances of cognitive function are considered to largely affect the outcome in patients with schizophrenia. There is much attention to the role of psychotropic compounds acting on serotonin (5-HT) receptors in ameliorating cognitive deficits of the disease. Among the 5-HT receptor subtypes, the 5-HT$_{1A}$ receptor is attracting particular interests as a potential target for enhancing cognition [31-33].

We have conducted a series of clinical trials to determine if 5-HT$_{1A}$ agonists improve cognitive function in patients with schizophrenia [34-36]. The addition of tandospirone (30 mg/kg), but not placebo, to typical antipsychotic drugs (mainly haloperidol) for 4-6 weeks, was found to improve verbal memory, memory organization, and executive function [34,36]. Furthermore, in a randomized, double-blind, placebo-controlled study, the addition of buspirone (30 mg/kg) outperformed placebo in improving attention/information processing in patients treated with atypical antipsychotic drugs, such as olanzapine and risperidone [35]. These findings provided the basis for the ability of 5-HT$_{1A}$ receptor stimulation to enhance cognition, a therapeutic approach that has promoted the development of novel antipsychotic drugs [7,31-33].

Studies on Animal Models of Schizophrenia

We measured extracellular lactate concentrations (eLAC) using *in vivo* micro dialysis technique with an enzyme reactor/fluoro metric detector, according to the methods of Korf *et al.* [37,38] with minor modification [39]. First, we investigated the acute effect of 5-HT$_{1A}$ agonists on lactate metabolism in the medial prefrontal cortex (mPFC) and basolateral amygdala (BLA) of naïve adult rats. For example, acute systemic administration of tandospirone (2.0 mg/kg, i.p.), a 5-HT$_{1A}$ partial agonist, caused a significant increase in eLAC in the mPFC, but not BLA [40].

Next, we examined the effect of chronic 5-HT$_{1A}$ receptor stimulation on lactate metabolism in the mPFC of rats administered MK-801, a non-competitive NMDA receptor antagonist, on postnatal days (PDs) 7-10. These animals have been shown to elicit (1) impairment of set-shifting test, a measure of PFC function, in early adulthood [41], (2) disruption of prepulse inhibition, a measure of sensorimotor gating [42,43], and (3) enhancement of methamphetamine-induced locomotor activity [43] after, but not before puberty. Moreover, the same model animals demonstrated a decreased number of parvalbumin-positive (PV-p) GABA interneurons in the mPFC and hippocampus [44]. These findings suggest that neonatal treatment with NMDA antagonist produces behavioral and histochemical abnormalities mimicking some of the features of schizophrenia. Especially, the decreased number of PV-p GABA neurons led to the hypothesis that this animal model would show altered lactate metabolism in the mPFC, because GABAergic neurotransmission regulates lactate production under basal conditions and during neural activation [39,45].

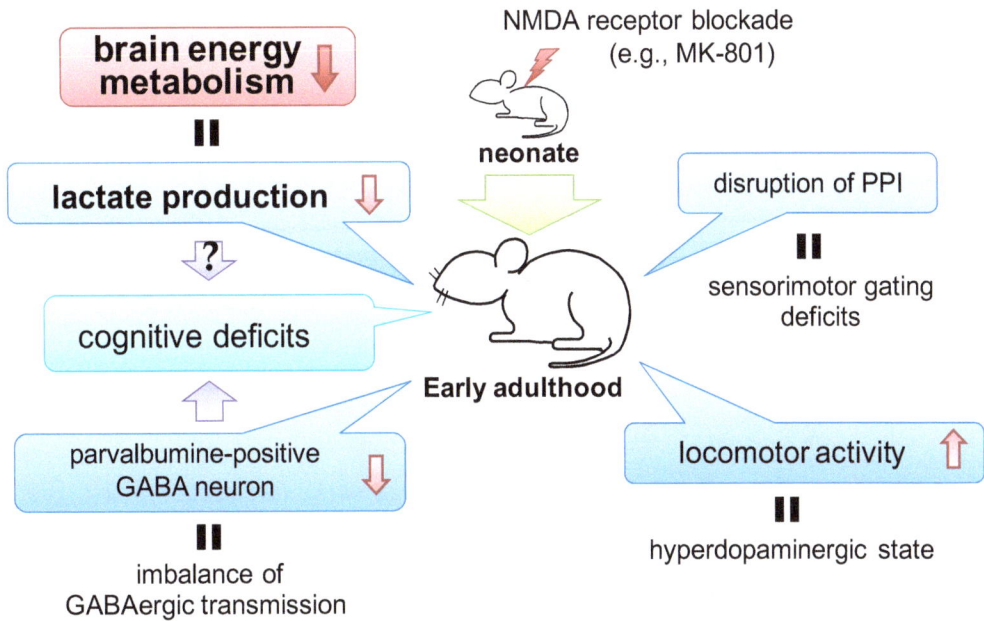

Figure 1: The schema of the behavioral, neurochemical and immunohistochemical abnormalities in an animal model of schizophrenia. Rats received MK-801 (0.2 mg/kg/day, s.c.) neonatally (postnatal days 7-10).

Transient exposure to MK-801 administrations in the neonatal stage suppressed foot-shock stress-induced lactate production in the mPFC, but not in the caudate-putamen (CPu) of rats in the young adult stage (PD 63), without changes in the basal levels [46]. MK-801-induced attenuation of lactate response to footshock was recovered by 14-day treatment with tandospirone (1.0 mg/kg/day) (starting on PD 49), which was antagonized by co-administration of WAY-100635, a 5-HT$_{1A}$ antagonist [46]. These results support the hypothesis that cognitive deficits of schizophrenia are produced, in part, by dysfunction of lactate metabolism in the mPFC.

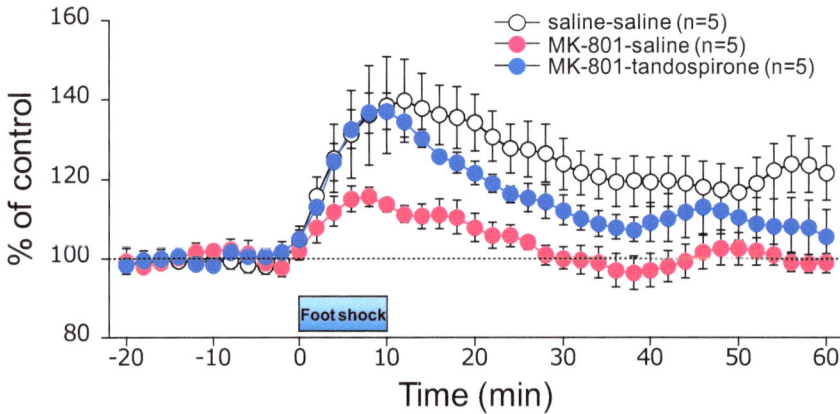

Figure 2: Time course of extracellular lactate concentrations (eLAC) induced by footshock stress in the medial prefrontal cortex (mPFC) of rats with (red circle) or without (open circle) neonatal MK-801 treatment. Neonatal administration of MK-801 led to suppression of footshock stress-induced eLAC increment on PD 63. Tandospirone (1.0 mg/kg/day), administered once daily for 14 days before the microdialysis examination, ameliorated MK-801-induced suppression of lactate production in the mPFC (blue circle).

The mechanisms by which tandospirone ameliorates lactate metabolism can be explained in several ways. For example, we proposed that $5\text{-}HT_{1A}$ receptor agonists stimulate glutamatergic neurotransmission through inhibitions of GABA neurons [7], as follows; $5\text{-}HT_{1A}$ receptors are located on pyramidal (glutamatergic) cells and GABAergic interneurons. Systemic administration of 8-OH-DPAT, a full $5\text{-}HT_{1A}$ agonist, to rats increases the discharge rate of pyramidal neurons in mPFC by inhibiting fast-spiking GABAergic interneurons through a preferential action on $5\text{-}HT_{1A}$ receptors on these latter neurons [33, 47].

Postmortem studies report reductions in the number of PV-p GABA interneurons in the mPFC of patients with schizophrenia [48]. Moreover, the concentration of cortical GABA and the activity of glutamate decarboxylase (GAD) 67, the enzyme that synthesizes GABA, have been shown to be decreased [49,50]. These observations have led to the "GABA hypofunction" theory [51]. However, a recent study with proton magnetic resonance spectroscopy demonstrated elevations of GABA concentrations in the mPFC of unmedicated patients with schizophrenia [52], while medicated patients elicited reductions or no change of GABA levels in the cortical regions and basal ganglia [53-55]. These findings are consistent with our hypothesis, discussed above, that the ability of 5-HT_{1A} stimulation to ameliorate cognitive disturbances of schizophrenia is mediated by enhancement of glutamatergic activity through inhibition of GABA neurons [33]. This concept is supported by clinical observations that augmentation therapy with tandospirone enhanced mismatch negativity, an electrophysiological cognitive marker of glutamate neuron activity, as well as neuropsychological performance in schizophrenia [6].

Lactate Metabolism as Protection Against Neurodegeneration

Schizophrenia and Neurodegeneration

Progressive pathophysiological processes possibly begins in the prodromal stage of schizophrenia, and continues after the onset of the illness [9,12,56]. The underlying mechanisms may include apoptosis (a.k.a. programmed cell death). The vulnerability of neurons to pro-apoptotic insults (pro-apoptotic triggers) would lead to selective dendritic and synaptic losses observed in subjects with schizophrenia [13]. The number of spines on the dendrites of pyramidal neurons in the frontal association cortex are largely reduced in schizophrenia, while neurons themselves are reduced in size and packed more density without any change in number [57,58]. While many factors have been found to promote apoptosis, several stimuli have also been suggested to play a role, including 1) glutamatergic excitotoxicity, 2) excess synaptic calcium flux, 3) oxidative stress, and 4) reduced neurotrophin levels (*e.g.*, brain-derived neurotrophic factor; BDNF, neurotrophin-3; NT-3) [13].

Neuroprotective Effect of Lactate

Accumulated evidence to date indicates that lactate plays a critical role in neuroprotection, as demonstrated by *in vitro* and *in vivo* experiments. Using hippocampal slices, the uptake of lactate into neurons was found to protect against the effects of incubation with glutamate [59]. An experiment conducted under culture conditions showed that over expression of GLUT1 in glial cells or MCT2 in neurons protected themselves against glutamatergic excitotoxicity [60]. It is note worthy that increased activity of GLUT1 or MCT2 enhances lactate concentrations in the extracellular space. MCT2 over expression in the hippocampal culture showed that glutamate increased lactate utilization, resulting in neuroprotection [60]. On the other hand, over expressed GLUT1 in the cortical glial culture also led to a significant increase in glucose uptake and enhanced extracellular lactate accumulation in the presence of glutamate. Moreover, glutamate toxicity was reduced in neurons in the presence of glial cells overexpressing GLUT1 [60].

The ability of lactate to protect against excitotoxic neurotoxicity has also been shown by *in vivo* examinations. Co-perfusion of glutamate and lactate into the fronto-parietal cortex of unanesthetized rats through microdialysis probe reduced the size of excitotoxic lesions compared to perfusion of glutamate alone [17]. Under this condition, perfusion of glutamate alone led to decreased extracellular glucose concentrations and increased lactate levels. These findings are consistent with the ANLS hypothesis predicting that glutamate enhances glucose uptake into astrocytes and lactate release from astrocytes into the extracellular space [15,19].

The neuroprotective effect of lactate is also explained by reduced production of reactive oxygen species (ROS) [18]. Using the rat hippocampal slice, the study showed that neural tissue activation was accompanied by aerobic lactate and nicotinamid adenosine dinucleotide diphosphate (NADH) production, the latter being produced when lactate is converted to pyruvate by mitochondrial lactic dehydrogenase (LDH) [61,62]. NADH, formed during lactate conversion to pyruvate in the neuronal mitochondrial membrane, neutralizes glutamate-induced neuronal damage due to production of ROS [18].

Lactate Metabolism as a New Target for Treatment of Schizophrenia

The neurodegenerative features of schizophrenia and the ability of lactate to protect neurons led us to hypothesize that lactate would provide a novel tool for the prevention of schizophrenia. Clinical and preclinical evidence predicts that 5-HT_{1A} agonists would improve the course of the disease. Fig. **3** shows the schema explaining this concept and the abnormalities of lactate metabolism in the brain, as demonstrated in our model animals.

Genetic and environmental factors, such as viral infection, may affect neural circuits during adolescence or young adulthood, leading to the emergence of positive and negative symptoms [63,64]. Cognitive impairment is present as early as the premorbid and prodromal stages of the illness [56]. Data from longitudinal studies suggest a progressive volume loss of discrete brain regions in patients even after the onset [65-68]. Enduring negative symptoms (deficit syndrome) and cognitive disturbances of schizophrenia may be caused by progressive neurodegeneration or the apoptosis procedure [13,56,69].

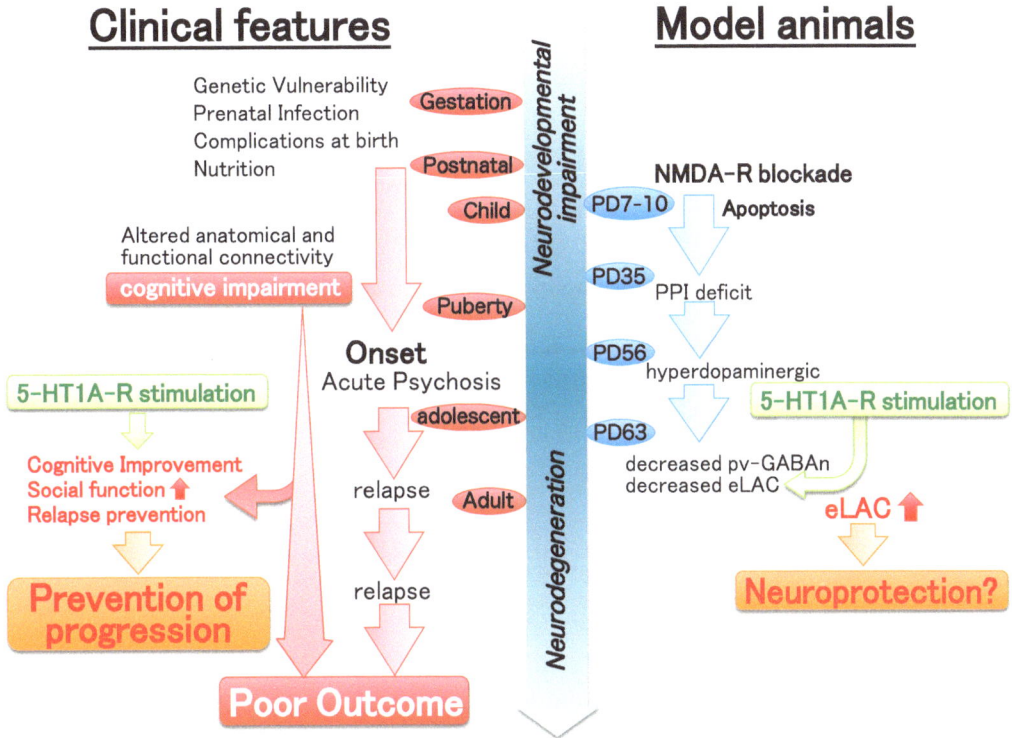

Figure 3: The relationship between clinical features of schizophrenia and behavioral and neurochemical manifestations in our animal model; PPI, prepulse inhibition; pv-GABAn, parvalbumin-positive GABA interneurons; eLAC, extracellular lactate concentrations.

The cognition-enhancing effect of 5-HT$_{1A}$ agonists in clinical trials is consistent with preclinical evidence for the ability of these agents to reverse suppression of lactate production in the mPFC of model animals [46]. Moreover, 5-HT$_{1A}$ agonists are expected to prevent progression to overt schizophrenia in vulnerable individuals through neuroprotective actions of lactate.

CONCLUSION

Although some AAPDs may elicit neuroprotective actions [70], about one third of patients are resistant to treatment with existing antipsychotic drugs [71]. We have proposed that lactate provides a novel probe for the development of therapeutics for schizophrenia. 5-HT_{1A} receptor agonists are one of the candidate agents to normalize lactate metabolism in the brain of subjects with the disease [72]. Further research, preferably of the translational nature, is needed to clarify this issue.

ACKNOWLEDGEMENTS

This study was funded by Takeda Science Foundation, Grants-in-aid for Scientific Research from Japan Society for the Promotion of Science, Grants-in-aid from the Ministry of Health, Labour and Welfare, Japan, and SENSHIN Medical Research Foundation.

CONFLICT OF INTEREST

The author(s) confirm that this chapter content has no conflict of interest.

REFERENCES

[1]   McGurk SR, Meltzer HY. The role of cognition in vocational functioning in schizophrenia. Schizophrenia research 2000; 45: 175-184.
[2]   Meltzer HY, McGurk SR. The effects of clozapine, risperidone, and olanzapine on cognitive function in schizophrenia. Schizophr Bull 1999; 25: 233-255.
[3]   Meltzer HY, Huang M. *In vivo* actions of atypical antipsychotic drug on serotonergic and dopaminergic systems. Prog Brain Res 2008; 172: 177-197.
[4]   Newman-Tancredi A. The importance of 5-HT1A receptor agonism in antipsychotic drug action: rationale and perspectives. Curr Opin Investig Drugs 2010; 11: 802-812.
[5]   Newman-Tancredi A, Kleven MS. Comparative pharmacology of antipsychotics possessing combined dopamine D2 and serotonin 5-HT1A receptor properties. Psychopharmacology (Berl) 2011; 216: 451-473.
[6]   Higuchi Y, Sumiyoshi T, Kawasaki Y, Ito T, Seo T, Suzuki M. Effect of tandospirone on mismatch negativity and cognitive performance in schizophrenia: a case report. J Clin Psychopharmacol 2010; 30: 732-734.
[7]   Sumiyoshi T, Uehara T. serotonin-1A receptors and cognitive enhancement in schizophrenia: role for brain energy metabolism. In: Schizophrenia in the 21st Century. Burne, THJ (Ed). In Tech, Rijeka, 2012) 127-140.
[8]   Lewis DA, Lieberman JA. Catching up on schizophrenia: natural history and neurobiology. Neuron 2000; 28: 325-334.
[9]   Lieberman JA. Is schizophrenia a neurodegenerative disorder? A clinical and neurobiological perspective. Biological psychiatry 1999; 46: 729-739.
[10]   Weinberger DR. Implications of normal brain development for the pathogenesis of schizophrenia. Arch Gen Psychiatry 1987; 44: 660-669.

[11]   Weinberger DR. Neurodevelopmental perspectives on schizophrenia. In: Psychopharmacology; The Forth Generation of Progress. Bloom, FE, Kupfer, DJ (Eds.) (Raven Press, New York, 1995) 1171-1183.

[12]   Lieberman JA, Jarskog LF, Malaspina D. Preventing clinical deterioration in the course of schizophrenia: the potential for neuroprotection. J Clin Psychiatry 2006; 67: 983-990.

[13]   Jarskog LF, Glantz LA, Gilmore JH, Lieberman JA. Apoptotic mechanisms in the pathophysiology of schizophrenia. Prog Neuropsychopharmacol Biol Psychiatry 2005; 29: 846-858.

[14]   Pellerin L, Magistretti PJ. Glutamate uptake into astrocytes stimulates aerobic glycolysis: a mechanism coupling neuronal activity to glucose utilization. Proceedings of the National Academy of Sciences of the United States of America 1994; 91: 10625-10629.

[15]   Pellerin L, Magistretti PJ. Sweet sixteen for ANLS. J Cereb Blood Flow Metab 2012; 32: 1152-1166.

[16]   Tsacopoulos M, Magistretti PJ. Metabolic coupling between glia and neurons. J Neurosci 1996; 16: 877-885.

[17]   Ros J, Pecinska N, Alessandri B, Landolt H, Fillenz M. Lactate reduces glutamate-induced neurotoxicity in rat cortex. Journal of neuroscience research 2001; 66: 790-794.

[18]   Schurr A, Gozal E. Aerobic production and utilization of lactate satisfy increased energy demands upon neuronal activation in hippocampal slices and provide neuroprotection against oxidative stress. Front Pharmacol 2011; 2: 96.

[19]   Pellerin L. Brain energetics (thought needs food). Curr Opin Clin Nutr Metab Care 2008; 11: 701-705.

[20]   Chih CP, Lipton P, Roberts EL, Jr. Do active cerebral neurons really use lactate rather than glucose? Trends in neurosciences 2001; 24: 573-578.

[21]   Chih CP, Roberts Jr EL. Energy substrates for neurons during neural activity: a critical review of the astrocyte-neuron lactate shuttle hypothesis. J Cereb Blood Flow Metab 2003; 23: 1263-1281.

[22]   Korf J. Is brain lactate metabolized immediately after neuronal activity through the oxidative pathway? J Cereb Blood Flow Metab 2006; 26: 1584-1586.

[23]   Dienel GA, Cruz NF. Neighborly interactions of metabolically-activated astrocytes *in vivo*. Neurochemistry international 2003; 43: 339-354.

[24]   Pellerin L. Lactate as a pivotal element in neuron-glia metabolic cooperation. Neurochemistry international 2003; 43: 331-338.

[25]   Pellerin L, Bouzier-Sore AK, Aubert A *et al.* Activity-dependent regulation of energy metabolism by astrocytes: an update. Glia 2007; 55: 1251-1262.

[26]   Gjedde A, Marrett S, Vafaee M. Oxidative and nonoxidative metabolism of excited neurons and astrocytes. J Cereb Blood Flow Metab 2002; 22: 1-14.

[27]   Aubert A, Costalat R. Compartmentalization of brain energy metabolism between glia and neurons: insights from mathematical modeling. Glia 2007; 55: 1272-1279.

[28]   Ramirez BG, Rodrigues TB, Violante IR *et al.* Kinetic properties of the redox switch/redox coupling mechanism as determined in primary cultures of cortical neurons and astrocytes from rat brain. Journal of neuroscience research 2007; 85: 3244-3253.

[29]   Smith D, Pernet A, Hallett WA, Bingham E, Marsden PK, Amiel SA. Lactate: a preferred fuel for human brain metabolism *in vivo*. J Cereb Blood Flow Metab 2003; 23: 658-664.

[30]   Wyss MT, Jolivet R, Buck A, Magistretti PJ, Weber B. *In vivo* evidence for lactate as a neuronal energy source. J Neurosci 2011; 31: 7477-7485.

[31]   Ohno Y, Tatara A, Shimizu S, Sasa M. Management of cognitive impairments in schizophrenia: the therapeutic role of 5-HT receptors. In: Schizophrenia Research: Recent Advances. Sumiyoshi, T (Ed). Nova Science Publishers, New York, 2012; 321-335.

[32]   Newman-Tancredi A, Albert PR. Gene polymorphism at serotonin 5-HT1A receptors: moving towards personalized medicine for psychosis and mood deficits? In: Schizophrenia Research: Recent Advances. Sumiyoshi, T (Ed. (Nova Science Publisher, New York, 2012) 337-358.

[33]   Sumiyoshi T, Higuchi Y. facilitative effect of serotonin 1A receptor agonists on cognition in patients with schizophrenia. Curr Med Chem, (in press).

[34]   Sumiyoshi T, Matsui M, Yamashita I *et al.* The effect of tandospirone, a serotonin(1A) agonist, on memory function in schizophrenia. Biological psychiatry 2001; 49: 861-868.

[35]   Sumiyoshi T, Park S, Jayathilake K, Roy A, Ertugrul A, Meltzer HY. Effect of buspirone, a serotonin1A partial agonist, on cognitive function in schizophrenia: a randomized, double-blind, placebo-controlled study. Schizophrenia research 2007; 95: 158-168.

[36]   Sumiyoshi T, Matsui M, Nohara S *et al.* Enhancement of cognitive performance in schizophrenia by addition of tandospirone to neuroleptic treatment. American journal of psychiatry 2001; 158: 1722-1725.

[37]   Kuhr WG, Korf J. Extracellular lactic acid as an indicator of brain metabolism: continuous on-line measurement in conscious, freely moving rats with intrastriatal dialysis. J Cereb Blood Flow Metab 1988; 8: 130-137.

[38]   Schasfoort EM, De Bruin LA, Korf J. Mild stress stimulates rat hippocampal glucose utilization transiently *via* NMDA receptors, as assessed by lactography. Brain research 1988; 475: 58-63.

[39]   Uehara T, Sumiyoshi T, Itoh H, Kurata K. Lactate production and neurotransmitters; evidence from microdialysis studies. Pharmacol Biochem Behav 2008; 90: 273-281.

[40]   Uehara T, Sumiyoshi T, Matsuoka T, Itoh H, Kurachi M. Role of 5-HT(1A) receptors in the modulation of stress-induced lactate metabolism in the medial prefrontal cortex and basolateral amygdala. Psychopharmacology (Berl) 2006; 186: 218-225.

[41]   Stefani MR, Moghaddam B. Transient N-methyl-D-aspartate receptor blockade in early development causes lasting cognitive deficits relevant to schizophrenia. Biological psychiatry 2005; 57: 433-436.

[42]   Uehara T, Sumiyoshi T, Seo T *et al.* Long-term effects of neonatal MK-801 treatment on prepulse inhibition in young adult rats. Psychopharmacology 2009; 206: 623-630.

[43]   Uehara T, Sumiyoshi T, Seo T *et al.* Neonatal exposure to MK-801, an N-methyl-d-aspartate receptor antagonist, enhances methamphetamine-induced locomotion and disrupts sensorimotor gating in pre- and postpubertal rats. Brain research 2010; 1352: 223-230.

[44]   Uehara T, Sumiyoshi T, Hattori H *et al.* T-817MA, a novel neurotrophic agent, ameliorates loss of GABAergic parvalbumin-positive neurons and sensorimotor gating deficits in rats transiently exposed to MK-801 in the neonatal period. J Psychiatr Res 2012; 46: 622-629.

[45]   Uehara T, Sumiyoshi T, Matsuoka T *et al.* Enhancement of lactate metabolism in the basolateral amygdala by physical and psychological stress: Role of benzodiazepine receptors. Brain research 2005; 1065: 86-91.

[46]   Uehara T, Itoh H, Matsuoka T *et al.* Effect of transient blockade of N-methyl-D-aspartate receptors at neonatal stage on stress-induced lactate metabolism in the medial prefrontal cortex of adult rats: Role of 5-HT1A receptor agonism. Synapse (New York) 2012; 66: 408-417.

[47]   Llado-Pelfort L, Santana N, Ghisi V, Artigas F, Celada P. 5-HT1A receptor agonists enhance pyramidal cell firing in prefrontal cortex through a preferential action on GABA interneurons. Cereb Cortex 2012; 22: 1487-1497.

[48]   Lewis DA, Hashimoto T, Volk DW. Cortical inhibitory neurons and schizophrenia. Nat Rev Neurosci 2005; 6: 312-324.

[49]   Perry TL, Kish SJ, Buchanan J, Hansen S. Gamma-aminobutyric-acid deficiency in brain of schizophrenic patients. Lancet 1979; 1: 237-239.

[50]   Bird JM. Computed tomographic brain studies and treatment response in schizophrenia. Canadian journal of psychiatry. Revue canadienne de psychiatrie 1985; 30: 251-254.

[51]   Lisman JE, Coyle JT, Green RW *et al.* Circuit-based framework for understanding neurotransmitter and risk gene interactions in schizophrenia. Trends in neurosciences 2008; 31: 234-242.

[52]   Kegeles LS, Mao X, Stanford AD *et al.* Elevated prefrontal cortex gamma-aminobutyric acid and glutamate-glutamine levels in schizophrenia measured *in vivo* with proton magnetic resonance spectroscopy. Arch Gen Psychiatry 2012; 69: 449-459.

[53]   Yoon JH, Maddock RJ, Rokem A *et al.* GABA concentration is reduced in visual cortex in schizophrenia and correlates with orientation-specific surround suppression. J Neurosci 2010; 30: 3777-3781.

[54]   Tayoshi S, Nakataki M, Sumitani S *et al.* GABA concentration in schizophrenia patients and the effects of antipsychotic medication: a proton magnetic resonance spectroscopy study. Schizophrenia research 2010; 117: 83-91.

[55]   Goto N, Yoshimura R, Kakeda S *et al.* No alterations of brain GABA after 6 months of treatment with atypical antipsychotic drugs in early-stage first-episode schizophrenia. Prog Neuropsychopharmacol Biol Psychiatry 2010; 34: 1480-1483.

[56]   Lieberman JA, Perkins D, Belger A *et al.* The early stages of schizophrenia: speculations on pathogenesis, pathophysiology, and therapeutic approaches. Biological psychiatry 2001; 50: 884-897.

[57]   Selemon LD, Goldman-Rakic PS. The reduced neuropil hypothesis: a circuit based model of schizophrenia. Biological psychiatry 1999; 45: 17-25.

[58]   Garey L. When cortical development goes wrong: schizophrenia as a neurodevelopmental disease of microcircuits. Journal of anatomy 2010; 217: 324-333.

[59]   Schurr A, Miller JJ, Payne RS, Rigor BM. An increase in lactate output by brain tissue serves to meet the energy needs of glutamate-activated neurons. J Neurosci 1999; 19: 34-39.

[60]   Bliss TM, Ip M, Cheng E *et al.* Dual-gene, dual-cell type therapy against an excitotoxic insult by bolstering neuroenergetics. J Neurosci 2004; 24: 6202-6208.

[61]   Schurr A. Lactate: the ultimate cerebral oxidative energy substrate? J Cereb Blood Flow Metab 2006; 26: 142-152.

[62]   Schurr A, Payne RS. Lactate, not pyruvate, is neuronal aerobic glycolysis end product: an *in vitro* electrophysiological study. Neuroscience 2007; 147:, 613-619.

[63]   Rapoport JL, Addington AM, Frangou S, Psych MR. The neurodevelopmental model of schizophrenia: update 2005. Mol Psychiatry 2005; 10: 434-449.

[64]   Fatemi SH, Folsom TD. The neurodevelopmental hypothesis of schizophrenia, revisited. Schizophr Bull 2009; 35: 528-548.

[65]   Cahn W, Hulshoff Pol HE, Lems EB *et al.* Brain volume changes in first-episode schizophrenia: a 1-year follow-up study. Arch Gen Psychiatry 2002; 59: 1002-1010.

[66]   Gur RE, Maany V, Mozley PD, Swanson C, Bilker W, Gur RC. Subcortical MRI volumes in neuroleptic-naive and treated patients with schizophrenia. American journal of psychiatry 1998; 155: 1711-1717.

[67]   Steen RG, Mull C, McClure R, Hamer RM, Lieberman JA. Brain volume in first-episode schizophrenia: systematic review and meta-analysis of magnetic resonance imaging studies. Br J Psychiatry 2006; 188: 510-518.

[68]   Kasai K, Shenton ME, Salisbury DF *et al.* Progressive decrease of left superior temporal gyrus gray matter volume in patients with first-episode schizophrenia. American journal of psychiatry 2003; 160: 156-164.

[69]   Carpenter WT, Jr., Arango C, Buchanan RW, Kirkpatrick B. Deficit psychopathology and a paradigm shift in schizophrenia research. Biological psychiatry 1999; 46: 352-360.

[70]   Lieberman JA, Bymaster FP, Meltzer HY *et al.* Antipsychotic drugs: comparison in animal models of efficacy, neurotransmitter regulation, and neuroprotection. Pharmacological reviews 2008; 60: 358-403.

[71]   Freedman R. Schizophrenia. N Engl J Med 2003; 349: 1738-1749.

[72]   Prabakaran S, Swatton JE, Ryan MM *et al.* Mitochondrial dysfunction in schizophrenia: evidence for compromised brain metabolism and oxidative stress. Mol Psychiatry 2004; 9: 684-697, 643.

Send Orders for Reprints on reprints@benthamscience.net

CHAPTER 6

Principles and Current Issues of Antiepileptic Drug Therapy

Laura Rosa Pisani, Vincenzo Belcastro, Giancarla Oteri and Francesco Pisani*

Department of Neurosciences, University of Messina, Messina, Italy

Abstract: The armamentarium to treat epilepsy includes today more than twenty drugs. These are classically distinguished as standard, traditional or first generation antiepileptic drugs (AEDs), which include phenobarbital, phenytoin, carbamazepine, valproic acid, ethosuximide and benzodiazepines, and new or second generation AEDs, which are vigabatrin, lamotrigine, felbamate, gabapentin, oxcarbazepine, tiagabine, topiramate, stiripentol, pregabalin, levetiracetam, rufinamide, zonisamide. More recently, other four compounds have been introduced, *i.e.,* lacosamide, retigabine, eslicarbazepine acetate and perampanel, also defined as third generation AEDs. The mechanism of antiepileptic action is mainly mediated by increase in inhibitory GABA activity and/or prolongation of sodium and/or calcium channel inactivation. From a pharmacokinetic point of view, the most relevant difference characterizing a number of the recent AEDs as compared to the traditional ones is the lack of or a milder induction of hepatic enzymes with consequent reduced risk of drug-drug interactions. Concerning therapeutic features, valproic acid exhibits the broadest spectrum of action and still remains the only AED which can be used to treat all types of seizures, from absences and other primarily generalized seizures to focal ones, and almost all syndromes. Among the recent compounds, lamotrigine, topiramate, levetiracetam and zonisamide have shown to be efficacious in primarily generalized seizures and in some idiopathic generalized epilepsies, but their effect against absence seizures is less potent than that of valproic acid or even irrelevant. All AEDs, except ethosuximide, exhibit similar efficacy against focal seizures and, apart from traditional drugs, lamotrigine, oxcarbazepine, gabapentin, topiramate and levetiracetam, and ZNS in Europe, have the indication of mono-therapy. Some of the new compounds have specific paediatric indications: vigabatrin against infantile spasms in West syndrome, felbamate and rufinamide against mixed seizures in the Lennox-Gastaut syndrome, and stiripentol against some types of seizures in Dravet syndrome. In spite of this variety of AEDs, the percentage of patients with refractory epilepsy has not changed over the last 50 years and is still stabilized around 30-40%. Adverse events are observed in one-third of patients on AED therapy. Frequent, unspecific and usually dose dependent CNS side effects occur with almost all AEDs and encompass sedation, somnolence, fatigue, and dizziness, usually attenuating or even disappearing over time. Acute idiosyncratic effects, such as skin rush, may be particularly troubling and may configure a hypersensitivity syndrome with rapid

---

*****Address correspondence to Pisani Francesco:** Department of Neurosciences, University of Messina, Polyclinic, Messina, Italy; Tel: +090 221207-34 8560 5758; Fax: +090+ 2212301; E-mail: pisanif@unime.it

degeneration to a severe and even life-threatening condition. Felbamate and vigabatrin are used exceptionally because of high incidence of aplastic anemia and liver failure occurring with the first and irreversible loss of visual field occurring with the second one. Subtle and slowly developing adverse effects, like bone mineral density reduction associated especially with traditional enzyme-inducing AEDs, require a continuous monitoring of the patient clinical condition. Antiepileptic therapy has special implications for women of child bearing age with regard to contraception, pregnancy and teratogenicity. There is some evidence that teratogenic effects, particularly frightening with valproic acid, with > 400mg/day dose of carbamazepine, and with polytherapies, are less frequent with lamotrigine. Given the large number of available AEDs, opportunities to tailor drug therapy on the individual patient are various. Treatment decisions, however, are complex and need to be individualised on the basis of careful evaluation of a number factors related to drug, disease and the patient. Choice of first-line therapy for a specific form of epilepsy, the time at which the drug should be started, and which strategy is most appropriate after failure of the first drug are key decision steps. Patient-specific factors include age, sex, childbearing potential, co-morbidities, and concomitant medications. Future directions include discovery of drugs with an improved safety profile, with more potent anti-seizure effect, able to prevent epileptogenesis and, possibly, to interact with specific genetic substrates.

Keywords: Traditional antiepileptic drugs (AEDs), second and third generation AEDs, neuronal membrane stabilization, ion channels modulation, GABA transmission, pharmacokinetics, drug interactions, pregnancy, adverse effects of AEDs, clinical indications of AEDs, refractory epilepsy.

HISTORICAL NOTES

The history of the modern treatment of epilepsy starts in the middle of the nineteenth century with the introduction of bromides. Previous approach was based on purely empirical attempts and magic beliefs.

From Bromides and Phenobarbital to the Third Generation Drugs: The Milestones

The first step in drug therapy of epilepsy was the result of an intuition of a skilled English obstetrician of the nineteenth century, Sir Charles Locock. In 1857 he, at that time President of the Royal Medical and Chirurgical Society in London, commented 52 cases of epilepsy presented by Edward H Sieveking and reported his successful personal experience in treating with potassium bromide 15 young women suffering from "hysterical epilepsy" [1]. On the basis of a German report that bromide had caused a reversible impotence, Locock tried the compound in young women troubled by sexual excitement obtaining an evident sedation. He thought that, given such an effect, bromide might result efficacious against epilepsy. Although it has been pointed out for historical correctness that two other distinguished English physicians, Bland Radcliffe and Samuel Wilks, independently made soon the same observation [2], classically the short Locock's dissertation remains the beginning of the history of the actual epilepsy treatment. Potassium bromide, therefore, was the choice treatment of epilepsy until 1912, year of publication of the paper by Alfred Hauptman on the antiepileptic effect of phenobarbital (PB) [3]. The drug was already largely used as a sedative. Hauptman, a young clinical assistant in Freiburg, gave PB to tranquillize epileptic patients who kept him awake in the ward during the night (at least so it was handed down) with their seizures. Surprisingly, he noted that, apart from sedation, epileptic fits were drastically reduced or completely disappeared. Thus, even this second milestone was a result of a serendipitous observation. The introduction of PB considerably decreased the popularity of potassium bromide in treating humans with epileptic seizures and still remains a strength of today's antiepileptic therapy. Twenty-six years thereafter Merritt and Putnam published their famous article on the antiepileptic effect of sodium diphenyl hydantoinate tested for the first time in a large population of epileptic patients [4]. Additionally, they underlined that the new drug was less sedative and hypnotic than phenobarbital. The compound had been tested as a tranquillizer in 1916, just few years after the introduction of PB [5]. Merritt and Putnam began their collaboration in 1935 at the Neurological Unit of the Boston City Hospital, screening 19 chemicals structurally similar to phenobarbital in an experimental model of epilepsy, namely electrically induced seizures in cats [6]. These phenyl derivatives had been provided by the pharmaceutical company Parke-Davis. The researchers found that diphenylhydantoin (phenytoin, PHT) displayed the strongest anticonvulsant effect with low or even absent sedative effects. As it has been described in detail [7, 8], some other compounds had been previously tested in experimental models of epilepsy and that the anticonvulsant effect can be separated from the sedative one had already been previously observed. However, the research of Merritt and Putnam is rightly considered a revolutionary and extraordinary shift in the therapeutic approach to epilepsy for the following reasons: a) for the first time a compound had been used in humans soon after it was tested in animals; b) for the first time it had been clearly stressed that the desired therapeutic action could be obtained without the coexistence of an unwonted effect; c) for the first time a fruitful and beneficial partnership between a scientific institution and a pharmaceutical industry had been made. After the paper of Merritt and Putnam, phenytoin soon became the best antiepileptic drug available. By the end of the World War II, pharmaceutical companies played an important role in antiepileptic drug discovery. At the beginning of 50s, another drug, primidone, the antiepileptic effect of which had been demonstrated by Yule Bogue in 1949, was introduced in the treatment of epilepsy [9]. The drug, a compound of the pyrimidinedione class, is largely converted into PB in the liver. The 50s are spectators of the marketing of another drug: ethosuximide (ESM) indicated for absence seizures [10, 11]. Carbamazepine (CBZ) had been developed during the 50s by the Swiss pharmaceutical J.R. Geigy AG in Basel, Switzerland. It was initially investigated as a drug for depression and psychosis due to its chemical structure similar to that of the antidepressant tricyclic compound imipramine [12]. Following the first historical trials in the first half of the 60s [13-15], CBZ was used as an anticonvulsant in the UK since 1965, and approved in the USA since 1974. Anticonvulsant properties of valproic acid (VPA), which is structurally unrelated to other antiepileptic drugs, were discovered by chance. The drug was first synthesized in 1882 by B.S. Burton as an analogue of valeric acid, found naturally in valerian [16]. It is a fatty acid that is a clear liquid at room temperature and, for many decades, its only use was in laboratories as a solvent for organic compounds. In 1962, Pierre Eymard, a research student at the University of Lyon (France) and colleagues used VPA as a solvent, investigating potentially anticonvulsant drugs with low aqueous solubility against pentylenetetrazol-induced convulsions in laboratory rats. A strong anticonvulsant activity was observed in all solutions where the compound was present, leading these authors to investigate VPA as a potentially useful agent for epilepsy [17]. It was approved as an antiepileptic drug in 1967 in France and has soon gained a worldwide use as a large spectrum antiepileptic drug [18]. The 60s were also the years of introduction of benzodiazepines (BZ). The first compound of this class was chlordiazepoxide, which, tested in a standard battery of animal experiments, showed very strong anticonvulsant effects.

The drug was speedy introduced throughout the world in 1960 by Hoffmann–La Roche pharmaceutical company under the brand name Librium [19, 20]. Following chlordiazepoxide, diazepam (DZ) was the second compound marketed under the brand name Valium in 1963 by the same company [21]. Soon it revealed its strong therapeutic effects against status epilepticus [22]. The more recent years are characterized essentially by three complementary and integrative development stages: a) synthesis of compounds with a chemical structure hypothesised to have a specific mechanism of action interfering with seizure activity; b) systematic screening against seizures in various experimental epilepsy models; c) careful initial evaluation tests in small numbers of subjects and then of patients aimed to explore tolerability and pharmacokinetic properties; these preliminary tests are followed by more large clinical trials with precise rules and designs, including randomization and control groups. The most explicative example of this policy is development of vigabatrin (VGB). The creation of this drug, marketed as Sabril, dates back to 1974 and was the result of a deliberate search for a molecule to increase γ -aminobutyric acid (GABA) levels in the central nervous system (CNS). Due to its chemical structure, very similar to the inhibitory neurotransmitter γ-aminobutyric acid (GABA), it acts as an irreversible inhibitor of GABA-transaminase, the enzyme responsible for GABA metabolism. This action results in increased levels of GABA in the central nervous system. Clinical trials began in the late 70s and early 80s [23-25], leading to first marketing approval in UK in 1989. It has now been approved in over 50 countries worldwide.

The actual Anticonvulsant Drug Development Program in the US resulted in the global licensing of a number of antiepileptic drugs (AEDs). These are, in alphabetical order, eslicarbazepine acetate, a pro-drug of eslicarbazepine (ESL), felbamate (FBM), gabapentin (GBP), lacosamide (LCS), lamotrigine (LTG), levetiracetam (LEV), oxcarbazepine (OXCZ), perampanel (PRP), pregabalin (PGB), retigabine, rufinamide (RFN), stiripentol (STP), tiagabine (TGB), topiramate (TPM), zonisamide (ZNS). Some of these drugs, for example ESL, has been approved only in Europe, but not in USA. In spite of increase in our knowledge in basic mechanisms of seizure activity and development of more accurate animal models of epilepsy, the last two decades are characterized by empirically, and sometimes serendipitously, discovery of new AEDs and this explains their diverse chemical structure (Fig. **1)**. More detailed information on the history of AEDs can be found in other excellent reviews [26-28].

MECHANISMS OF ACTION OF ANTIEPILEPTIC DRUGS

An epileptic seizure is the consequence of an abnormal and excessive neuronal activity in the brain. This implies that a drug intended to counteract seizures should be able to potentiate inhibitory neurotransmission and/or inhibit neuronal excitability with the final result of stabilizing neuronal membrane. These effects can be obtained trough different mechanisms. The main targets for AED action are voltage-dependent sodium (Na^+), calcium (Ca^{2+}), potassium channels, $GABA_A$ receptors, h-channels, excitatory amino acid receptors, some enzymes and synaptic proteins [29-33]. Blockade of voltage dependent sodium and calcium channels and the GABA mimetic effects prevail over the other mechanisms. On the basis of the present knowledge, some drugs act through a single mechanism of action, some others with two or three, while at least ten out of 21 available drugs exhibit multiple mechanisms of actions. (Table **1**) illustrates the mode of action of individual drugs. Although continuous effort has been spent to establish a link between a given mechanism of action or an association of different mechanisms and clinical effects in terms of efficacy and toxicity, this still remains a goal to be achieved [34]. In general, AEDs with a broad spectrum of clinical activity usually exert their action through multiple mechanisms (Table **1**) [34]. However, although BZs act through a single mechanism of action, *i.e.,* affinity to GABA-A receptor (Table **1**) [29-33], they exhibit a broad spectrum of action, being effective drugs against both partial and generalized seizures.

Increase in Inhibitory Transmission: The GABA Target

GABA is one of the main inhibitory neurotransmitters in the brain and interacts with three specific receptors, $GABA_A$, $GABA_B$ and $GABA_C$. PB and BZ bind to sites associated to $GABA_A$ receptors in the form of a receptor complex, which controls opening of the chloride channel. When GABA binds to the receptor complex, the channel is opened and chloride anions enter the neuron, inducing a condition of hyperpolarization. $GABA_B$ receptors are metabotropic and, as so, they are associated to a cascade of second messengers. Ionotropic $GABA_C$ receptors are mainly located in the retina.

VPA potentiates GABA transmission through different effects on GABA metabolism. It stimulates the GABA synthesizing enzyme, glutamate-decarboxylase, and inhibits the GABA metabolizing enzymes, GABA transaminase, succinic-semialdehyde-dehydrogenase, and aldehyde reductase [29-33].

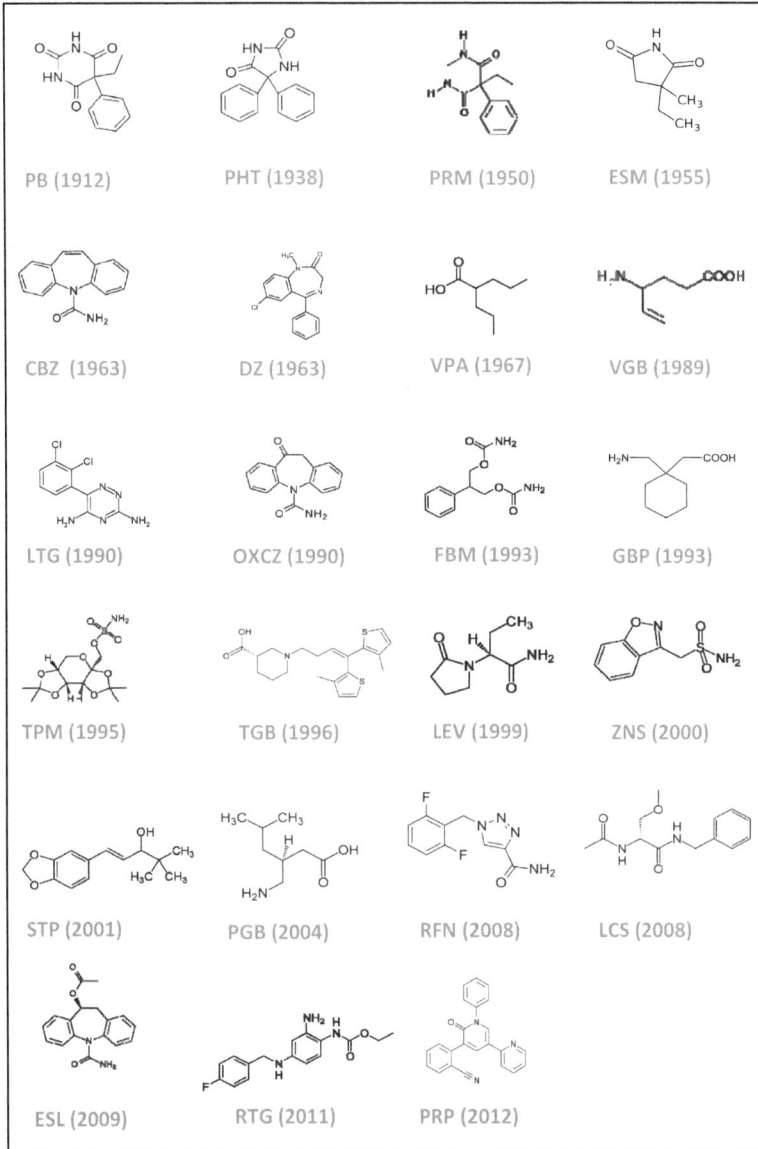

Figure 1: Chemical structure of antiepileptic drugs (AEDs) with in parentheses their earliest year of use. PB= phenobarbital, PHT= phenytoin, PRM= primidone, ESM= ethosuximide, CBZ= carbamazepine, DZ= diazepam, VPA= valproic acid, VGB= vigabatrin, LTG= lamotrigine, OXCZ= oxcarbazepine, FBM= felbamate, GBP= gabapentin, TPM= topiramate, TGB= tiagabine, LEV= levetiracetam, ZNS= zonisamide, STP= stiripentol, PGB= pregabalin, RFN= rufinamide, LCS= lacosamide, ESL= eslicarbazepine acetate, RTG= retigabine, PRP= perampanel.

Among the recent AEDs, VGB and TGB have been specifically and successfully synthesized to act selectively through the GABAergic system. The former increases the synaptic concentration of GABA by irreversible inhibition of the main GABA metabolizing enzyme GABA-aminotransferase (GABA-transaminase), whilst the latter inhibits neuronal and glial uptake of GABA occurring through specific GABA transporters (Fig. **2**) [29-33]. GBP was also

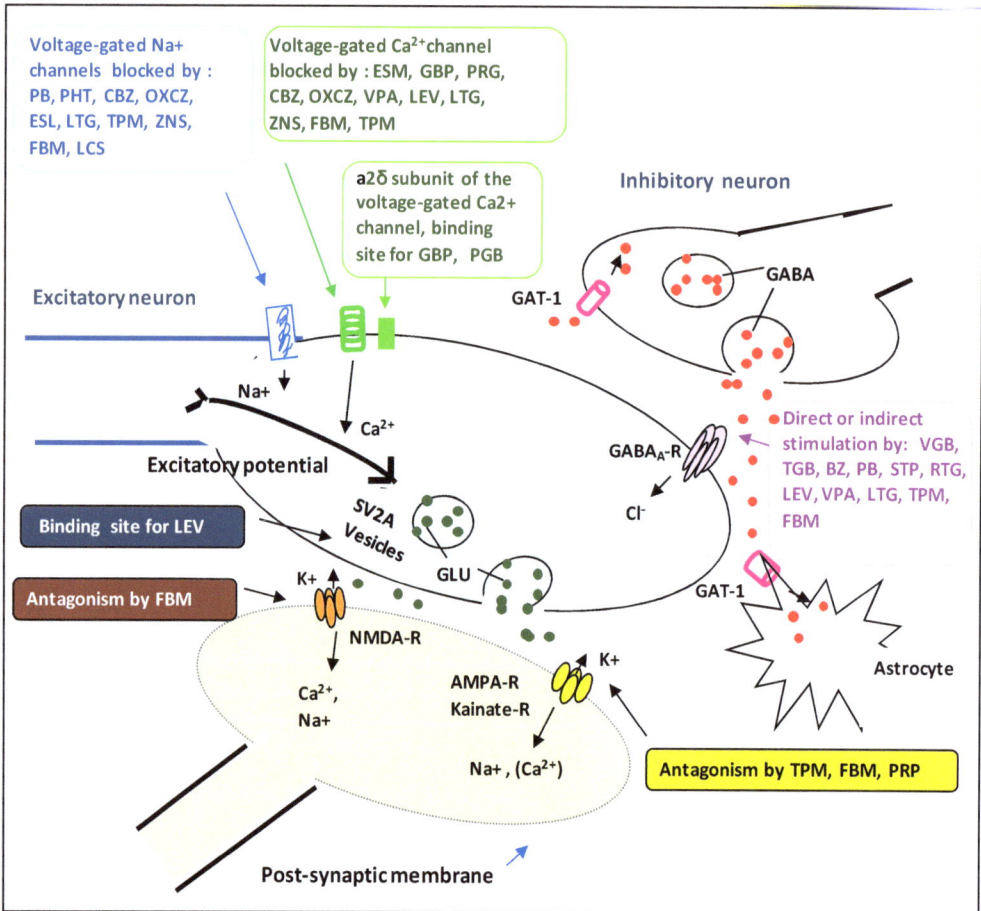

Figure 2: Main mechanisms of action of AEDs. These include: (a) increased inhibitory GABA transmission through increased GABA production by inhibition of its catabolism (VGB), its transport into the cell (TGB) or receptor stimulation (other AEDs); (b) modulation of ion channels with consequent reduced excitatory potential; (c) antagonism of glutamate receptors. New targets, *e.g.*, binding to SV2A vesicles of LEV, have been recently identified for some third generation AEDs. Primary action of RTG is missing. Actions of anti-absence drugs on T-type calcium channels are also missing. For details see the text and references [29-33].

designed to act stimulating GABA transmission and, as a precursor of GABA easily entering the brain, it increases brain synaptic GABA. Additionally, its mechanism of action is mediated by a decreased influx of calcium ions into neurons *via* a specific subunit of voltage-dependent calcium channels. Most AEDs act through an increase in GABA levels and/or binding to $GABA_A$ receptor in association to other mechanisms (Table **1**) (Fig. **2**) [29-33].

Blockade of Sodium (Na^+), Calcium(Ca^{++}) and Potassium(K^+) Channels

Genes mutations may lead to abnormal neuronal activity and determine inherited epilepsies which are today included in the big chapter of channelopathies. An ever increasing number of channelopathies due to the mutation of subunits of voltage-dependent sodium, potassium, calcium and chloride neuronal channels has been identified with relevant implications for antiepileptic therapy and development of new AEDs [35]. In particular, voltage-gated sodium channels are proteins responsible for the generation of action potentials and are the main targets for most conventional and recent AEDs (Table **1**) (Fig. **2**). Following an action potential, voltage-dependent sodium channels enter an inactive state.

AEDs exert their effect by stabilizing an inactivated form of the channels. One fast- and different slow-inactivated states have been identified [36]. CBZ and PHT block preferentially the excitation of cells that are firing repetitively, and the higher the frequency of firing, the greater the block produced (use-dependency effect). Available data indicate that both drugs specifically stabilize the fast-inactivated state [29-33]. The consequence of this block is prevention from returning of the channels to the resting state, and thus reducing the number of functional channels available to generate action potentials. LTG, like PHT and CBZ, decreases sustained high-frequency repetitive firing of voltage-dependent sodium action potentials and this effect has been seen to result in a preferential decreased release of pre-synaptic glutamate [29-33]. Although considerable differences among drugs does exist (Table **1**), also TPM, OXCZ, ESL and ZNS exhibit this primary mechanism of action (Table **1**) (Fig. **2**). A novel mechanism has been recently demonstrated for LCS. This drug is the only AED known to reduces sodium channels availability by selective enhancement of slow inactivation but without apparent interaction with fast inactivation gating [36].

Slow inactivation is different from fast-inactivation in that it does not produce complete channel blockade. LCS, therefore, affects neurons which are depolarized or active for long periods of time like it happens at an epileptic focus. This specific mechanism open innovative and promising opportunities to develop a new class of AEDs [37].

Table 1: Mechanisms of action of individual AEDs

		Increase in GABA levels	Affinity to GABA$_A$ Receptor	Blockade of Sodium Channels	Blockade of Calcium Channels	Inhibition of Glutamate Excitation	Activation of Potassium	Others Mechanisms Channels
One Mechanism	VGB	+						
	TGB	+						
	BZs*		+					
	ESL			+				
	LCS			+				
	ESM				+ (T-type)			
	GBP				+ (N-, P/Q-type)			
	PGB				+ (N-, P/Q-type)			
	PRP					+		
At least 2 mechanisms	STP	+	+					
	PHT			+				+
At least 3 mechanisms	RTG	+	+				+ Kv7.2-7.5	
	LEV		+		+ (N-type)			+
	CBZ			+	+ (L-type)			+
	VPA	+			+ (T-type)			+
	OXCZ			+	+ (N-, P-type)			+
4-to-6 mechanisms	ZNS			+	+ (N-, P-, T-type)	+		+
	PB	+	+	+		+		+
	LTG	+		+	+ (N-, P/Q-, R-, T-type)	+	+	
	TP M	+	+	+	+ (L-type)	+		+
	FBM	+	+	+	+ (L-type)	+		+

Data from references [29-31]. Controversial mechanisms are not included. PRM is not included in this and other tables because of its nearly complete transformation into PB. *= BZs used in epilepsy include, apart from DZ, nitrazepam, clonazepam, clobazam, lorazepam, and midazolam.

Similarly to sodium channels, voltage-gated calcium channels play a key role in regulating neuronal excitability and, therefore, it is not surprising that several AEDs exert their action by interfering with them (Table **1**) (Fig. **2**). Calcium channels are divided into different subtypes, L, N, P/Q, T and R, according to their electrophysiological characteristics [38]. The low-voltage channel T is involved in the modulation of thalamo-cortical oscillatory activity and in the generation of spike-wave discharges in the electroencephalogram, which both are main features of absence epilepsy [39, 40]. VPA and ESM, first choice drugs against absence seizures, reduce T-type channel currents [29-33] and this is an example of a fruitful combination of an identified mechanism of drug action with key pathogenetic steps of an epileptic disorder [34]. GBP and PGB are structurally related compounds which are derivatives of GABA (Fig. **1**). The early hypothesis of a predominant GABAergic effect has not been sufficiently supported by experimental data and their complete mechanism of action has still to be elucidated. Modest actions on the GABAergic system [41] and on voltage-gated potassium channels [42] have been proven. However, the primary mechanism of action which explains their pharmacological profile is the inhibition of calcium currents *via* high-voltage-activated channels containing the a2d-1 subunit [43] (Table **1**) (Fig. **2**). Inhibition of calcium channels results in turn in a reduction of neurotransmitter release of noradrenaline, serotonin, dopamine and substance P with an attenuation of postsynaptic excitability [44]. This most likely explains their beneficial effects on a variety of clinical disorders, including, apart from epilepsy, neuropathic pain, fibromyalgia, some types of headache and anxiety disorders. Potassium channels have been less explored as target for AEDs as compared to sodium and calcium channels. To date, a lot of potassium + channel related genes have been identified in the human genome, and experimental evidence strongly suggests that they play a role in influence neuronal excitability. Apart from some evidence indicating that LTG and TPM modulate potassium currents [29-33], RTG seems to differ from all currently approved AEDs. Its primary mechanism of action, in fact, consists, as a positive allosteric modulator of KCNQ2-5 (K(v) 7.2-7.5), in opening potassium channels. In particular, KCNQ2-5 channels are predominantly expressed in neurons and a human genetic mutations relating to KCNQ2/3 underlies the inheritable syndrome of benign familial neonatal convulsions. RTG increases the number of KCNQ channels that are open at rest and thus amplifies the inhibition in the brain [45].

Other Mechanisms of Action

Among the different mechanisms exhibited (Table **1**), both FBM and TPM have been found to limit glutamate-mediated excitatory neurotransmission. FBM blocks NMDA receptors and TPM acts on AMPA receptors (Table **1**) (Fig. **2**) [29-33]. As above mentioned, LTG also is an antagonist of glutamate transmission *via* block of sodium channels. On the basis of our present knowledge, LEV exhibits a unique mechanism of action not involving the classical targets of the other AEDs. It binds to an isoform of SV2A (Synaptic Vescicle protein A), a class of pre-synaptic proteins involved with synaptic vesicle exocitosis and neurotransmitter release. The binding of LEV to this site is reversible, saturable, and highly stereoselective with the final result of modulating neuronal excitability [46]. Although different steps need to be still elucidated, this target for AEDs action may reveal a new strategy research for epilepsy therapy. A recent study has demonstrated a low distribution of vesicle SV2A in the cerebral cortex and in hyppocampus of spontaneously epileptic rats exhibiting both tonic convulsions and absence seizures [47]. This suggests that SV2A could play a role in the process of epileptogenesis and seizure propagation. Another target of AED action is the H-channel (cyclic nucleotide gated), which has a high permeability to potassium ions and tends to stabilize the membrane potential toward the resting potential against both hyperpolarizing and depolarizing inputs. LTG and GBP target the H-channels by increasing the hyperpolarization-activated cation current [48, 49].

AED Action in Experimental Models of Epilepsy

Since the era of Merritt and Putnam, who in the late '30s firstly investigated PHT in animal models of epilepsy prior to introduce it in clinical practice [4, 6], putative antiepileptic drugs are screened in animals, most frequently in rodents. Models of inducing seizures in normal animals essentially include: (a) maximal electrical shock induced seizures in mice or rats (MES), (b) seizures induced by s.c. administration of pentylenetetrazol (PTZ) in mice or rats, and (c) amygdala-kindling focal seizures in rats [50]. These classical models have important limits, the main of which is that they cover fits triggered by artificial means in normal animals. Human epilepsy is characterized by spontaneous recurrent seizures, a very complex pathological disorder to which genetic, molecular, cellular, biochemical and system neuronal pathways abnormalities concur. Thus, other animal models of epilepsy have been developed over time, which include: (a) a number genetic models (animals with spontaneous recurrent seizures, *e.g.,* rats/mice with spike-wave discharges and transgenic mice, and animals with reflex seizures, *e.g.,* DBA/2 mice, photosensitive baboons) and (b) animals with chronic spontaneous recurrent seizures consequent, for example, to status epilepticus electrically or chemically provoked [51]. (Table **2**) summarizes the effect of individual AEDs against the main animal models in relation to the main mechanism of AED action. In general, drug positive action in MES model is considered to be predictive of efficacy against generalized tonic-clonic seizures and an effect against PTZ-induced seizure is thought to be predictive of drug efficacy against non-convulsive generalized seizures (absence, myoclonic seizures). The amygdala-kindling model, obtained by repeated applications of electrical stimuli through a depth electrode in the amygdala (or hippocampus) is considered a model of chronic focal seizures and drug efficacy in this model is predictive of efficacy against partial epilepsies. This differentiation, however, although of some value, is too simple and imperfect and, as it can be seen in (Table **2**), most findings deriving from these models are not applicable to human epilepsy. ESM, for example, is efficacious against PTZ-induced seizures and, as expected, it is efficacious against absences, but VGB, TGB and PB, although efficacious against PTZ-seizure, are all ineffective against absences and may aggravate them.

More recently, since introduction of new drugs has not reduced the percentage of patients with refractory epilepsy, which is stabilized at around 30%, the thought has gained room that animal models should explore novel mechanisms of drug actions and new pathophysiological steps of epilepsy. Animal models of refractory epilepsy have been developed in more recent years to explore drugs with a different mechanism of action as compared to existing AEDs. An interesting model is that of MAM-exposed rats [52]. MAM is the acronym of the antimitotic agent MethylAzoxyMethanol acetate which in rats in utero produces a neuronal migration disorder similar to the cortical dysplasias seen in human brain. As known, epileptic seizures secondary to cortical dysplasias are very resistant to AED therapy. In this model, seizures provoked by either kainate or the cholesterol biosynthesis inhibitor AY-9944 are refractory to valproate, ethosuximide or carbamazepine. Thus, these rats seem to have refractory seizures.

Table 2: Anticonvulsant effect of individual AEDs in rodent models of epilepsy. GCS= Generalized Convulsive Seizures; GNCS= Generalized Non Convulsive Seizures; NE=not effective. Adapted from ref. [5, 53]

Drug and main Mechanisms of Action	MES (GCS)	s.c. PTZ (GNCS)	Amygdala-Kindling (focal seisures)
Blockade of Na$^+$ (and Ca^{2+}) channels			
PHT	+	NE	+
CBZ	+	NE	+
OXCZ	+	NE	+
LTG	+	±	+
ZNS	+	±	+
Blockade of Ca^{2+} Channels			
ESM	NE	+	NE
Increase of GABA Transmission			
BZs	+		+
VGB	NE	+	+
TGB	NE	+	+
Multiple Mechanisms			
VPA	+	+	+
FBM	+	+	+
TPM	+	NE	+
PB	+	+	+
New Mechanisms			
GBP	±	±	+
PGB	+	NE	+
LEV	NE	NE	+
LCS	+	NE	+
RTG	+	+	+
PRP	+	+	+

An alternative approach to develop animal models of drug resistant epilepsy is the treatment of large group of kindled or epileptic rats with AEDs followed by a selection of subgroups of animals that either respond or do not respond to this treatment. An animal model of epilepsy allowing selection of subgroups of animals with drug-refractory and drug responsive seizures could be a valuable tool to study why and how seizures become intractable and to develop more effective treatment strategies. Details of animal models of epilepsy can be found in excellent comprehensive reviews in the literature [53, 54]. Drug toxicity tests include the rotorod test, the position head test, and the gait and posture test [51, 53, 54]. Animal models for seizures and epilepsy have played a fundamental role in advancing our understanding of basic mechanisms underlying ictogenesis and epileptogenesis and, although major limits are implicit in these models, they are continuing to be of very relevant assisting help in the development of novel AEDs [51, 53, 54].

PHARMACOKINETICS AND SERUM LEVEL MONITORING OF AEDS

Drug pharmacokinetics include time- and dose-related processes, namely absorption, distribution, metabolism and elimination, which ultimately determine the ability of a given drug to achieve and maintain in the body concentrations that optimize efficacy and minimize toxicity. Pharmacokinetic parameters can be largely influenced by both physiological (*e.g.,* age and pregnancy) and pathological (*e.g.,* liver and kidney diseases) conditions and by the presence of other drugs and, therefore, they are susceptible of relevant variations over time. Since the majority of patients with epilepsy require chronic therapy, knowledge of the pharmacokinetic features of a drug is a key factor to make a correct drug choice and to use it properly. Selected pharmacokinetic parameters of AEDs are illustrated in (Table **3**).

Absorption

Most AEDs undergo a rapid and complete or nearly complete absorption when given orally, with exclusion of CBZ, and GBP, which show a slow and/or incomplete absorption (Table **3**) [55-57]. RTG has complete absorption, but incomplete bioavailability due to a first pass effect. CBZ is absorbed slowly and erratically, with peaks after single doses occurring at 4-24 hours. In particular, absorption is faster with oral suspension (peaks at 1-2 hours), of intermediate speed with standard tablets (peaks at 4-5 hours) and long with extended-release formulations (peaks at 5-24 hours) [55]. The causes may be: scarse CBZ solubility in water (a parenteral formulation is in fact not available), and a slow dissolution rate. Despite the irregular absorption rate, oral bioavailabilty of CBZ, *i.e.,* the extent of drug absorbed, is approximately 80% of the given dose [58]. GBP exhibits a low oral bioavilabilty with values ranging from approximately 60% at a dose of 400 mg to 35% at a dose of 1600 mg when administered three times daily [59]. GBP absorption, in fact, occurs through a saturable aminoacid transport system and as dose is increased, biavailability decreases [59]. RTG exhibits an oral bioavailability, as compared to an i.v. administration, of 60% [56, 57]. Most often, administration of AEDs with food slows absorption but not the extent of the drug absorbed. Calcium containing antacids may reduce absorption of a number of AEDs (see below).

Elimination Half-life

This parameter is influenced by drug distribution and clearance processes and has relevant practical implications. It in fact determines (a) the time to reach steady-state conditions, *i.e.,* when the amount of drug intaken and that eliminated are the same, and (b) the number of daily drug administrations. As an accepted rule, time to achieve steady-state is four-to-six times the value of half-life. Clinically, evaluation of the efficacy of a drug, once AED treatment is chronically started, has to be made under steady-state conditions. Consequently, for example, clinical effects of VPA or CBZ can be evaluated within few days after the established daily dose has been achieved and maintained, while 3-4 weeks are needed to evaluate the effects of PB or ZNS [55]. Similar differences do exist among AEDs in relation to the number of daily administrations. VPA or CBZ need to be administered three-four times a day to lessen excessive daily fluctuations in serum drug levels during a dose interval and hence to reduce drug toxicity possibly occurring at concentration peaks or the risk of breakthrough seizures in conjunction with excessively low serum drug levels. Conversely, one daily administration is largely sufficient with PB. This concept is surely applicable to drugs exhibiting a close correlation between serum concentrations and clinical effects, but it has a less value for drugs, like VGB, the mechanism of action of which implies induction of an enzymatic activity and thus clinical effects are not directly correlated with drug concentration in blood [55].

Metabolism and Excretion

Although some AEDs are predominantly or exclusively excreted unchanged by the kidneys (VGB, GBP, PGB, LCS), the majority of AEDs are metabolized in the liver by hydroxylation or conjugation (Table **3**) [55-57]. Most AEDs exhibit linear enzyme kinetics, in which clearance remains constant and changes in daily dose are paralleled to proportional changes in serum concentration. Some drugs exhibit non-linear (Michaelis-Menten) kinetics because of changes in their metabolic rate. PHT show a progressively saturable metabolism [55]. Saturation can already occur at serum therapeutic PHT levels and the consequence is that small increments of dose result in a disproportionate rise in serum drug levels and drug toxicity. CBZ shows an opposite phenomenon. It induces its own metabolism in a dose-dependent manner and, although the dose is kept stable, serum levels undergo a decrease with possible loss of seizure control [55]. Among the newer AEDs, STP exhibits also non-linear pharmacokinetics with decrease in clearance secondary to dosage increments [60]. Oxidative biotransformation of AEDs occurs predominantly through the cytochrome P450 (CYP) isoenzyme system in the liver, which includes three main families of enzymes: CYP1, CYP2, and CYP3. In particular, the CYP2C9, CYP2C19, and CYP3A4 isoenzymes are commonly involved in the metabolism of AEDs [61]. At this level, important drug interactions can occur (see below) and induce large variations in the serum drug levels among patients taking the same dose. Additionally, genetic factors can induce relevant variation in the metabolism rate. The CYP2C9 and CYP2C19 do exist in the form of genetic (allelic) variants causing different degrees of enzymatic activity [62, 63]. Both are involved in the PHT biotransformation which is at least two fold decreased in subjects who are poor or slow metabolisers (*i.e.,* carriers of mutated alleles of both isoenzymes) as compared to rapid or extensive metabolisers. Slow metabolisers are frequent, *i.e.,* approximately 10%, in Asians and relatively rare, 1%, in the Caucasians [62, 63]. CBZ is converted to the active metabolite CBZ-10,11-epoxide through the microsomal enzyme epoxide hydrolase, which also is polymorphic [62, 63]. The role of this polymorphism in the determining variations in CBZ clearance has to be still established. Another main metabolic pathway of AEDs is conjugation *via* the enzyme uridine diphosphate glucuronosyltransferase (UGT), which include VPA, LTG and some BZs These and other metabolic pathways of AEDs are illustrated in (Table **3**).

Table 3: Pharmacokinetic parameters of AEDs. *= non-linearity of some pharmacokinetic parameters (see the text for details). MHD=mono-hydroxy-derivative. R.e.= renal excretion

Drug	Oral Bioavailability (%)	Elimination Half-Life (h) °	Main Routes of Elimination	Protein Binding (%)
PB	~ 100	75-125	oxidation, conjugation, r.e. (25%)	50
PHT*	~ 100	10-100	oxididation	70-95
ESM	~ 100	40-60	oxididation	<5
CBZ*	< 85	10-25	oxididation	75-80
DZ	~ 100	40-100	demethylation, hydroxilation, glucuronidation	> 95
VPA*	~ 100	10-20	oxidation, glucuronidation	85-95
VGB	60-80	5-8	r.e.	0
LTG	~ 100	15-30	glucuronidation	55
OXCZ	pro-drug of MHD (95%)	8-15 (MHD)	keto-reduction > 4(MHD > glucuronidation, conjugation	40
FBM	~ 100	14-23	oxidation, r.e.(>40%)	25-35
GBP*	< 60	5-7	r.e.	0
TPM	~ 100	20-30	oxidation, r.e. (30-55%)	15
TGB	~ 100	4-13	oxidation	96
LEV	~ 100	6-8	hydroxilation, r.e. (>60%)	<10
ZNS	~ 100	50-70	glucuronidation, conjugation, acetylation oxidation, r.e. (~30%)	40-60
STP	n.e.^	4.5-13	glucuronidation, demethylation	99
PGB	~ 100	5-7	r.e.	0
RFN*	85	6-12	hydroxilation, acyl-glucuronidation	23-34
LCS	~ 100	~ 6	r.e.	0

Table 3: contd.

ESL°°	> 90	20-24	glucuronidation, renal excretion (>60%)	30
RTG	~ 60	8-10	hydrolysis./N-acetylation, glucuronidation	80
PRP	~ 100	70-120	oxidation, glucuronidation	95

°= Elimination half-life is subject to large variability in consequence of drug-interactions, auto-induction (like CBZ) or auto-inhibition (like PHT) of metabolism, age and disease states (for more information see the text and specific references [55-57]). *= these AEDs exhibit non-linearity of some pharmacokinetic parameter: PHT exhibits metabolic saturation within therapeutic levels with consequent marked prolongation of its elimination half-life; CBZ exhibits metabolic auto-induction within therapeutic levels with remarkable reduction in its elimination half-life; VPA exhibits a saturable protein binding at high serum levels with an increase in free unbound fraction; this, however, has little therapeutic relevance if concomitantly an inhibition of metabolism does not occur; GBP shows a saturable absorption at usual doses with progressive reduction in its bioavailability; STP also exhibits a nonlinear kinetic profile (*i.e.,* Michaelis-Menten type) with multiphasic elimination curve; its clearance diminishes with increased doses; RFN shows a decreased bioavailability at high doses; °°=ESL is eslicarbazepine acetate that is rapidly and extensively transformed into eslicarbazepine; bioavailability and half life values refer to eslicarbazepine; ^= oral bioavailability of STP has not been established clearly and variable values have been reported; possible saturation of first-pass metabolism has also been suggested.

Age-, Pregnancy-, and Disease-Induced Changes in AED Pharmacokinetics

Age influences markedly AED pharmacokinetics [64-66]. Elimination rates and protein binding of most drugs are diminished in the neonatal period, the first 6 months of infancy, because liver and kidneys undergo maturation. In children drug clearance progressively increases and exceeds that of adult patients. Increments achieve values of 20-40% in mean as compared to adults, but values of >150% have been also reported [65]. Translated into clinical practice, this means that children require higher doses than adults, even twice the adult mg/kg dosage, and more frequent drug administrations over the day because of large fluctuations in serum drug levels. Other pharmacokinetic parameters, *e.g.,* absorption and volume of distribution, show also differences in paediatric age as compared to adults [64-66]. Therefore, the approach to paediatric drug dosing is complex and needs a carefull evaluation. As general recommendation, drugs that are extensively biotransformed in the liver should be administered with extreme care until the age of 2 months [66]. At the age of 2-6 months, a choice of drug dosage based on bodyweight may be used [66]. After 6 months of age, bodyweight in conjunction with body surface area have been suggested as good line for drug dosing [66]. For drugs primarily excreted by kdneys, in the first 2 years of life drug dosing should be selected on the basis of markers of renal function, such as serum creatinine and p-aminohippuric acid clearance [66]. In the elderly drug elimination is almost invariably reduced as compared to adult patients as a consequence of physiological reduction in liver and kidney blood flow [64-65]. For the majority of AEDs extensively metabolized in the liver, prediction of pharmacokinetic changes cannot be made with accuracy in an individual patient. Conversely, with kidney eliminated drugs, determination of creatinine clearance may be useful in predicting individual changes in drug clearance and hence dosage adjustment [64].

In addition to changes in drug clearance, in elderly drug protein binding is reduced with consequent increase in the free (pharmacologically active) serum drug amount. This implies that, especially with highly bound drugs, like PHT, VPA, DZ and STP, clinical drug effects occur at lower total serum drug levels than those usually considered [64]. Drug disposition is significantly altered during pregnancy. Increased renal elimination, and altered hepatic enzyme activity usually induce a reduction in serum drug levels [67, 68]. This reduction has been observed with the majority of older AEDs. In particular, changes in protein binding lead to a greater reduction in total than free (active) drug concentrations and monitoring of free drug concentrations has been recommended specifically for phenytoin and valproic acid [67, 68]. Data on pregnancy-induced pharmacokinetic changes of more recent AEDs are particularly detailed for LTG which has been extensively studied. Its apparent clearance shows a marked increase, up to 300%, in the late pregnancy as compared to baseline values. with some changes occurring early. This clearance increase is most probably due to an enhanced metabolism and can be associated to a loss of seizure control. LTG levels may then rise precipitously after delivery, leading to symptoms of lamotrigine toxicity [68]. More limited data indicate that a similar decline in plasma concentrations of the active monohydroxy derivative (MHD) of OXCZ may occur in late pregnancy [68].

Renal or hepatic diseases may affect markedly AED kinetics and require careful considerations in prescribing dosage to epileptic patients suffering from these diseases. Reduced renal function lead to accumulation of those AEDs predominantly or exclusively excreted by kidneys, like GBP, PGB, VGB, TPM, LEV, LCS [69, 70]. Determination of renal clearance can help in dosage adjustments. Low protein-bound AEDs, like PB, LEV, LTG, TPM, the active metabolite of OXCZ MHD, LCS and others (Table **3**), are extensively removed by hemodialysis and supplemental doses are required for dialysed patients [69, 70]. Hepatic dysfunction reduces, as expected, enzymatic metabolism of AEDs and causes hypoalbuminemia. However, because of concomitant changes in hepatocellular activity, biliary excretion and hepatic blood flow, calculation of hepatic impairment cannot be made with sufficient accuracy and appropriate dosage adjustments are very difficult [69]. VPA and FBM are potentially hepatotoxic and should be avoided in patients with hepatic disorders. Febrile illnesses may enhance drug metabolism and during fever higher AED dose may be required to maintain therapeutic efficacy.

Serum Level Monitoring of AEDs

From a historical perspective, introduction of this practice in the routine clinical setting in the '70s has been of essential help in identifying, taking awareness and giving a correct value to a number of aspects, most of which previously underestimated, characterizing AED therapy. These aspects include:(a) large variability of serum drug levels among patients taking the same dose, (b) more close relationship between serum drug levels and clinical events, (c) need of optimizing drug dosages and regimen on an individual basis, (d) patient noncompliance more common as it was expected , (e) frequent drug interactions [71]. The correct progressive acquisition of the advantages of mono.-therapy over poly-therapy and the introduction of a growing number of new AEDs have undoubtedly contributed to reduce the use of AED serum monitoring. However, clinical experience gained over the last four decades and data deriving from nonrandomized investigations have confirmed the usefulness of monitoring serum levels of most old and recent AEDs if used appropriately [72,73]. According to a recent report of an ad hoc Commission of the ILAE(International League against Epilepsy) [74], determination of serum AEDs levels is indicated in the following situations: a) when an unexpected and unexplained clinical event occurs, *e.g.* sudden loss ofseizure control ina patient previously well controlled or, even, sudden gain of seizure control in a patient with apparentrefractory seizures, or onset of toxic manifestations, (b) to verify patient compliance especially in populations of patients at risk, *e.g.* children, elderly patients often relied on careers, and psychiatric patients, (c) during pregnancy, especially for drugs, like LTG, showing marked serum level variations (see above), d) patients who developed concomitant diseases, especially renal and hepatic diseases, known to alter markedly AED disposition (see above), e) with drugs, like PHT, exhibiting non-linear pharmacokinetics and, thus, at risk to achieve high serum levels following small dose increases, f) to have, after starting treatment, a baseline steady-state concentration which will be useful for future evaluations. In the same report, it has been stressed that the predefined "therapeutic range" as intended in the past years,i.e. an interval where most patients are expected to show an optimal response, has to be substituted with the conceptual term of "individual therapeutic concentration". In clinical practice, in fact, it is common experience that many patients tolerate well and/or need serum concentrations above the usual therapeutic range, while others gain complete seizure control, or even experience adverse effects, at concentrations below it. To summarize, determination of serum AED levels can be appropriated for all drugs when a clinical manifestation of any type is open to various and complex interpretations. In line with this concept, therefore, although the serum level/response relationship of more recent AEDs has been so far less investigated as compared to conventional AEDs [73, 74], in certain specific situations determination of serum levels of these drugs may be appropriate. Reference levels for most old and new AEDs derived from literature data are given in (Table **4**).

Table 4: Serum target range of AEDs

	Serum Level (mg/l)
PB	10-40
PHT	10-20
ESM	40-100
CBZ	4-12
VPA	50-100
LTG	5-18
OXCZ (MHD)	12-24
GBP	6-21
TPM	4-25
LEV	10-40
ZNS	7-40

AED INTERACTIONS

Drug-drug interactions are one of the most relevant aspects of antiepileptic therapy since they occur frequently and often alter unexpectedly the clinical picture [75-77].

Drug interactions may occur at any step of the kinetic processes (pharmacokinetic interactions) and at any level of the mechanisms of drug action (pharmacodynamic interactions). To have a clinical impact, however, changes in drug concentration into the body and/or interferences at the intrinsic drug activity level have to be of sufficient magnitude to give rise to clinical manifestations [75-77]. Pharmacokinetic drug interactions are by far the most studied, defined and documented as compared to the pharmacodynamic ones. Although, in fact, the latter can be easily identified in experimental animal models, they are rarely proven with sufficient accuracy in humans. A complete list of the large number of interactions involving AEDs can be found in specific article reviews of the literature [75-77, 80]. A list of more clinically relevant interactions is given in (Tables **5-7**).

Table 5: Clinically relevant AED interactions resulting in reduced serum levels of the affected drugs. Dose increments of these drugs may be required to achieve and/or maintain their effects

AEDs Affecting Other Drugs	AEDs	Various Drugs	Oral Contraceptives	Antidepressants and Antipsychotics	Antimicrobal Drugs
CBZ, PB, PHT	BZs, ESM,	Warfarin,	Estrogen	Clomipramine	Doxycycline
	LTG, OXCZ	Antineoplastics,		Imipramine	Indinavir
	RFN, STP,	Immunosuppressants		Chlorpromazine	Itraconazole,
	TGB, TPM,	Cortisol derivatives		Haloperidol	Metronidazol
	ZNS VPA	Dextropropoxyphene		Aripripazol	Praziquantel
	PRP	Calcium antagonists		Clozapine,	
		Statines, Methadone,		Olanzapine	
		Theophylline, Tyroxine		Quetiapine,	
				Risperidone	
				Ziprasidone,	
TPM		Digoxin, Metformin Hydrochlortiazide Carboanhydrase inhibitors Pioglitazone	Estrogen*		
ESL, OXCZ	PHT, PB		Estrogens		
	LTG, PRP				
FBM	CBZ, Clobazam				

*TPM induces the estrogen component of the contraceptive pill at daily doses of > 200 mg. The list of interactions is not exhaustive; for more detailed data see references [75-78]. Modified from reference [77].

Table 6: Clinically relevant AED interactions resulting in increased serum levels of the affected drugs. Dose reductions of these drugs may be required to avoid the risk of toxic effects

AEDs Affecting Other Drugs	AEDs	Various Drugs	Oral Contraceptives	Antidepressants and Antipsychotics	Antimicrobal Drugs
VPA	PB, CBZ,	Cisplatin, etoposide		Amitriptyline	Carbapenem
	LTG, ESM			Nortriptyline	Imipenem
	RFN,				Meropenem
					Panipenem
FBM	PB, PHT,	Warfarin	Estrogen		
	VPA,				
	Clonazepam				
RFN	PHT, PB	Triazolam	Estrogens		
	LTG,CBZ				
STP	CBZ, PB				
	PHT, VPA				
	Clobazam				
PRP	OXCZ				

The list of interactions is not exhaustive; for more detailed data see references [75-78]. Modified from reference [77].

Table 7: Clinically relevant drug-drug interactions resulting in reduced or increased serum levels of the affected AEDs. Dose adjustments of AEDs may be required to optimize AED therapy

Drug classes and Related Compounds	Affected AEDs	Effect on Serum AED Levels
Antidepressants and Antipsychotics		
Haloperidol	CBZ	Increase
Risperidone	VPA	"
Chlorpromazine	CBZ	"
Clomipramine	CBZ, PHT, PB, VPA	"
Sertraline	CBZ, PHT, LTG, VPA	"
Oral Contraceptives	LTG, VPA	Reduction
Antimicrobal Drugs		
Macrolides (Clarithromycin, Erythromycin, Troleandomycin)	CBZ	Increase
Rifampicin	LTG	"
Isoniazid	CBZ, PHT, VPA, ESM	

Table 7: contd….

Others		
Probenecid	CBZ	Increase
Ketoconazole	PRP	"

The list of interactions is not exhaustive; for more detailed data see references [75-78]. Modified from reference [77].

Absorption

Calcium containing antacids can significantly reduce PHT, PB, CBZ, GBP absorption by decreasing the acidity of the stomach and also by the precipitation of these AEDs with formation of insoluble complexes difficult to be absorbed [75-77]. A particular situation is the concomitant administration of PHT with nutritional formulations, resulting in a marked reduction in PHT absorption [75-77]. Epileptic patients requiring continuous feedings need higher PHT doses and monitoring of serum PHT levels may help in dosage adjustments. Co-administration of other AEDs such as VPA, LTG, GBP or LEV with enteral feeding supplements is less problematic and usually an increase in drug dose is not required. [75, 77].

Drug Distribution

Drug interactions occurring at this level are rarely of clinical relevance and only drugs exhibiting a high protein binding (*i.e.,* >80-90%), like PHT and VPA, may be involved. When a drug is displaced by another drug from plasma proteins, the free fraction increases, but the free drug concentration, *i.e.,* the active drug which diffuses across tissues, does not change due to a concomitant increase in drug elimination consequent to a contemporary increase in drug diffusion into the elimination organs. To be of clinical significance, therefore, displacement of a drug from plasma proteins must occur in parallel with a reduced drug metabolism. This is, for example, the case of VPA-PHT interaction. VPA is able to displace PHT from plasma proteins and inhibits its metabolism with consequent risk of PHT toxicity [75-77].

Metabolism

The most important pharmacokinetic interactions with AEDs occur at this level, in which induction or inhibition effects on drug metabolising enzymes take place. As above mentioned, enzyme activity is under the control of genes which determine inter-individual variability in the magnitude of a given interaction [78]. CBZ, PHT and PB are the classical AEDs which stimulate many drug metabolizing enzymatic pathways, including CYP and glucuronyl-trasferase enzymes. These drugs, therefore, lower drastically the concentration and the clinical effect of a large number of concomitant drugs (warfarin, steroidal, anti-neoplastic, psychotropic, cardiovascular, immunosuppressant drugs, *etc.*), including various AEDs (VPA, LTG, TPM, TGB, RFN, STP, ZNS, BZs, PRP) [75-77]. Conversely, VPA is an enzyme- inhibitor and slows the elimination rate with risk of toxicity precipitation of PB, LTG, CBZ-epoxide, among AEDs, and of a number of other drugs, such as amitriptyline, nortriptyline, some antibiotics, cisplatin, and others [75-77]. If the affected drug has an active metabolite, enzyme induction/inhibition can result in increased/decreased metabolite concentration and related clinical effects. In some cases, AED-AED interactions are bi-directional. The interaction between CBZ and VPA is a clear example. VPA inhibits the enzyme epoxide hydrolase, which is responsible for the metabolism of the active metabolite of CBZ, CBZ-epoxide, and induces an increase of this metabolite. Concomitantly, CBZ causes a fall in serum VPA levels. Thus, in patients treated with VPA-CBZ co-medication monitoring of both compounds and of CBZ-epoxide may result extremely useful to guide dosage adjustments [79]. Overall, the newer AEDs are less interacting because of their main renal elimination or of absent or negligible plasma protein binding (*e.g.*, GBP, PGB, LEV, LCS, VGB and others (Table **3**) [80]. Of the new AEDs, LTG, TPM and OXCZ are the most interacting. The metabolism of LTG is, as above mentioned, susceptible to both enzyme inhibition (by VPA, rifampicin) and enzyme induction (by CBZ, PHT, PB and oral contraceptives) [80]. OXCZ, TPM, FBM, RFN, ESL stimulate the elimination of oral contraceptives and compromise contraception control [80]. Here below are highlighted interactions occurring between AEDs and other classes of drugs of common use. Because of their frequency and the fact that they cause the need of dosage adjustments, these interactions are of particular relevance in clinical practice.

AED Interactions with Anticoagulants

Warfarin, administered as a racemic mixture of the R- and S- enantiomers, is metabolized by CYP2C9 (S-enantiomer) and by different CYP isozymes (R-enatiomer). CBZ and PB enhance warfarin metabolism, reduce significantly its anticoagulant effect and a warfarin increased dosage is required [75-77]. PHT initially induces an increase in the anticoagulant effect of warfarin because of a possible combined protein-binding displacement and metabolism inhibition of the drug. During chronic treatment, the PHT-induced enhancement of warfarin biotransformation probably prevails with the implicit consequence of a need to increase warfarin dosage [75-77]. An opposite effect occurs when PB, CBZ or PHT are discontinued and, in this case, a risk of hemorrhagic events is high. Clearly, careful monitoring of international normalized ratios is crucial in these situations. Whenever possible, these drug combinations should be avoided as fatal cases have been also reported [81]. AEDs devoid of hepatic inducing properties or those eliminated by kidneys (Table **3**) should be preferred in patients requiring warfarin treatment [80].

AED Interactions with Anti-Hypertensive Drugs and Statins

These interactions complicate treatments especially of elderly patients. Nimodipine, felodipine, nicardipine, nifedipine, and amlodipine, and the beta antagonists propanolol and metoprolol are extensively metabolized by CYP3A4. CBZ and PHT induce a substantial reduction in the bioavailability of these medications and, therefore, larger doses of these medications might be required to maintain their therapeutic effect [75-77]. Some other beta-antagonists, such as atenolol and labetalol, diuretics and ACE inhibitors are eliminated by the kidneys or are not extensively metabolized, and thus would not be expected to interact with AEDs. Most commonly prescribed lipid-lowering medications, *i.e.,* statins, such as lovastatin, simvastatin, atorvastatin, and fluvastatin are extensively metabolized in the liver and, therefore, concomitant administration of enzyme-inducing AEDs may lower significantly their therapeutic effects [75-77]. Other statins, namely pravastatin and rosuvastatin are not extensively metabolized, so significant pharmacokinetic interactions with AEDs would not be anticipated.

AED Interactions with Oral Anti-Contraceptives

A number of old and newer AEDs, including PB, PHT, CBZ, OXCZ, ESL, PRP, and TPM at > 200mg daily doses, stimulate the metabolism of estrogen components contained in the contraceptive pills with consequent risk of abolishing their effects [75-77]. Conversely, oral contraceptives lower serum LTG levels by more than 50% and, specifically, ethinyl estradiol induces LTG metabolism *via* uridine-glucuronyl-transferase activity [82, 83]. In view of the changes in LTG kinetics during menstrual cycle and menopause, this LTG-hormone interaction is of particular relevance [84].

AED Interactions with Psychotropic Drugs

Psychiatric disorders are very common in patients with epilepsy [85, 86]. Thus, interactions of psychotropic drugs with AEDs have to be constantly taken into consideration in clinical practice as they may require dosage adjustments [87]. CBZ, PHT and PB lower a large number of drugs, including tricyclic antidepressants such as amitriptyline, nortriptyline, imipramine, and desipramine, non-tricyclics such as sertraline, paroxetine, mianserin, citalopram, and nefazodone, and both the older typical and newer atypical antipsychotics haloperidol, chlorpromazine, risperidone, clozapine, olanzapine, quetiapine, ziprasidone, and aripiprazol [87-90]. VPA causes significant increases (50%-60%) in serum concentrations of amitriptyline and nortriptyline [88]. Given that both tricyclics facilitate seizures especially at high dosages [91], it is important in adjusting drug doses to evaluate, besides mood changes, also possible changes in seizure control [87]. Fluoxetine has been shown to inhibit the metabolism of PHT and CBZ and risperidone, and possibly haloperidol have been shown to increase the carbamazepine serum concentration [89-90]. Given the frequency and the complexity of most of these interactions, whenever possible preference should be given to the choice of less interacting newer AEDs [87].

AED Interactions with Antibiotic and Antiviral drugs

The antibiotic macrolides clarithromycin, troleandomycin and erythromycin increase serum CBZ concentrations markedly [75-77]. Isoniazid has the same effect on serum PHT, CBZ, ESM and VPA levels and rifampicin on LTG levels [75-77]. PB, PHT and CBZ stimulate the metabolism of doxycicline, metronidazol, indinavir, itraconazole and, conversely, VPA lowers elimination of imipenem, panipenem, and meropenem [75-77].

AED Interactions with Antineoplastics, Immusuppressants and Other Drugs

As expected, CBZ, PHT, and PB induce the biotransformation of a number of antineoplastics (methotrexate, cyclophosphamide, irinotecan, tamoxifen) and immunosuppressants (ciclosporin, tacrolimus) [75-77]. TPM has been seen to stimulate elimination of digoxin, metformin, and hydrocholtiazide [80]. Conversely, VPA increases serum levls of cisplatin with risk of toxicity [75-77].

A Complex, Highly Relevant AED-AED Interaction: VPA-LTG

It is common experience in the clinical practice that a minority of patients with difficult-to-treat epilepsy show beneficial effects by taking combination therapy with two or more AEDs [92]. At the present time, however, although various investigative approaches have been made, no clear indications can be drawn from available literature data on which drug combination can be chosen against a specific form of epilepsy [92]. Some report has indicated that VPA in association with ESM is useful against atypical absences not responding to mono-therapy [93]. By far, the most documented association resulting in synergistic effects is that of VPA and LTG against focal seizures. The first report dates back to 1992 when a drastic response to VPA-LTG co-medication was observed in some patients with focal seizures refractory to other therapies [94]. Two subsequent studies performed in a larger groups of epileptic patients gained strong evidence that combining VPA with LTG results in synergistic effects against focal seizures [95, 96]. In particular, in one of these studies [96] it was possible to optimize the therapeutic response at serum levels of both drugs lower than those of individual drug used in mono-therapy. These findings would exclude that the clinical effects are due to an elevation of serum LTG levels caused by VPA through a metabolic inhibition [97]. A number of studies performed in adult and in paediatric populations have confirmed the efficacy of LTG-VPA combination in different forms of both generalized and partial forms of epilepsy [92]. In addition to these beneficial effects, there is evidence that the VPA-LTG pharmacodynamic interaction leads also to toxicity. A severe and disabling tremor has been reported following treatment with the two drugs [98], which has been observed not to be consequent to high serum levels of VPA or LTG [97]. Conversely, the VPA induced rise of serum LTG levels, *i.e.,* their pharmacokinetic interaction, might contribute to the onset of skin rash and other more serious consequences, *i.e.,* Stevens-Johnson syndrome and toxic epidermal necrolysis, occurring during treatment with LTG especially at high initial doses [99, 100].

ADVERSE EVENTS OF AEDS

Since antiepileptic therapy has a very long, often lifetime, duration, toxicity of AEDs is a key element for its success. Epidemiological studies have indicated that approximately 40% of patients suffer from AED toxicity which can obliges physicians to change medication [101-107]. Apart from efficacy in controlling seizures, therefore, retention rate of a given drug, namely its effectiveness, is also determined by the development of adverse events (AEs). Given such a relevance, in recent years AE are systematically and intensively investigated in controlled trials both in paediatric populations and in adults [104-106]. A consequential consideration is that whilst knowledge of the toxicity of the older AEDs derives mainly from observational data but their use over many years has allowed a large identification of long AEs, the newer AEDs have been more systematically studied, but their potential long-term AEs are at present time not completely known. Undoubtedly, a correct knowledge of the pharmacological profile of individual drugs and the physician's experience help in predicting toxicity manifestations. This in its turn influences drug choice, given that efficacy differences have not been substantially demonstrated among AEDs (see the specific section below). A common strategy to avoid initial intolerance is, for example, mono-therapy in conjunction with an appropriate dose titration period [108]. Starting with small doses and proceeding with small dose increments is especially recommended for CBZ, PHT, TPM, LTG, OXCZ, ESL, ZNS, FBM, LCS and others [108]. A close follow-up with a careful evaluation of all the signs and symptoms complained by patients are additional correct attitudes which may contribute to minimize or even avoid the consequences of AED toxicity. In spite of these efforts, however, AE cannot be always prevented since multiple factors and especially individual susceptibility on genetic basis are fundamental factors in determining the development and severity of AEs [107]. Recently, an important exception in AE prevention has been achieved with pharmacogenomic studies regarding idiosyncratic reactions to CBZ [106, 107]. The most common and most clinically relevant AE are briefly illustrated below. An overview of AEs related to individual drugs and of target organs possibly affected by AED toxicity is given in (Tables **8** and **9**).

Table 8: Main adverse events of individual AEDs

Drug	Dose-Related	Idiosyncratic	Long-Term
CBZ	diplopia, dizziness, hyponatremia, mild leucopoenia, gastrointestinal discomfort	rash/exfoliation, hepatitis, aplastic anemia, Stevens-Johnson syndrome in HLA-B 1502	osteomalacia
PHT	ataxia, dizziness, diplopia, tremor	rash/exfoliation, hepatitis, marrow aplasia, lympho-proliferative disorders	osteomalacia, gum hyper-plasia, facial coarsening/hirsutism, cerebellar syndrome, peripheral neuropathy, folate deficit
VPA	weight gain, alopecia, tremor, thrombocytopenia, gastrointestinal discomfort	encephalopathy, hepatic failure, pancreatitis polycystic ovary syndrome (causality not yet proven)	
PB	sedation (adults), hyperactivity (children) ataxia, cognitive dysfunction, depression	rash, hepatitis	osteomalacia, connective tissue disorders, sex dys-function, folate deficiency
ESM	headache, photophobia, nausea	rash, blood dyscrasias	
BZs	sedation, somnolence, ataxia,	rare renal, hematologic and behavioural disorders hepatic toxicity	
VGB	behavioural disorders		visual field loss/blindness
LTG	dizziness, headache, diplopia, ataxia, nausea, sleepiness,	rash, especially when combined with VPA, rare hepatic failure	
LEV	sedation, fatigue, dizziness, depression	leucopoenia	
OXCZ	sedation, fatigue, headache, dizziness, diplopia, hyponatremia gastrointestinal discomfort	rash/exfoliation	
TPM	cognitive slowing, nephrolithiasis, fatigue paresthesias, weight loss, word finding difficulty, dizziness, rare glaucoma		
GBP/PGB	sleepiness, dizziness, ataxia, weight gain	rare rash	
ZNS	sedation, dizziness, headache, fatigue distress, paresthesias, agitation, irritability, behavioural disorders	rash, marrow aplasia, rare hepatic damage aplastic anemia,, agranulocytosis, hyperthermia	nephrolithiasis

Table 8: contd….

FBM	headache, insomnia, weight loss, gastrointestinal discomfort, agitation	rash, hypersensitivity, fatal aplastic anemia, fatal hepatic failure	
TGB	dizziness, somnolence, "abnormal thinking"		
STP	nausea, vomiting		
RFN	somnolence, dizziness, headache, nausea shortened QT interval		
ESL*	dizziness, somnolence, headache, nausea		
LCS*	dizziness, nausea, gastrointestinal discomfort prolongation of PR interval at high doses		
RTG*	drowsiness, dizziness, slurred speech, diplopia		
PRP*	dizziness, somnolence, weight gain, irritability		

*= These drugs, also defined as third generation AEDs, are the most recent introduced and, therefore, their toxicity profile might be not sufficiently known. For more information see references [101, 111].

Table 9: Systems/organs which can be affected by AED toxicity and which should be monitored during AED therapy

Nervous System	Endocrine System	Skin
Sedation, tiredness, cognitive dysfunction, behavioural disorders, suicidality, cerebellar atrophy, peripheral neuropathy	Decreased serum thyroxine, cortisol and sex hormone levels	Acne, hirsutism, alopecia, chloasma
Blood	**Immune system**	**Liver**
Anaemia, folate deficiency Thrombocytopenia, lymphoma	IgA deficiency, systemic lupus erythematosus	Various hepathopathies
Bone	**Connective tissue**	**Pregnancy**
Osteomalacia teratogenicity	Gum hypertrophy, coarsened facial features Dupuytren's contracture	Obstetric complications,

For more information see references [101, 111].

Common Unspecific, Early Onset CNS Effects

Since AEDs act on neuron excitability, it is not surprising that the most frequent adverse events AEs are by far those at the Central Nervous System (CNS) level

[101, 102, 109]. They are usually observed early during dose escalation, often subtle, frequently considered by patients detrimental in terms of quality of life since they cause subjective sensations, but in most cases tending to diminish or even to disappear over time [101, 102, 109]. These effects include typically dizziness, sedation, distractibility, somnolence, ataxia, nausea, fatigue, diplopia and may occur invariably with almost, if not all, AEDs [101, 102, 106, 109].

Idiosyncratic Reactions and Immunological Effects

Idiosyncratic drug reactions occur at any dose, usually within the first weeks of treatment, in genetically predisposed individuals, and are unpredictable [110, 111]. Because the initial manifestation may deteriorate and prelude to a life-threatening condition, they require immediate attention and, possibly, drug suspension. These reactions, approximately 10% of all AE, are of different types [110, 111]. Most frequently, a benign skin rash is the first and only manifestation of an immune-mediated hypersensitivity [111]. This, however, can evolve toward serious, life-threatening, systemic conditions, like the Steven Johnson's syndrome and the toxic epidermal necrolysis [112-114] Although potentially all AEDs may trigger both these conditions, Steven Johnson's syndrome and toxic epidermal necrolysis have been particularly documented with CBZ, PHT, PB and LTG [112-114]. Other rare idiosyncratic reactions include those involving non-immune-mediated individual susceptibility with abnormal production or defective detoxification of cytotoxic metabolites [107, 110, 111]. These may lead to liver cell necrosis, acute liver failure, possibly leading to liver transplantation and even death. Production of reactive metabolites may also cause neo-antigen formation triggering immuno-allergic mechanisms. These may undergo rare cases of severe hepato-toxicity associated most frequently to PHT, but observed also with CBZ and more rarely with VPA [107, 110]. Other more rare idiosyncratic effects are those occurring at the CNS level but in different target tissues. Example of these effects include AED-induced dyskinesias, the mechanism of which has not yet been elucidated [107]. General strategies to minimize the consequences of idiosyncratic reactions include evaluation of risk factors, a careful dose titration, and careful monitoring of clinical response. An important advance in understanding idiosyncratic reactions has been made recently. Large investigations, firstly performed in East populations, in fact, have identified the relevant role of human leukocyte antigen-related genes as predictors of the risk of serious CBZ-induced cutaneous reactions [115-117]. It has been therefore recommended that particularly Asian patients be tested for the HLA-B*1502 allele, in order to identify those at high risk of developing Stevens-Johnson syndrome and toxic epidermal necrolysis after administration of CBZ [115]. Future pharmacogenomic research will increase our knowledge on the relationship among genes, individual drugs and idiosyncratic drug reactions and, most probably, will indicate incisive strategies to prevent all these severe and potential mortal AEs.

Cognitive Impairment, Suicidal Behaviour and Other Psychiatric Disorders

Changes in mood, onset of behavioural and other psychiatric disorders and cognitive impairment are frequently observed in patients undergoing AED therapy [118-121]. These effects can be detrimental in relation to the quality of life and well-being. Cognitive impairment, in particular, penalize patients who work or study. Cognitive effects include diminished attention and memory, slowing of the executive function, and language [115-117]. A particular effect is world finding difficulty more frequently observed with TPM especially at high doses and especially in patients with a left-sided EEG focus [119]. PB is typically associated to hyperactivity in school- and preschool-aged children. While a general agreement does exist on a contribution of AED treatment to these AE, onset, degree and course over time is also influenced by different factors including the type of epilepsy, concomitant diseases, co-existence of learning disability, history of previous psychiatric disorders, familial predisposition [118-121]. Given such a multiplicity of causative factors, it is difficult to extrapolate clear information from available literature data on the contribution of individual drugs. A favourable psychotropic profile has been indicated for CBZ, OXCZ, VPA, GBP, PGB, LCS and, particularly, for LTG especially with regard to affective disorders [121]. PB, TPM, VGB, LEV, TGB, ZNS, PRP and FBM have been associated more frequently with adverse mood and cognition effects [118-121]. Recent papers, however, have highlighted the methodological limits of the studies already made and have stressed the need for more specific clinical trials [121]. In general terms, complex poly-therapies and high doses of AEDs cause a number of these AE. The risk of overtreatment of epilepsy which easily leads to an imbalance between benefits and drug toxicity with predominance of the latter should be constantly taken into consideration by clinicians and avoided [108]. A concern of particular clinical relevance is that AEDs may increase the risk of suicidal though and behaviour [121, 122]. This risk has been recognized since many years in patients with epilepsy [123] and has been evaluated on an epidemiological basis as being 5-fold higher as compared to that of the general population [123]. This value has been found to be even higher (up to 25%) in temporal lobe epilepsy [122]. Additionally, co-morbidities like depression and cognitive impairment seem to be the main risk factors for suicide in epilepsy [121]. Only recently, however, the AED-induced suicidal thought and behaviour has re-gained great attention. In 2008, the US Food and Drug Administration (FDA), after a meta-analysis of data deriving from AED clinical trials, published a warning for this class of drugs in relation to suicidal risk [124]. This risk has been found to be 0.43 per 1000 treated patients as compared to the rate observed in the placebo population which was approximately an half, *i.e.,* 0.22 [124]. Following the FDA alert, an abundant literature has expressed doubts on the conclusions and the action of FDA [121, 124-126]. An interesting, observational, large subsequent study has concluded that the use of AEDs was associated with an increased risk of suicide-related events among patients with or without depression, those who did not have epilepsy, or bipolar disorder, but not in patients with epilepsy [127]. Given the clinical relevance of the matter and in the hope that future studies will clarify as soon as possible all the factors involved in suicidal behaviour, it is important that clinicians be careful and do not underestimate mood disorders and emotional experience complained by patients.

Teratogenic Effects and Other AE on Women of Reproductive Age

Treatment of epilepsy in women of childbearing age is a very complex challenge. On the one hand, the majority of women approaching pregnancy or already pregnant cannot discontinue AED therapy to keep seizures under optimal control. Recurrence of fits, in fact, is associated with the risk of health problems for both the mother, through, for example, a direct influence on the function of the hypothalamic-pituitary axis and consequent alteration of the release of sex steroid hormones, and/or her product of conception, through, for example, direct injuries and possible brain ischemia [128, 129]. On the other side, AED treatment can double, in comparison with untreated women, the risk of major congenital malformations and may be even higher with VPA and with poly-therapies [128-130]. Teratogenic effects include congenital malformations, intrauterine growth retardation, fetal loss, impaired postnatal development, and behavioural problems [130]. As a general rule, avoidance of VPA and AED poly-therapies, especially LTG in combination with VPA, have been recommended to minimize the risks of major birth defects [128-130]. Assumption of at least 0.4 mg per day of folic acid for some months before pregnancy and vitamin K supplements to minimize possible bleeding risk have been also indicated as a potential beneficial therapeutic strategies [131]. More recently the results of large European, US and Australian investigations have been published [132-135]. In particular, the data so far accumulated in an European registry indicate that the malformation rate with VPA and PB at any dose and with CBZ at a dose of >400mg per day is significantly higher than that of LTG at a dose of <300mg per day and CBZ at a dose less than 400mg (2.0% and 3.4%, respectively) [132]. It derives from this study that, apart differences in teratogenic potential of individual drugs, the dose plays also a fundamental role. Similar conclusions have been achieved by other authors in a large US study, in which it was observed that VPA and PB were associated with a higher risk of major malformations than LTG and LEV [133]. Additionally, it has been seen that TPM was associated with an increased risk of cleft lip in comparison with that of a population of control. Experience with the more recent AEDs is of course limited. Two recent investigations, one Danish and the other Australian, have reported the first that exposure to LTG, LEV, TPM, GBP, and OXCZ, was not associated with major malformation as compared to neonates not exposed and the second that the new drugs are not more teratogenic than the traditional ones used in mono-therapy [134-135]. Future research is needed to consolidate the whole of the aforementioned findings and, in particular, the responsibility of individual drugs. Sufficient evidence, however, has been accumulated indicating that VPA, especially in combination with LTG, and AED poly-therapies exert detrimental effects on the fetus [128-130, 134]. Additionally, recent studies strongly suggest that prenatal exposure to AEDs, in particular PB and PHT, may be complicated by negative effects on long-term neurodevelopment and cognitive performance of the infants [136-138] .

A number of sex hormone alterations has been reported during AED therapy. The clinical consequences of these alterations may be infertility and development of a polycystic ovary syndrome [139-141]. The latter has been seen to be two-fold higher with VPA as compared to other AEDs [141].

Given the complexity of the matter on the whole, it is very important that physicians inform women with epilepsy about possible hormonal disorders, fertility, planning a pregnancy, teratogenicity of AEDs, distribution of AEDs in the milk and other important aspects.

Haematological Abnormalities

Mild alterations of various blood parameters are commonly observed during AED therapy, including thrombocytopenia, leukopenia, erythrocytopenia, fall in haemoglobin concentration and others [102, 107]. These alterations can be potentially associated to all AEDs. A mild anemia, for example, is encountered in clinical practice in approximately 5% of patients treated with CBZ, OXCZ or PHT and usually is transient and does not require drug discontinuation [102, 107]. Especially OXCZ, but also CBZ, have been associated to hyponatriemia and, if this remains within a change of 20%, no medication change has to be made [102, 107]. PB, CBZ and PHT, as enzyme inducing drugs, may cause folate depletion with development of megaloblastic anemia [102, 107]. In these case, folate supplementation is sufficient to bring values within the normal ranges. Folate depletion is probably also the cause of hyper-homocysteinemia observed especially in patients treated with enzyme-inducing drugs, but sometimes reported also in patients treated with VPA and with some of the newer drugs, like OXCZ and TPM [142, 143]. Rarely, a severe anemia with systemic involvement has been observed with a number of AEDs and in this case a drug suspension is imperative [102, 107]. Bleeding disorders have been reported with VPA which can induce a dose-dependent thrombocytopenia, low fibrinogen, and inhibition of platelet aggregation [144, 145]. As a rule, when an increased bleeding tendency is observed after VPA exposure, all coagulation factors should be monitored and, if deemed clinically advisable, VPA should be discontinued. Among AEDs, FBM has been associated to severe aplastic anemia in 1:6,000 [102, 107]. Because of this effects and because of its association with severe liver failure, FBM is today used only in selected cases. Asymptomatic elevation of liver enzymes, *i.e.,* alanine aminotransferase (ALT), aspartate aminotransferase (AST), and/or gamma-glutamyl transferase (gamma-GT) is of common observation, especially elevation of gamma-GT, in clinical practice without this implying drug discontinuation [146]. An increase in total cholesterol, low-density lipoprotein cholesterol, and high-density lipoprotein cholesterol has been also described in patients treated with CBZ, PB and VPA [147]. Increase in lipoproteins and/or homocysteine might have clinical significance. Literature data, in fact, indicate that epileptic patients chronically treated with CBZ, PHT or VPA in mono-therapy exhibit increased common carotid arterial intima- media thickness (CCA IMT) correlated with the duration of AED treatment and suggest that patients with epilepsy are at higher risk as compared to non-epileptic patients to develop earlier atherosclerosis with vascular disorders and, possibly, brain atrophia [148-151]. To summarize, alterations of blood parameters are commonly observed during AEDs therapy and, in the great majority of cases, are mild in magnitude, asymptomatic, sometime reversible and usually do not require change in medication. Although all AEDs can potentially induce haematological alterations, these are more frequently associated with some of them, like CBZ, PHT and VPA, among traditional AEDs, and with OXCZ and, especially for the severity of manifestation, with FBM among the newer drugs. In any case, since blood abnormalities, even if common and mild, may in rare occasion deteriorate into serious and also life-threatening conditions, they require an immediate and careful evaluation on an individual basis and a close monitoring over time.

Slowly Developing, Late Appearing AE : Weight, Cosmetic, Hormonal, and Bone Mineral Density Changes

Changes in body weight are frequently observed during AED therapy [152]. Weight gain is commonly observed with VPA, GBP, and PGB, but it has been reported also with CBZ and VGB. Conversely, weight loss and anorexia have been associated to TPM, FBM, and ZNS. Other cosmetic changes, commonly developing after chronic exposure to PHT, include hirsutism, gingival hyperplasia, and coarsening of facies [102, 107]. AED therapy may also affect the hormonal system. Apart from the aforementioned alterations in women, AED induced changes in male hormones have been also described. These include, for example, VPA induced decrease in follicle-stimulating hormone (FSH) and luteinizing hormone (LH) and, *vice versa*, an enhancement of dehydroepiandrosterone sulfate (DHEAS) concentrations and CBZ induced decrease in testosterone/sex-hormone binding globulin (SHBG) ratio [153]. Other changes has been reported for thyroid hormones [154]. Enzyme inducing AEDs, like PB, PHT, CBZ and OXCZ have been reported to decrease the levels of free and protein bound thyroxine (T4) without affecting thyrotropin (TSH). VPA increases the serum concentrations of free triiodothyronine (FT3). All these changes have been observed to reverse, even after years, following AED discontinuation [154]. The real impact of all these changes on the patient clinical condition requires more investigation to be better defined.

Another important issue is the AED induced detrimental effects on bone health. Bone mineral density is lower after chronic exposure to AEDs and epileptic patients exhibits a higher fracture risk as compared to non-AED-exposed subjects [155-157]. The risk has been seen to be higher for women and for those with longer duration of treatment. Drugs involved include PB, CBZ, OXCZ, VPA, LEV and others. The exact mechanisms underlying this effect have not yet been fully clarified, but recent evidence suggests that AED induction of some metabolic pathway involved in vitamin D catabolism might play a pathogenetic role [158].

PRINCIPLES OF ANTIEPILEPTIC DRUG THERAPY

AED therapy is characterized by a long-term duration which lasts, in many cases, a lifetime. This characteristic markedly influences key step decisions, like the time at which therapy should be started, which drug should be chosen for first-line treatment, which change in medication should be made in the case of initial treatment failure. The long duration also implies the need of a dynamic approach and a constant attention to ever possible changes in the clinical situation over time. Seizure control may be affected, for example, by (a) change in drug disposition consequent to age, drug interactions, pregnancy, concomitant diseases, (b) direct influence of physiological conditions, like pregnancy with its hormonal changes, or pathological conditions, like those, for example, associated with fever which lowers seizure threshold, (c) changes in lifestyle with, for example, excessive intake of alcohol, which facilitate seizures, or sleep deprivation caused by stress (d) development of drug tolerance, (e) modification of disease severity, (f) noncompliance [159-161]. The number of AEDs today available allows various possibilities to adapt treatment to a given patient and to a given clinical condition. Available data have shown that no or little difference does exist among the various AEDs in terms of efficacy. Up to 60-70% of people with epilepsy may have seizures well or completely controlled with a proper AED treatment, while a percentage of

30-40% of patients exhibit forms of epilepsy difficult to treat [159-161]. The use of a single drug has different advantages over drug combinations, including fewer AE, better tolerability, avoidance of drug-drug interactions, better adherence to treatment and minor costs [162, 163]. Initial treatment, therefore, should always be mono-therapy both in adults and children. Apart from seizure control, the other aspect requiring close monitoring is development of drug AE. As afore indicated in the specific section, AED therapy is frequently associated to AE, fortunately mild and irrelevant in the majority of cases, but always requiring attention since a manifestation, *e.g.,* skin rush, initially interpreted as benign, might degenerate into more severe and systemic conditions. Drug toxicity may develop slowly over time, be initially not evident but increasing subtly in magnitude, not easily imputable to chronic therapy, but potential cause of devastating effects on quality of life [164]. Examples of this kind include (a) development of psychiatric disorders, (b) cognitive impairment with detrimental effects on study and/or work activities, (c) osteomalacia with consequent risk of fractures, and (d) changes in body weight with, for example in case of an increase, possible consequences on cardiovascular system [159]. Identifying drug toxicity in clinical practice may result not always easy, since the cause/effect relationship is open to different interpretations, the patient does not complain of all symptoms because he ascribes them as disease-related manifestations, or simply because a given sign is considered irrelevant and an insufficient time is dedicated to find an explanation. In case of persistence of seizures, a particular attention has to be placed to avoid the risk of overtreatment. In some cases, seizures of low frequency and mild in their manifestations, *e.g.,* brief focal seizures or brief myoclonic jerks, might be better accepted by the patients than certain AE of drugs. To summarize, drug treatment of epilepsy requires thorough knowledge of a number of key aspects, including AED-, disease- and patient-related factors (Fig. **3**) which all need to be evaluated and balanced on an individual basis. Sharing treatment decisions with the patient after a careful and detailed information is mandatory to gain patient compliance which is the basis to optimize therapy.

Starting AED Therapy

Several studies have been made with the aim of identifying the risk rate of further seizure after a first seizure. In a large and deep investigation, the average risk, calculated on the basis of a meta-analysis, was approximately 50% [165]. In another study, this value was more than 70% after a second seizure and even if the patient had initiated treatment [166] These risk evaluations led to the widely accepted consensus that treatment should be started after at least two unprovoked seizures [161]. It is to be emphasized, however, that that several factors contribute

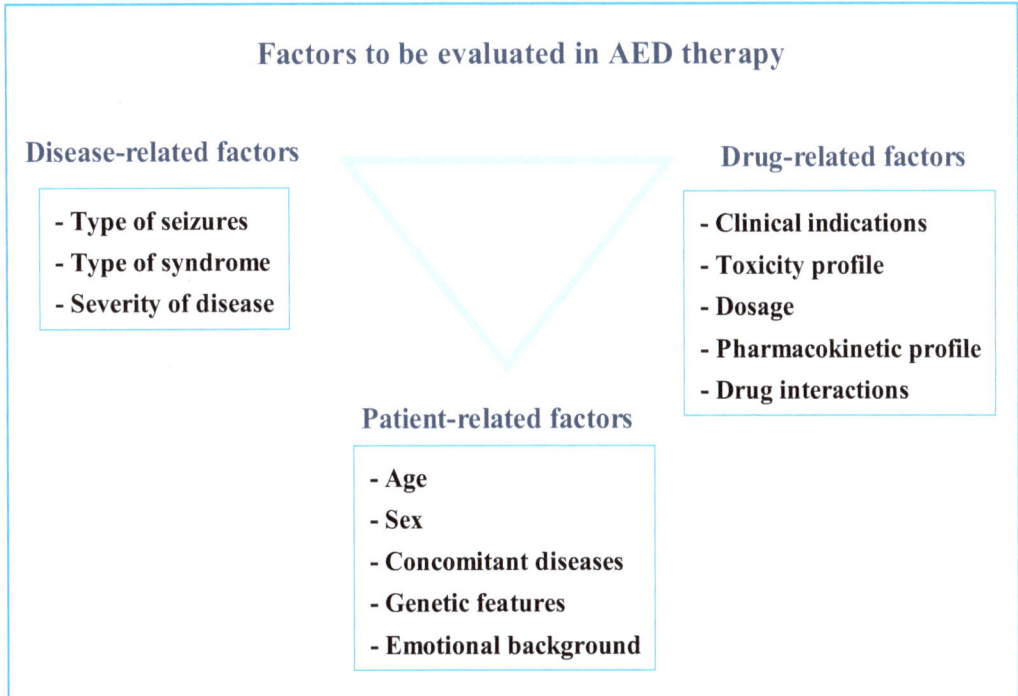

Factors to be evaluated in AED therapy

Disease-related factors

- Type of seizures
- Type of syndrome
- Severity of disease

Drug-related factors

- Clinical indications
- Toxicity profile
- Dosage
- Pharmacokinetic profile
- Drug interactions

Patient-related factors

- Age
- Sex
- Concomitant diseases
- Genetic features
- Emotional background

Figure 3: Factors regarding the features of epilepsy, the drug profile, and the patient characteristics, and requiring careful evaluation in commencing and conducting AED therapy.

in clinical practice to make difficult the evaluation of the real risk of further seizures: (a) type of epilepsy (syndrome, symptomatic seizure, *etc.*) (b) presence of brain pathologies, which might be the direct cause or, even, facilitate seizures (c) abnormalities of the electric brain activity as revealed by the electroencephalogram (EEG), (d) genetic predisposition, (e) unidentified causes. Various possible combinations of these and other variables determine the large and often unpredictable fluctuations observed in the natural history of epilepsies commonly observed in clinical practice [167]. These factors also justify the large range of seizure risk values that can be found across studies in the literature [161].

Another consequence of such a great variability, is that, when facilitating factors are present, the risk of further seizures after a first seizure is not different from that after a second seizure [168]. On these grounds, the more recent definition of epilepsy of the ILAE, stressing that the disease is characterized by "an enduring predisposition to generate seizures", implies that also a first seizure may require treatment when aggravating factors occur [169]. Two large studies have investigated the influence of time of starting treatment on the long-term prognosis of epilepsy [170, 171]. These studies came to the same conclusion, namely that deferral of treatment as compared to immediate medication has no effect on long-term prognosis both in terms of seizure recurrence and in terms of seizure consequences (*e.g.,* status epilepticus or injuries). To summarize, starting AED therapy is often a medical dilemma since it frequently regards situations not clinically well defined and open to different possible evaluations. Therefore, the decision to commence AED treatment is the result of a strictly individualised process requiring a compromise among physician's experience and expertise, evaluation of the most consolidated literature data on the matter, and the emotional approach of the patients towards the newly arisen condition, *i.e.,* the risk of seizures. This risk with its implications on driving, work, social discomfort, and even on the rare but possible occurrence of sudden unexpected death (SUDEP, see below) has to be balanced with the development of AE. As afore mentioned, an emotional reaction of the patient, easily understandable but frequently neglected, should be always taken into a proper consideration to minimize noncompliance. Some types of epilepsy, *e.g.,* benign childhood epilepsy with centrotemporal spike, febrile seizures, benign occipital epilepsies, and reflex epilepsy, may not always require treatment. In conclusion, although indications can be drawn from available literature, starting AED treatment remains a step with not clear-cut edges, it should rely on a careful evaluation of the strictly individual clinical situation and should be shared with the patient after a detailed information of all potential future facets.

Selection of the First Treatment: Spectrum of AED Action and Clinical Indications of AEDs

Selection of the first drug, once the decision to commence therapy has been taken, has to be based on evaluation of the triad of factors regarding the type of epilepsy, the patient characteristics and the drug profile [159-161]. These factors, illustrated in Fig. **3**, are indissoluble linked among them. The initial consideration deals with the evidence of drug efficacy and effectiveness, *i.e.,* retention rate, for a given type of seizure or epileptic syndrome. In adult and elderly patients focal seizures undoubtedly predominate over the primarily generalized forms. The number of drugs to treat focal seizures is more than double as compared to that available for primary generalized seizures (Table **10**). The large body of data so far accumulated indicate that no or little differences do exist among AEDs in terms of efficacy [161]. In particular, a large un-blinded pragmatic study has shown that LTG is the most effective as compared to CBZ, OXCZ, TPM and GBP [172]. On the other hand, an head-to head blinded controlled and randomised comparison fails to reveal any substantial difference between LTG and CBZ and concluded that the two drugs exhibit comparable effectiveness against focal seizures [173]. Similar results, namely substantial equivalence or non-inferiority in seizure control, were achieved in other head-to-head first class studies, regarding CBZ and LEV [174], and CBZ and ZNS [175]. A relatively small difference in efficacy has been demonstrated in favour of LTG as compared to PGB, with 68% of seizure free newly diagnosed patients *vs.* 52% [176] In this study, both drugs exhibited similar tolerability. LTG, GBP and CBZ have exerted similar efficacy in elderly patients with newly diagnosed epilepsy, with a better toxicity profile of LTG and PGB [177]. Selection of a drug as first choice treatment against primarily generalised seizures and syndromes regards mainly paediatric populations and this implies a number of features different from those of drug selection against focal seizures [178-180]. These features include : (a) the number of controlled trials is obviously smaller than that of adults and methodological aspects often hamper the relevance of the results; (b) consequently, AEDs licensed for mono-therapy and initial mono-therapy are fewer; (c) diagnosis of syndrome, a problem encountered by far more commonly in paediatric age, is often difficult after the first seizure and requires time; (d) a correct diagnosis, on the other hand, should be made to avoid drug induced paradoxical seizure aggravation, a phenomenon observed in generalized epilepsies when a wrong drug is used; (e) regardless of paediatric age, the number of AEDs exhibiting a spectrum of action which is extended to primary generalized epilepsy is small as compared to the number of AEDs efficacious against focal seizures (Table **10**).

Table 10: Clinical indications of AEDs in epileptic seizures

Focal Seizures with or Without Secondary Generalization	
Adults	
All	PB, PHT, CBZ, VPA, LTG, VGB, FBM, TPM, GBP, PGB, TGB, OXCZ, LEV, ZNS, ESL, LCS, RTG, PRP
Initial mono-therapy	PB/PRM, PHT, CBZ, VPA
Possible options for Initial mono-therapy *	OXCZ, LEV, TPM, LTG, GBP, ZNS
Used in add-on only	TGB, PGB, LCS, ESL, RTG, PRP
Not routinely used	VGB, FBM
Children	
All	PB/PRM, PHT, CBZ, VPA, LTG, TGB, GBP, TPM, OXCZ, LEV
Initial mono-therapy	PB, PHT, CBZ, VPA
Possible options for Initial mono-therapy*	LTG, TPM, OXCZ, LEV
Use in add-on only	TGB, GBP
Not routinely used	/
Primary Generalized Seizures	
Adults	
All	PB/PRM, VPA, LTG, FBM, TPM, LEV
Initial mono-therapy*	PB, VPA,
Possible options for Initial mono-therapy	LTG, TPM, LEV
Used in add-on only	LEV
Not routinely used	FBM
Children	
All	PB/PRM, VPA, ESM, LTG, FBM, TPM, LEV
Initial mono-therapy*	PB/PRM, VPA, ESM
Possible options for Initial mono-therapy	LTG, TPM, LEV
Used in add-on only	/
Not routinely used	FBM, VGB, RFN, STP

*Some new drugs have been approved for use as initial mono-therapy; although conventional AEDs are formally recommended in new-onset epilepsy patients, new drugs offer a larger possibility to tailor therapy to patients in terms of no or fewer drug interactions and/or different toxicity profiles; °° VGB and FBM can be used only in selected patients when all other drugs have resulted ineffective, because of their severe AE (progressive and irreversible reduction of visual field for VGB and aplastic anemia and liver failure for FBM); VGB can be used as initial mono-therapy in infantile spasms and FBM as adjunctive therapy in the Lennox-Gastaut syndrome in children (> 4 years) and adults. RFN has been approved for the use of Lennox-Gastaut syndrome and STP for the Dravet's syndrome. BZs are not included in the list because of their primary sedative effects; they can be virtually used in all types of seizures/ syndromes, with some caution in the Lennox-Gastaut syndrome for possible aggravation of tonic seizures; as previously mentioned, PRM has not been included in the list because of its nearly complete biotransformation into PB. This table is merely indicative and summarizes the main

clinical indications and recommendations found in the literature, deriving from specific scientific Associations (ILAE= International League against Epilepsy, ANN= American Academy of Neurology, NICE= UK National Institute for Health and Clinical Excellence) and from specific Organisms (FDA= Food and Drug Administration, EMA= European Medicines Agency, and Others), some differences can be found among these Societies; additionally, this table cannot be considered all-comprehensive since licensed indications of AEDs exhibit some differences among Countries and it cannot be considered definitive since an ever increasing accumulation of literature data can lead to change of drug indications especially for AEDs of more recent introduction.

With these premises, it is not surprising that VPA is the drug of first choice in the majority of these forms and especially in paediatric patients [181]. VPA resulted more efficacious than LTG and better tolerated than TPM in a very large, un-blinded, randomised, controlled UK study in a population of patients with generalized and unclassifiable seizures [182]. Based on the most relevant data of the literature, a recent review has synthesised the most consolidated lines of treatment of primary generalized seizures and syndromes: VPA and ESM are the fundamental drugs against childhood absence seizures, TPM as mono-therapy and LTG and LEV as adjunctive therapy can be used to treat primarily generalized tonic-clonic seizures, and LEV shows to be efficacious in myoclonic seizures [183]. As emphasized many times throughout all the text, also in the case of AEDs used in generalized seizures and syndromes any decision should balance drug efficacy with toxicity profile and drug interaction potential [184]. In conclusion, the choice of a given drug for initial treatment should rely on the basal consideration that an evidence-based guideline indicating the overall optimal recommended initial mono-therapy has not yet been developed [185, 186].

This is the consequence of a lack of a sufficient number of rigorous, well-designed, properly conducted first class trials, *i.e.,* those producing results of level A (the highest) of evidence, regarding especially generalized epilepsies and children [185]. Similarly, drug toxicity has not been investigated systematically and with adequate methodology and differences found across the literature data are often of a too small magnitude to allow any substantial drug characterization in quantitative terms [178-180, 184-186]. As also shown by in-depth studies reviewing the most relevant trials regarding refractory epilepsy, clear differences among conventional and newer AEDs cannot be drawn from available results and this has been attributed to a series of methodological factors [187-189]. One of these factors is the daily dose. This, as often stressed throughout the text, should be optimized on individual bases since large inter-individual differences in therapeutic response and drug tolerability do exist. (Table **11**) gives general indications on the most common initial and maintenance doses of individual AEDs.

Table 11: Dose-related aspects of AEDs

AED	Initial target maintenance dose (mg/day)	Usual maintenance doses (mg/day)	Administrations/day (n)
PB	50-100	20-200	1
PHT	200-300	200-400	1-2
ESM	500-750	500-1500	2-3
CBZ	400-600	400-1600	2-3*
VPA	500-1000	500-2500	1-3*
VGB	1000	1000-3000	1-2
LTG	100-300°	100-500°	2
OXCZ	600-900	600-2400	2-3
FBM	1800-2400	1800-3600	3-4
GBP	900-1800	900-3600	2-3
TPM	150-200	150-400	2
TGB	15-30°°	15-50°°	2-4°°
LEV	1000	1000-3000	2
ZNS	200	200-600	2
STP	50 (per kg)^	50-4000 (per kg)^	2-3
PGB	150-300	150-600	2-3
RFN	400-1800	400-3200	2
LCS	200-300	200-400	2
ESL	800	800-1200	1
RTG	300	300-1200	3
PRP	2	8-12	1

Values indicated refer to adults (with the exception of STP indicated in severe myoclonic epilepsy in infancy, also known as SMEI or Dravet's syndrome) and are purely indicative as they may be very different, this depending on age, individual response, individual susceptibility to toxic effects, interactions with other drugs. *= for these drugs extended release formulations are commercially available and the lower number of daily administrations refers to the use of these formulations; °= dosage of LTG may vary greatly, being the lowest in presence of VPA which strongly inhibits its metabolism with consequent marked increase in serum LTG levels, and the highest in presence of conventional enzyme-inducing AEDs which stimulate LTG metabolism with consequent marked decrease of serum LTG levels; intermediate doses can be used when LTG is combined with VPA and one or more conventional AEDs. ^= as above mentioned, STP is only licensed as adjunctive therapy in infancy and the dose is more strictly adjusted on the basis of the body weight. Dose titration suggestions are intentionally not given in this Table. Apart from some AEDs which strictly require small increments weekly, like LTG and TPM, because of their AEs (see the text), it is opinion of the Authors that dose titration schedules cannot be based on the pharmacological properties of a given drug only, but they may vary greatly on the basis of psychological patient acceptance and drug tolerability, both showing remarkable and unpredictable inter-individual variability. Modified from reference [161].

Paradoxical AED-Induced Seizure Aggravation

The phenomenon of the so called "paradoxical aggravation" of seizures refers classically to worsening of seizures in patients suffering from idiopathic generalized epilepsy (IGE), caused by an AED indicated to treat focal seizures, typically CBZ and PHT [190-192]. Drugs facilitating seizures by lowering the seizure threshold are known and include especially some psychotropic drugs belonging to the class of antidepressants and antipsychotics [193, 194]. Thus, the term "paradoxical" most probably reflects the implicit conceptual contradiction that an anti-seizure drug may facilitate seizures. Seizure exacerbation may manifest as an aggravation in frequency and/or severity of pre-existing seizures, induction of new types of seizures or even of status epilepticus [190-192]. The mechanisms underlying this phenomenon are multiple, complex, not clearly identified and, on the whole, poorly understood. Specific pharmacodynamic actions likely mediate the most common type of paradoxical aggravation, which regards exacerbation of primarily generalized seizures, *e.g.,* absences, myoclonic jerks, tonic-clonic seizures in patients with IGE, induced by drugs which typically are anti-focal-seizure drugs, *i.e.,* CBZ, PHT, OXCZ, GBP, PGB, TGB, VGB, and others [190-192]. This effect is common, easily predictable and, as a logical consequence in clinical practice, it implies, after a first seizure of unclear interpretation, especially in paediatric age, the choice of a drug, like VPA, with a very wide spectrum of action [180-184]. Probable more specific mechanisms might underlie the detrimental effect of LTG on myoclonic seizures in the Dravet syndrome [195], juvenile myoclonic epilepsy [196], and, in general, in IGE in which this drug can trigger *de novo* myoclonus [197]. Conceptually, these effects of LTG are of more difficult interpretation, given the spectrum of action of the drug extending also to some types of primary generalized seizures, including tonic-clonic seizures and, to a lesser extent, absences [161, 180, 183, 184, 198]. Another example characterized by a great specificity is aggravation of tonic seizures and even induction of a tonic status by some BZs, namely DZ and clonazepam, in patients with Lennox-Gastaut syndrome [199, 200]. In this case, it can be hypothesized that the sedative effect of these drugs may contribute to trigger specifically tonic seizures, known to occur especially during the night and under the influence of the vigilance level. AED-induced seizure precipitation may be also mediated by detrimental effects of a given AED on a basal pathology from which a patient is suffering. VPA has been associated to seizure exacerbation to a lesser extent as compared to other AEDs [201]. In particular conditions, however, the drug, which is known to may exert toxic actions on the liver and cause hyperammoniemic encephalopathy, may facilitate seizures. An example of this type include a VPA-induced increase in seizures in a case of nonketotic hyperglycinemia recently described [202]. Paradoxical aggravation of seizures requires attention in clinical practice. In theory, all AEDs may be responsible of such an effect and the literature is abundant of reports suggesting a cause-effect relationship for many AEDs. A careful evaluation is needed to consider other factors potentially involved in seizure control deterioration, *e.g.*, noncompliance, spontaneous fluctuations of the disease, development of tolerance to treatment, development of a co-morbidity, genetic factors [190-192].

Drug Resistant Epilepsy: Current Therapeutic Strategies and Future Prospects

Although in recent years the armamentarium of AED therapy has been enriched by a number of new drugs, a substantial proportion of patients, namely 30-40%, suffers from unremitting seizures which do not respond or do show an unsatisfactory response even to carefully optimized treatments [203]. These patients are exposed to the risk of developing chronic drug toxicity, undergoing negative effects consequent to drug-drug interactions, sudden unexpected death, and, ultimately, all detrimental consequences of an useless overtreatment [108]. A crucial point to take into consideration is whether or not a timely and appropriate treatment may contribute to avoid development of drug unresponsiveness and/or, at least, to contain and minimize risk factors facilitating the shifting process of a given form of epilepsy to a drug resistant one. This point, however, implies a premise, namely that epilepsy or at least some specific forms of epilepsy are not drug resistant at the beginning but they do become drug resistant over time and under the influence of a number of factors. Several clinical and experimental excellent studies were performed to address this question, but the complexity of the matter has not yet allowed to achieve clear conclusions. Clinical investigations have identified a number of factors which may reveal early drug resistance and which include: (a) elevated number of seizures before starting treatment, (b) unsatisfactory response to initial treatment, (c) structural brain lesions, (d) age of onset (very early and very late), (e) aetiology (symptomatic *vs.* idiopathic/cryptogenic seizures), (f) type of seizures (secondary generalized *vs.* focal seizures only; syndromic *vs.* nonsindromic forms), (g) EEG abnormalities, (h) genetic determinants [203-209]. Other data support the concept that, even if epilepsy exhibits initially features of drug resistance and does not respond satisfactorily to the first two appropriate drugs at correct dosages, changes in drug therapy may lead to a sizeable proportion of patients (approximately 20-to-50% as reported in various studies) who achieve a condition of prolonged (1-5 years) seizure improvement or, in a minority (~ 17%) of patients, remission [210-214]. A detailed study aimed at quantifying the effects of drug change on seizures previously unresponsive to the first two AEDs has shown that drug resistance is a graded process requiring failure of six AEDs to be labelled as absolute [211]. Recently, the ILAE has formulated a consensus definition and has proposed that drug resistant epilepsy is defined as "failure of adequate trials of two tolerated, appropriately chosen and used antiepileptic drug schedules (whether as monotherapies or in combination) to achieve sustained seizure freedom" [215]. This definition, as emphasised by the same authors, has not to be considered as a rigid conceptual fact, but rather a tool and an attempt to make work in progress more homogeneous than that previously conducted. As recently emphasized, clinical trials to asses therapeutic effects of AEDs should be better designed to improve quality of results and make them more finalized and useful for clinical purposes [216]. Apart from the pharmacological approach, however, other factors may contribute to determine such a great and unpredictable variability in drug intractability. One of the most important is surely the natural course of a given form of epilepsy [217] From an experimental point of view, two types of pathogenetic mechanisms have essentially gained sufficient evidence to explain, at least in part, drug resistance. One occurs at a cellular level and is related to an excessive and abnormal functional activity of multidrug transporter proteins in the epileptogenic focus. This abnormal activity leads to an impaired and reduced access of AEDs into the tissue targets through the blood-brain barrier [218-220]. The other mechanism is mediated by a reduced sensitivity of specific receptors to AEDs (so called target hypothesis as compared to the previous one called transporter hypothesis) [218-219]. Particular attention has been recently devoted to genetic determinants. Two large studies suggest that screening for copy number variations and gene enrichment using microarray-based comparative genomic hybridization could be useful to identify gene mutations especially in epileptic patients with drug resistant epilepsy associated with mental retardation and neuropsychiatric features [221-222]. Future directions in the matter of drug resistant epilepsy, therefore, include genetic features and a series of new targets at cellular and sub-cellular levels (see also the section dedicated to the mechanisms of action of AEDs) [223]. Future pre-clinical studies should take into consideration all the identified targets to explore which of them may be better pursued to have relevant therapeutic implications in clinical practice and, thus, to resolve or even reduce consistently the problem of drug resistant epilepsy [54, 224].

AED Treatment of Status Epilepticus

Status epilepticus (SE) is a medical emergency condition associated to high degree of mortality and morbidity, the latter including substantially detrimental effects on cognitive function and on seizure propensity [225-227]. It is classified, on the basis of its electro-clinical manifestations, in generalized or partial SE or in convulsive/non-convulsive SE, on the basis of the presence or not of convulsions [225-227]. Effective treatment has to be initiated as soon as possible once the diagnosis has been made to prevent permanent neuronal damages. Since aetiology is a primary prognostic factor, it is mandatory to look for an underlying cause of SE to treat this concomitantly with AED medication. Classically, first-line therapies for SE include rapid i.v. administration of BZs, slow infusion of PHT or, in case of failure, PB, and more recently the usefulness of i.v. VPA has been also suggested [225-227]. In a large, randomized, double-blind, multicenter study, which is a landmark in the treatment progress of SE, a comparison has been made of i.v. medication with DZ (0.15 per kilogram of body weight) followed by PHT (18mg), PHT alone (18mg), PB (15mg) and lorazepam (0.1mg) [228]. Lorazepam resulted more effective than PHT in generalized convulsive SE and of equal efficacy as compared to the other treatments. Because of its easier use, the suggestion has been derived that lorazepam is the first choice drug against SE [228]. In a more recent in-depth review [229], lorazepam was confirmed a first line medication, better tolerated than DZ, in controlling SE. Although precise guidelines universally accepted have not yet been produced, there is general consensus that in early/initial SE lorazepam can be considered the first choice drug [229-231]. Buccal midazalam is a valid alternative to lorazepam, especially in children in the cases in which the intravenous access is difficult or unavailable [229, 231]. If BZs fail to control SE and it is at a second stage, defined as established SE, i.v. PHT, but even PB and VPA, can be used [230].

At this stage, more recently introduced drugs, namely LEV and LCS, can be also taken into consideration on he basis of the available positive observations [231, 232]. Despite lack of evidence from controlled studies, both drugs exhibit a favourable pharmacokinetic profile and good tolerability, especially lack of sedating effects, which make them promising potential alternatives to the standard anti-SE drugs [231, 232]. SE at the third stage, namely refractory SE, needs treatments in an intensive care unit, where a variety of anaesthetics, like propofol and thiopental, and midazolam can be administered through continuous i.v. infusion in association to a close monitoring of vital functions and cerebral electrical activity [231, 232]. In cases of very refractory SE, a number of possible pharmacological and non-pharmacological treatments have been considered, including ketamine, isoflurane, desflurane, topiramate, pregabalin, hypothermia, magnesium, pyridoxine, immunotherapy, ketogenic diet, vagal nerve stimulation, deep brain stimulation and others [233]. A positive outcome with recovery to baseline was observed in 35% of patients [233]. To summarize, i.v. lorazepm, i.v. DZ, buccal and, as recently demonstrated [234]; intramuscular midazolam can be all considered first-line medications [228-234]; i.v. PHT, PB, and VPA are to be considered second choice AEDs; LEV and LCS can be also included at this stage. Third-line drugs are anaesthetics, like propofol and thiopental, and a number of alternative treatments, in case of very refractory SE, listed above. As it has been frequently emphasised, SE, especially generalized convulsive SE, is most often manifestation of an acute brain insult, direct o secondary to systemic disorders, and that early identification of prognostic factors may result relevant to avoid or minimize the risk of under- or over-treatments [226, 227, 231, 233]

Ongoing research is finalized to identify specific risk factors influencing prognosis of SE, *e.g.,* age, duration, aetiology, degree of consciousness, and others, to assess and validate scales suitable to evaluate the severity of SE, and to formulate clear therapeutic guidelines, sufficiently consolidated by scientific evidence, and still lacking [226, 235-238].

Therapeutic Features of Specific Patient Populations

Treatment of epilepsy in specific patient populations requires knowledge of a number of aspects which may interfere each with the other and complicate the clinical picture. Clinicians should be familiar with: (a) the features of the specific condition, (b) the pharmacokinetic profile of AEDs and other classes of drugs used in combination, and its possible changes induced by that condition, (c) drug interactions, (d) the possible detrimental effects of AEDs on the co-morbid disorder and those of the associated drugs on epilepsy. An overall view of these aspects allow to make the correct therapeutic choices. (Table **12**) summarizes the most common treatment features of patients populations affected by disorders/conditions of frequent observation in clinical practice.

Psychiatric Patients

Psychiatric co-morbidity, including depression with suicide thoughts, anxiety disorders and psychosis, is very common among patients with epilepsy and percentages even superior to 40% have been reported in the previous and more recent literature [239-242]. A so high co-morbidity and the observation that these disorders often precede the onset of seizures strongly suggest that these pathologies share common pathogenetic mechanisms with epilepsy [240-243]. In particular, episodes of major depression and suicide attempts have been observed to be an independent risk factor of forthcoming development of unprovoked seizures, possibly through shared underlying neurochemical pathways [243]. Other recent data have drawn attention on another possible pathogenetic mechanism which causes onset of depression in patients with chronic epilepsy. Major depressive disorders, in fact, have been observed to be significantly associated to the development of mesial temporal sclerosis. In this case, progression of such a sclerosis and recurrence of seizures provoke functional and structural changes in limbic circuitries with consequent development of mood depression [244].

Irrespective of the hypothesised mechanisms, the term "bidirectional" used to describe the pathogenetic association of depression and epilepsy is most appropriated. [240]. Such a complex pathogenetic ground is paralleled by a similar complexity of the therapeutic management, the most important points of which have been recently highlighted by an ad-hoc Commission of the ILAE [245]. The majority of epileptic patients with psychiatric disorders require adequate treatment not only for the manifestations typical of these disorders, but also for additional subtle and not easily recognizable disturbances. It has been recently observed, for example, that depression worsen significantly the perception of AED-induced AEs with detrimental effects on quality of life [246]. This situation may lead to misinterpretation of the clinical picture and to an erroneous treatment [246]. One of the most important therapeutic concern is that both antidepressants and antipsychotics may facilitate seizures because of their lowering effects on seizure threshold [87, 91, 194, 242]. This aspect is largely known among physicians. For this reason, they are often reluctant to use these classes of drugs and tend to underestimate the importance of treating psychiatric co-morbidities in patients with epilepsy [247, 248]. Seizure incidence rate reported in the literature range from 0.1 to 1.5% with the use of antidepressants or antipsychotics, these values being clearly higher as compared to the incidence of the first unprovoked seizure in the general population [87, 91, 194, 242]. The extent of the seizure risk increases dramatically up to 30% in patients treated with high doses [87, 91, 194, 242]. The proconvulsant potential differs among psychotropic drugs A recent large analysis made by FDA in US on a total population of 75,873 patients in phase II and III trials has pointed out that the incidence of seizures is significantly higher than placebo in patients treated with clomipramine, bupropion immediate release, alprazolam, clozapine, olanzapine, quetiapine, and lower with the use of second-generation antidepressants [249]. Evidence has been derived from this analysis that (a) some second-generation antidepressants may exert anticonvulsant effects, as it was clearly demonstrated also for tricyclic compounds *in vitro* and in experimental models of epilepsy in animals [193]; (b) depression, psychotic disorders and also obsessive-compulsive disorder are associated with reduced seizure threshold. On the basis of these and other indications [245-249], psychiatric disorders should be carefully evaluated and adequately treated in patients with epilepsy. The unspecific fear of using psychotropic drugs is unjustified with most compounds [248, 249]. A careful selection of drugs is also required in treating seizures in patients suffering from psychiatric co-morbidities. Depressed patients, for example, should not be treated with AEDs known to affect negatively mood, like PB, LEV and TPM and, conversely, LTG, CBZ or OXCZ should be preferred [250-252]. Pharmacokinetic interactions between AEDs and psychotropic drugs, which occur frequently [240], have been illustrated before and listed in (Tables **5-7**).

Women

Treatment of women with epilepsy exhibits a number of specific, multifaceted and unique implications which are particularly relevant during childbearing age and even more relevant in the case of planning a pregnancy or during pregnancy (Table **12**) [253, 254]. Teratogenic effects of AEDs and pregnancy-induced changes in AED pharmacokinetics have been afore-illustrated. In synthesis: (a) VPA, especially at high doses and used in combination with other AEDs, and CBZ at doses higher than 400mg/day are to be avoided because of cause of an approximately two/three-fold increase in birth defects; (b) serum levels of many AEDs, especially LTG and OXCZ, show a progressive decline, typically more pronounced in the last period of pregnancy, with consequent risk of breakthrough seizures and possible need of drug dose adjustment; (c) mono-therapy and a careful pursuit of identifying the lowest effective dosage to prevent particularly generalized tonic-clonic seizures, which are potentially harmful for both the mother and her conception product, should be mandatory [128, 253, 254]. Apart from these aspects, other critical issues require to be contemplated. Treatment of juvenile myoclonic epilepsy (JME), catamenial epilepsy, and post-partum-related conditions are some of the most relevant examples. JME, or Janz syndrome, is of frequent observation in young women, being the age of onset between 12 and 18 years, and is substantially characterized by the co-existence of myoclonic jerks, absences and generalized tonic-clonic seizures. The most effective drug is VPA which has a response of approximately 80% of the cases. Given the aforementioned detrimental effects of this drug in women, however, other drugs have been suggested as alternative to VPA [255]. These drugs include LEV, LTG, TPM, and ZNS, which are especially efficacious against tonic-clonic seizures and to a lesser extent against absences and myoclonus [256, 257]. Clonazepam may be added to one of the above drugs when these fail to control satisfactory myoclonic jerks [257]. It is to emphasize that no first class trials or head-to-head drug comparisons exist and treatment options for JME are based primarily on observational studies [185, 256, 257]. Catamenial epilepsy is defined as a condition in which seizure exacerbation shows a periodicity related to the menstrual cycle and affects approximately one third of women [258]. A variety of treatments have been proposed, including acetazolamide, cyclic use of BZs, different AEDs and, as hormonal changes play a crucial role in determining seizures periodicity [259], hormonal therapy [258, 260]. Evidence for the effectiveness of these treatments comes from small, uncontrolled studies or even anecdotal reports. Interestingly, a recent double-blind, randomized, placebo-controlled trial aimed at testing cyclic progesterone as a possible useful medication in women with drug resistant focal epilepsy, has identified a subset of women with evident perimenstrual seizure exacerbation that were responsive to this treatment [261]. As known, estrogens facilitate generation of seizures.

Table 12: Common subtle adverse drug effects and/or aspects to be discussed with patients an/or their relatives characterizing specific patient populations treated with AEDs

Patient Population	Characterizing Therapy-Related Aspects
Children	hyperactivity, nervousness, reduced scholastic performance induced by some AEDs.
Adults	difficulties in working activity, driving, and sexual activity induced by some AEDs.
Elderly patients	impairment of cognitive function (*e.g.,* pseudo-dementia), psychiatric disorders, bone fractures consequent to falls.
Women	hormonal disorders (*e.g.,* polycystic ovary syndrome with VPA), pregnancy complications, teratogenicity of some AEDs, catamenial epilepsy, decreased efficacy of the contraceptive pill caused by enzyme-inducing AEDs, progressive reduction of serum AED levels during pregnancy, especially occurring with LTG and OXCZ and in the last period of pregnancy, with possible need of dosage adjustments to minimize the risk of breakthrough seizures, low IQ of offspring with maternal exposure to VPA.
Psychiatric Patients	compliance, seizure threshold lowered by some psychotropic drugs, worsening of psychic conditions, especially depression with possible suicide thoughts, induced by some AEDs, worsened perception of toxic effects of AEDs
Patients with concomitant diseases	drug interactions causing changes in disease picture and leading to possible misinterpretation of the clinical manifestations; possible changes in the pharmacokinetic features of AEDs and/or other associated drugs with possible consequent need to use alternative drugs; possible changes in pharmacodynamic responses.

The post-partum period also is a complex state for women with epilepsy. Changes of the hormonal condition, body weight, drug metabolism rate, sleep periodicity, seizure frequency may require re-adjustments of the antiepileptic therapy and concern regarding breastfeeding and specifically drug transfer into milk is a constant [262]. A strict collaborations of neurologist, gynaecologist and, if necessary, psychologist and a valid preconception counselling may be helpful and can contribute to minimize the above indicated critical issues regarding women with epilepsy.

Elderly Patients

The incidence of epilepsy in patients aged >60 years, the most rapidly growing segment of the population, is higher than in any other period of life [263]. Its treatment requires specific considerations. Firstly, elderly patients have a progressive, physiologic, age-related reduction in liver and kidney functions. This, as above described, may alter drug pharmacokinetics, causing, as a final result, a slow drug elimination with the risk of an increasing drug accumulation into the body [64, 65]. Mono-therapy and very low initial drug dose with subsequent gradual upward titration until seizures are controlled, therefore, are general and correct recommendations to start AED therapy in elderly patients [263-266]. These recommendations are particularly valid on the basis of other aspects: (a) epileptic seizures usually show a satisfactory and prompt response to medication at low doses in this segment of the population, (b) multiple medications for concomitant diseases, like hypertension, atherosclerosis, diabetes, osteoporosis, hearth diseases, behavioural disorders, are an almost necessary constant and facilitate extremely relevant drug-drug interactions (Tables **5-7**), (c) it is common experience that older patients exhibit a particular vulnerability to toxic drug effects, especially those at the CNS [263-266]. Exacerbation induced by some AEDs of pre-existing problems such as cognitive difficulties, mood depression, motor hamper, tremor and ataxia with consequent easy falls and bone fractures are of common observation once AED therapy is started [263-266]. The possible usefulness of comparing quantitative measures of balance to obtain more information on toxicity drug profile has been explored in older patients stabilized on treatment with CBZ, LTG or GBP [267]. Interestingly, the study was carried out in asymptomatic patients, namely in patients without complaint of dizziness or imbalance. The conclusion was that LTG is less probable to cause disequilibrium than does CBZ in older people [267]. This potential advantageous profile of LTG over CBZ derives also from specific double-blind, randomized, controlled trials performed in elderly patients with newly diagnosed epilepsy [268-269]. Overall these studies indicate that LTG and also GBP are equally efficacious as CBZ in controlling seizures but they are better tolerated and can be considered first-choice drugs in elderly patients. Post-stroke seizures and seizures occurring in patients with dementia account for a noticeable percentage of epilepsy in older people. Unfortunately, there is a paucity of high-level clinical trials and the few available data cam from observational studies [270-276]. LTG, GBP and the more recently introduced LEV appear to be more advantageous as compared to conventional drugs in terms of lack of interactions with drugs largely used in elderly patients, like salicylates and anticoagulants, and lack of detrimental effects on recovery, bone health, and cognition as induced PB, PHT and on blood sodium levels and hearth conduction as caused by CBZ [263-266, 270, 274]. Despite of this evidence and the general recommendation deriving from available literature, the use of conventional AEDs continues to predominate remarkably over that of the newer drugs and little attention is given to the above illustrated issues typical of AED therapy in older age [277, 278].

DISCONTINUATION OF AED THERAPY

It is known that AED therapy is purely symptomatic, *i.e.,* it exerts a control on seizures and prevents seizure recurrence, but has no effect on the natural course of epilepsy [279]. In paediatric age, in which several syndromes have been identified, sufficient knowledge has gained to predict how a specific syndrome responds to therapy and how it evolves when drugs are withdrawn [279]. As typical examples, benign childhood epilepsy with centrotemporal spikes, also defined benign rolandic epilepsy, exhibits no or little risk of relapse after therapy discontinuation; on the other side, seizures reappear in a percentage of >80% in patients affected by juvenile myoclonic epilepsy, *i.e.,* Janz syndrome [279]. However, in the majority of patients, especially in adults and old patients, seizures cannot be incorporated in a well defined syndromic picture and the outcome both during treatment and after drug discontinuation in the case of achievement of a seizure freedom condition is largely unpredictable. The few available literature data indicate that: (a) at least 2 years of seizure freedom for children and 3-to-5 years for adults are the periods suggested before discontinuing AEDs; (b) the relapse risk is 20-25% at 1 year and 25-30% at 2 years after AED withdrawal; (c) worse prognostic factors include partial epilepsy, an underlying neurological condition, an abnormal neurological examination, adolescent onset of epilepsy, abnormal EEG, a longer time to achieve seizure control, and longer duration of epilepsy [280-284]. These are of course general indications and neither the exact time duration of seizure freedom before withdrawing AEDs nor the optimal rate of tapering of AEDs have been clearly established and codified into clear guidelines. In daily clinical practice, given that AED therapy is associated with both short and long term toxic effects, patients press to discontinue drugs when epilepsy is in remission. Decision on the time to start drug withdrawal and on the rhythm at which drugs are tapered off are largely based on empirical grounds. Careful information on potential consequences on driving, job, leisure activity especially over the first months of drug discontinuation together with a proper consideration of the emotional state of the patients and his family are all important recommendations which contribute to optimize this very important, complex and not yet well-defined step of the antiepileptic therapy [280-284].

ESSENTIAL CURRENT-KNOWLEDGE-BASED REMARKS AND FUTURE DIRECTIONS OF AED THERAPY

Drug therapy of epilepsy is a very complex matter which is characterized by a number of unique aspects as compared to other therapies. The unpredictable nature of the episodes of impaired or loss of consciousness makes the patient particularly insecure in his daily life activities and particularly vulnerable as far as his emotional state is concerned [85, 239, 245] . It derives that he is dependent on treatment and, contemporary, he is obliged to accept that it is of long term duration and induces toxicity. Some toxic effects consequent to chronic AED administration are of difficult interpretation and not easily distinguishable from disorders partly induced by shared mechanisms underlying epilepsy. This is the case, for example, of mood depression or cognitive dysfunction, which may be both caused by some AEDs or by some functional/structural changes of neural circuitries which support epilepsy itself. As above discussed, some AEDs may induce depressive effects, hyperactivity especially in children, agitation and even psychosis, and cognitive dysfunction [251, 252, 285-287]. On the other hand, mesial temporal sclerosis, recurrence of seizures, some abnormal EEG patterns have been seen to be significantly associated with mood disorders and/or impaired cognition [240-244, 288, 289]. It is obvious that these causative factors may occur concomitantly and make the clinical picture and its treatment of very difficult task. Epilepsy is a dynamic pathologic process with unpredictable changes in severity over time and, consequently, changes in drug response [218]. Apart from

Table 13: Key points in AED therapy

1) The long duration of AED therapy implies a dynamic approach and a careful evaluation over time of possible: a) changes in pharmacokinetic properties of AEDs due to age, pregnancy, concomitant diseases (especially of kidneys and liver), and concomitant use of additional drugs; b) development of subtle drug toxicity (*e.g.,* osteomalacia, mood and/or cognition disturbances); c) development of unexpected patient non-compliance; d) development of drug tolerance; e) changes of disease severity.
2) Some AEDs exhibit non-linear pharmacokinetics and, consequently, plasma drug concentrations can become unexpectedly disproportional in relation to the dose: a) PHT serum levels can rise suddenly with possible toxicity precipitation because of metabolism saturation; b) CBZ shows over time a dose-dependent auto-induction of its own metabolism with consequent fall of plasma levels and possible loss of efficacy; c) GBP absorption decreases markedly up to 50% as the dose is increased because of a saturable transport system across the gastrointestinal tract; d) STP shows a decreased clearance with increasing dosage.
3) Given that little or no efficacy difference has been found among AEDs, drug selection has to be based on the toxicological profile adapted to the clinical condition of an individual patient; as examples: patients with severe hepatic diseases should given AEDs eliminated by kidneys (*i.e.,* LEV, LCS, GBP, PGB) and, *vice versa*, these drugs should be avoided in patients with kidney diseases; in women planning a pregnancy, VPA and CBZ (>400mg/day) should be avoided for the risk of major neonatal malformations.
4) Determination of plasma AED levels has not to be routinely made, but only on the basis of a specific and finalized question.
5) Drug resistant epilepsy facilitates the risk of overtreatment with detrimental effects on quality of life; high drug doses and combinations of AEDs should be avoided; only VPA+LTG against partial and generalised seizures and VPA+ESM against atypical absences have gained sufficient evidence of beneficial effects.
6) The long duration of AED therapy requires a continuous and in-depth dialogue with the patient and his family to obtain and maintain patient compliance; emotional components and easily occurring misinterpretations of the information and recommendations given by the physician might weak over time patient compliance.
7) The progress made over years in terms of scientific data acquisition and hence in terms of rational antiepileptic therapy does not exclude the old principle of "try and see"; the clinical dimension, in fact, is characterized of an endless amount of variables often undefined, elusive and misleading. In particular, individual genetic variations and individual changes in drug response consequent to age and concomitant diseases imply unpredictability of response to AED therapy and dose optimization requires a constant balance of seizure control with adverse drug effects.
8) There are not guidelines to discontinue AED therapy; in the case of a well defined syndrome, it is essential to possess an in-depth knowledge of the syndrome features; in other conditions, at least 2 years of seizure freedom for children and 2-5 years for adults are needed before discontinuing AEDs; factors such as duration of epilepsy, age at disease onset, persistence of EEG abnormalities, and symptomatic seizures are predictive of a negative prognosis.

Points indicated in the table are derived from references [159-164].

spontaneous changes of he natural course of the disease, loss of efficacy may be a consequence of development of tolerance [290, 291]. This is an adaptive process of the body to repeated drug administration, is reversible after drug discontinuation and is different depending on individual characteristics and on the type of drug [290, 291] The practical implication of these factors is that adjustments of medication over time are required to maintain optimal AED treatment. (Table **13**) gives a list of the most important and common challenges and features characterizing AED therapy, based on current knowledge and frequently encountered in daily clinical practice. Of course, the list is not all-comprehensive and other aspects may be related to AED therapy. One of these is, for example, SUDEP, an acronym referring to a specific phenomenon not yet elucidated in its pathogenesis an defined as sudden unexpected death in epilepsy.

Sudden Unexpected Death in Epilepsy (SUDEP) and its Potential Relationship with AED Therapy

SUDEP is a devastating epilepsy outcome occurring especially in patients with very refractory epilepsy, the incidence rate of which has been evaluated as up to nine per 1,000 patient-years in a population of epileptic patients candidate for epilepsy surgery [292, 293]. Many conditions have been reported to be closely associated with this devastating epilepsy outcome, including nocturnal seizures, cardiac arrhythmias, possibly related to a sympatho-vagal imbalance, frequency of generalized tonic-clonic seizures, long-duration epilepsy, male sex, serotonin system pathology [292-296]. Concerning treatment-related factors, a recent paper has found a positive association of SUDEP with females treated with LTG [297]. Although at the current stage of research the majority of these data cannot be considered definitive, sufficient evidence has been gained strongly suggesting that recurrence of generalized tonic-clonic seizures is one of the most relevant risk factors of SUDEP [298]. In support of this, it has been derived from an in-depth analysis of the most solid data of the literature that adjunctive therapy efficacious to prevent this type of seizures reduces by seven times the risk of this fatal event [299].

Therapy Optimization as Result of Integration between Evidence Based Medicine and Clinical Practice

As it has been stressed many times throughout all this chapter, optimization of AED therapy is a complex process requiring adaptation to patients on an individual bases. As so, the more effective results can be achieved only through an integration between the most qualified scientific evidence, *i.e.,* the evidence based medicine (EBM), and the individual clinical expertise gained through the daily medical practice [300]. EBM gives information on the most reliable and scientifically qualified data of the literature trough systematic reviews, meta-analysis, and critical elaboration of the literature results. This work is even more fruitful in the case of rapid changes of a given scenery, like that of AED therapy, which is currently characterized by the introduction of many drugs in a short period of time. A continuous up-dating of clinicians is required to the end not only of an appropriate therapeutic approach for the patient care but also for formal legal aspects of drug prescription. Treatment of epilepsy has deeply changed over the latter 100 years [301, 302]. In the previous century, AEDs were introduced just on the basis of a serendipitous intuition or of observation of few cases. Modern controlled clinical trials follow rigorous rules to test the real efficacy/effectiveness of a given compound. In recent years, however, awareness has been gained that clinical studies exhibit some limits and still require further methodological refinements [303-306]. Some of these limits include mainly substantial differences among studies which make them not comparable: (a) different and generally too short duration of the various study periods (baseline, active treatment, *etc.*), which hampers, for example, acquisition of the true drug effectiveness, (b) control groups not homogeneous, (c) use of different drug doses, (d) rare head-to-head comparisons (e) different efficacy parameters, (f) toxic effects of medication not formally investigated and quantified, (g) differences in definition of drug resistance [303-306]. In this sense, for example, a recent proposal has been formulated by the ILAE in the attempt to uniform worldwide the mean of drug resistant epilepsy [215]. In synthesis, the results deriving from clinical trials are rarely comparable and not always they are applicable in a clinical practice setting. As a consequence, beside a general awareness that clinical trials have to improve in methodology [303-306], observational pragmatic studies, *i.e.,* those reflecting daily clinical practice, have been reappraised and recognized to be useful in giving information complementary to that derived from controlled clinical trials [307, 308]. In other world, the old principle "try and see" is partially valid for a number of aspects of AED therapy, especially those related to long-term efficacy and long-term tolerability of AEDs (see also Table **13**).

Future Directions: Disease-Modifying Drugs and Gene-Related Therapeutic Approach

A current central limit of the treatment of epilepsy is that all AEDs currently available are only symptomatic. In other terms, they are unable to modify the course of disease, but only exert a partial or total control of seizure recurrence. Similarly, AEDs do not prevent development of chronic epilepsy. Following a brain injury, an early treatment may be effective against acute provoked seizures, but fails to stop the epileptogenic process leading, for example, to post-stroke epilepsy, post-traumatic epilepsy, post-surgery epilepsy, and so on [309]. One important line of research is identification of compounds able to protect neurons from the detrimental consequences of an injury and hence to avoid the occurrence of epileptogenesis [310, 311]. Progress has been made in understanding some steps following brain damage. Functional changes occur with shift to a condition of imbalance between inhibitory, mainly GABA-mediated, neuronal transmission in favour of the excitatory glutamate-related transmission [310, 311]. These functional changes may precede or even be from the start associated to structural tissue modifications. These include essentially neuronal loss, caused by necrosis/apoptosis, followed by compensatory plasticity processes of neo-synapto-genesis/neuron-genesis with consequent inflammatory and immune responses [310, 311]. More recent research has also identified some genes involved in these network rearrangements and potential promising targets for drug intervention [310-312]. The experimental data on the matter are abundant and tools to investigate the basic mechanisms of epileptogenesis include kindling and post-status epilepticus models of temporal lobe epilepsy [313]. Transfer of these data into human epilepsy with the aim of applying novel therapeutic strategies, however, has been so far disappointing and no drug has clearly proven to exert neuro-protective and/or antiepileptogenic and disease-modifying effects in man [309]. A recent meta-analysis aimed at evaluating the consequences of AED withdrawal given to prevent post traumatic seizures did not find any evidence of effectiveness of the treatment in preventing late seizures [314]. In the future it is hoped that identification of new epileptogenesis-related targets will contribute to synthesize new compounds able to exert true disease-modifying effects, a goal generating great interest in clinical practice [315]. This remains, therefore, an attractive and fascinating area of research.

In recent years, a surprisingly enormous progress has been made in identifying genes and understanding the genetic influx underlying epilepsy. This progress has led to revolutionary views of the classification of epilepsy and epileptic syndromes with a parallel extraordinary intensification of the genetic research relating treatment of drug resistant epilepsy [316-318]. Genetic variability in drug metabolism of PHT was reported more than 25 years ago and patients are, on the basis of the metabolic rate of some liver isoenzymes, *i.e.,* CYP2C and CYP2C19, classified as poor or extensive metabolizers [62, 63]. Slow metabolizers are more exposed to PHT accumulation and consequent toxicity. Pharmacogenomic studies aimed at predicting drug efficacy and drug toxicity, on the basis of individual genetic makeup, are complex and did not yet led to homogeneous and conclusive results. Current investigations are oriented to identify genes encoding drug target, *e.g.,* sodium channel-related SCN1A, drug transport (ABCB1), human leukocyte antigen (HLA) proteins, and novel biologically plausible candidate genes through microarray analysis of samples of brain tissue deriving from epilepsy surgery [316-318]. One of the most experimented approach in gene therapy of epilepsy is the use of recombinant viral vectors to reinstate the altered balance between inhibitory and excitatory transmission [319-320]. Promising results, for example, have been obtained through transduction of neuropeptide genes such as galanin and neuropeptide Y (NPY) in specific brain areas in experimental models of seizures [319, 320]. Therefore, gene therapy is a promising new approach for the treatment of epilepsy and efforts are currently made to improve our knowledge on cell transplantation techniques and the development of recombinant viral vectors for gene delivery [319, 320]. Four years ago a very interesting and stimulating debate has been published focusing on whether or not genetic information may be useful to treat patients with epilepsy [321]. Very recently, it has been found that patients carrying genetic variants associated with decreased dopaminergic activity exhibit a high propensity to develop psychiatric AEs, such as irritation and aggression, induced by LEV [322]. On the basis of this acquisition and that demonstrating that Asian patients with a particular HLA allele, HLA-B*1502, are at a higher risk for Stevens-Johnson syndrome when using CBZ [115-117], we have to stop the previously mentioned debate [321] in favour of a real therapeutic usefulness of genetic knowledge [323].

In conclusion, AED therapy has made an extraordinary progress over time [301, 302] even if the way to solve all related problems, first of all drug resistance, still remains a long way. Epilepsy is one of the most common, serious neurological disorders, affecting an estimated 50 million people worldwide. The hope is that the directions currently taken to increase our knowledge on the therapy of this disease are right and will allow to apply soon innovative strategies to alleviate suffering of the patients and improve remarkably their quality of life.

ACKNOWLEDGEMENTS

Our sincere gratitude to Emilio Perucca, Ted Reynolds and Alan Richens who directly and through their papers have deeply influenced our approach to the field of antiepileptic drug therapy.

CONFLICT OF INTEREST

The authors confirm that this article content has no conflicts of interest.

REFERENCES

[1]   Locock C. Discussion of a paper by Sieveking, EH. Analysis of fifty-two cases of epilepsy observed by the author. Lancet 1857; 1(1760): 527-8.
[2]   Friedlander WJ. Who Was 'the Father of Bromide Treatment of Epilepsy'? Arch Neurol. 1986; 43(5): 505-7.
[3]   Hauptmann A. Luminal bei epilepsie. Munch Med Wochenshr 1912;59:1907–9.
[4]   Merritt HH, Putnam TJ. Sodium diphenylhydantonate in the treatment of convulsive disorders. JAMA 1938;111:1068–75.
[5]   Werdnecke E. Phenyläthylhydantoin (Nirvanol), ein neues Schlaf- und Beruhigungsmittel. Deutsch Med Wochenschr 1916;42: 1193.
[6]   Putnam TJ, Merritt HH. Experimental determination of the anticonvulsant properties of some phenyl derivatives. Science. 1937 May 28;85(2213):525-6.
[7]   Friedlander WJ. Putnam, Merritt, and the discovery of Dilantin. Epilepsia. 1986;27 Suppl 3:S1-20.
[8]   Glazko AJ. Addendum to "Putnam, Merritt, and the discovery of Dilantin".Epilepsia. 1987; 28(1): 87-8.
[9]   Handley R, Stewart AS. Mysoline; a new drug in the treatment of epilepsy. Lancet. 1952;1(6711): 742–44.
[10]   Zimmerman FT. Use of methylphenyl-succinimide in treatment of petit mal epilepsy. AMA Arch Neurol Psychiatry. 1951;66:156–162.
[11]   Millichap JG. Milontin: a new drug in the treatment of petit mal. Lancet. 1952;2(6741):907-10.
[12]   Schindler W, Hafliger F Über Derivate des Iminodibenzyls. Helv Chim Acta 1954;37: 472-83.
[13]   Lorge M. Klinische Erfahrungen mit einem neuen Antiepilepticum, Tegretol (G 32 883) mit besonderer Wirkung auf die epileptische Wesenveränderung. Schweiz Med Wochenschr 1963; 93: 1042-7.

[14]   Lustig B. On the treatment results with the new anti-epileptic agent G 32,883. Med Welt. 1964;57: 203-4.

[15]   Sigwald J, Bonduelle M, Sallou C, Raverdy P, Piot C, Van Steenbrugghe. A new anti-epileptic agent: carbamyldibenzazepine or 5-carbamoyl-5-H-Dibenz(B,F)azepine (G-32,883). Presse Med. 1964;72:2323-4.

[16]   Burton B.S. On the propyl derivatives and decomposition products of ethylacetoacetate. Am Chem J 1882; 3: 385–395.

[17]   Meunier H, Carraz G, Meunier Y, Eymard P, Aimard M (1963). "Propriétés pharmacodynamiques de l'acide n- dipropylacetique". Therapie 18: 435–438.

[18]   Jeavons PM, Clark JE. Sodium valproate in treatment of epilepsy. Br Med J. 1974;2(5919): 584-6.

[19]   Brodie RE, Dow RS. Chlordiazepoxide in epilepsy. Northwest Med. 1962 Jun;61:513-6.

[20]   Brock JT, Dyken M. The anticonvulsant activity of chlordiazepoxide and Ro 5-2807. Neurology. 1963;13:59-65.

[21]   Barreraperez S. Clinical study of valium in 50 epileptic patients. Medicina (Mex). 1964;44: 123-6.

[22]   Gastaut H, Naquet R, Poire′ R, Tassinari AC. Treatment of status epilepticus with diazepam (valium). Epilepsia, 1965;6:167–82.

[23]   Gram L, Lyon BB, Dam M. Gamma-vinyl-GABA: a single-blind trial in patients with epilepsy. Acta Neurol Scand, 1983;68(1): 34-9.

[24]   Rimmer EM, Richens A. Double-blind study of gamma-vinyl GABA in patients with refractory epilepsy. Lancet. 1984; 1(8370): 189-90.

[25]   Schechter PJ, Hanke NF, Grove J, Huebert N, Sjoersdma A. Biochemical and clinical effects of gamma-vinyl GABA in patients with epilepsy. Neurology. 1984;34(2): 182-6.

[26]   Smith M, Wilcox KS, White HS. Discovery of antiepileptic drugs. Neurotherapeutics 2007;4: 12–7.

[27]   Brodie MJ. Antiepileptic drug therapy the story so far. Seizure. 2010;19(10): 650-5.

[28]   Bialer M, White HS. Key factors in the discovery and development of new antiepileptic drugs. Nature Rev 2010;9: 68–82.

[29]   Rogawski MA, Loscher W. The neurobiology of antiepileptic drugs. Nat Rev Neurosci 2004;5:553-564.

[30]   Ben-Ari Y, Holmes GL. The multiple facets of gamma-aminobutyric acid dysfunction in epilepsy. Curr Opin Neurol 2005;18:141-145.

[31]   Meldrum BS, Rogawski MA. Molecular targets for antiepileptic drug development. Neurotherapeutics. 2007;4(1): 18-61.

[32]   White S, Smith DM, Wilcox KS. Mechanism of action of antiepileptic drugs. Int Rev Neurobiol 2007;81: 85– 110.

[33]   Lasoñl W, Dudra-Jastrzêbska M, Rejdak K, Czuczwar SJ. Basic mechanisms of antiepileptic drugs and their pharmacokinetic/pharmacodynamic interactions: an update. Pharmacol Reports 2011;63:271-292.

[34]   Brodie MJ, Covanis A, Gil-Nagel A, Lerche H, Perucca E, Sills GJ, White HS. Antiepileptic drug therapy: does mechanism of action matter? Epilepsy Behav. 2011;21(4):331-41.

[35]   Mantegazza M, Rusconi R, Scalmani P, Avanzini G, Franceschetti S. Epileptogenic ion channel mutations: from bedside to bench and, hopefully, back again. Epilepsy Res. 2010;92(1):1-29.

[36] Escayg A, Goldin AL Sodium channel SCN1A and epilepsy: mutations and mechanisms. Epilepsia 2010; 51(9): 1650-8.

[37] Errington AC, Stöhr T, Heers C, Lees G. The investigational anticonvulsant lacosamide selectively enhances slow inactivation of voltage-gated sodium channels". Molecular Pharmacology 2008 73 (1):157–69.

[38] Curia G, Biagini G, Perucca E, Avoli M. Lacosamide: a new approach to target voltage-gated sodium currents in epileptic disorders. CNS Drugs. 2009;23(7):555-68.

[39] Perez-Reyes E. Molecular characterization of T-type calcium channels. Cell Calcium 2006;40 (2):89–96.

[40] Broicher T, Seidenbecher T, Meuth P, Munsch T, Meuth SG, Kanyshkova T, Pape HC, Budde T. T-current related effects of antiepileptic drugs and a Ca^{2+} channel antagonist on thalamic relay and local circuit interneurons in a rat model of absence epilepsy. Neuropharmacology. 2007 Sep;53(3):431-46.

[41] Weiergräber M, Stephani U, Köhling R. Voltage-gated calcium channels in the etiopathogenesis and treatment of absence epilepsy. Brain Res Rev. 2010;62(2):245-71.

[42] Errante LD, Williamson A, Spencer DD, Petroff OA. Gabapentin and vigabatrin increase GABA in the human neocortical slice. Epilepsy Res. 2002;49:203–210.

[43] McClelland D, Evans RM, Barkworth L, Martin DJ, Scott RH. A study comparing the actions of gabapentin and pregabalin on the electrophysiological properties of cultured DRG neurones from neonatal rats. BMC Pharmacol. 2004;4:14.

[44] Fink K, Dooley DJ, Meder WP, Suman-Chauhan N, Duffy S, Clusmann H, Göthert M.Inhibition of neuronal Ca(2+) influx by gabapentin and pregabalin in the human neocortex. Neuropharmacology. 2002;42:229–236.

[45] Gunthorpe MJ, Large CH, Sankar R. he mechanism of action of retigabine (ezogabine), a first-in-class K+ channel opener for the treatment of epilepsy. Epilepsia. 2012;53(3): 412-24.

[46] Lynch BA, Lambeng N, Nocka K, Kensel-Hammes P, Bajjalieh SM, Matagne A, Fuks B. The synaptic vesicle protein SV2A is the binding site for the antiepileptic drug levetiracetam. Proc Natl Acad Sci 2004;101:9861–9866.

[47] Hanaya R, Hosoyama H, Sugata S, Tokudome M, Hirano H, Tokimura H, Kurisu K, Serikawa T, Sasa M, Arita K. Low distribution of synaptic vesicle protein 2A and synaptotagimin-1 in the cerebral cortex and hippocampus of spontaneously epileptic rats exhibiting both tonic convulsion and absence seizure. Neuroscience 2012.

[48] Poolos NP, Migliore M, Johnston D. Pharmacological upregulation of h-channels reduces the excitability of pyramidal neuron dendrites. Nat Neurosci 2002;5:767–774.

[49] Surges R, Freiman TM, Feuerstein TJ. Gabapentin increases the hyperpolarization-activated cation current Ih in rat CA1 pyramidal cells. Epilepsia 2003;44:150–156.

[50] White HS. Clinical significance of animal seizure models and mechanism of action studies of potential antiepileptic drugs. Epilepsia. 1997;38 Suppl 1:S9-17.

[51] Pitkanen A, Schwartzkroin PA, Moshe' SL. Models of seizures and epilepsy. Amsterdam: Elsevier; 2006.

[52] Smyth MD, Barbaro NM, Baraban SC. Effects of antiepileptic drugs on induced epileptiform activity in a rat model of dysplasia. Epilepsy Res 2002;50:251–64.

[53] Löscher W. Critical review of current animal models of seizures and epilepsy used in the discovery and development of new antiepileptic drugs. Seizure. 2011;20(5):359-68.

[54] Galanopoulou AS, Buckmaster PS, Staley KJ, Moshé SL, Perucca E, Engel J Jr, Löscher W, Noebels JL, Pitkänen A, Stables J, White HS, O'Brien TJ, Simonato M; American Epilepsy Society Basic Science Committee and the International League Against Epilepsy

Working Group on Recommendations For Preclinical Epilepsy Drug Discovery. Identification of new epilepsy treatments: issues in preclinical methodology. Epilepsia. 2012;53(3):571-82.

[55]   Perucca E. An introduction to antiepileptic drugs. Epilepsia 2005;46(suppl.4):31-37.

[56]   Bialer M, Johannessen SI, Levy RH, Perucca E, Tomson T, White HS. Progress report on new antiepileptic drugs: a summary of the Tenth Eilat Conference (EILAT X). Epilepsy Res. 2010;92(2-3):89-124.

[57]   Patsalos PN, Berry DJ. Pharmacotherapy of the third-generation AEDs: lacosamide, retigabine and eslicarbazepine acetate. Expert Opin Pharmacother. 2012;13(5):699-715.

[58]   Marino SE, Birnbaum AK, Leppik IE, Conway JM, Musib LC, Brundage RC, Ramsay RE, Pennell PB, White JR, Gross CR, Rarick JO, Mishra U, Cloyd JC. Steady-state carbamazepine pharmacokinetics following oral and stable-labeled intravenous administration in epilepsy patients: effects of race and sex. Clin Pharmacol Ther 2012;91(3):483-8.

[59]   Gidal BE, DeCerce J, Bockbrader HN, Gonzalez J, Kruger S, Pitterle ME, Rutecki P, Ramsay RE. Gabapentin bioavailability: effect of dose and frequency of administration in adult patients with epilepsy.Epilepsy Res 1998; 31(2):91-9.

[60]   Levy RH, Loiseau P, Guyot M, Blehaut HM, Tor J, Moreland TA: Stiripentol kinetics in epilepsy: nonlinearity and interactions. Clin Pharmacol Ther, 1984;36:661–669.

[61]   Levy RH. Cytochrome P450 isozymes and antiepileptic drug interactions. Epilepsia. 1995;36 Suppl 5:S8-13.

[62]   Mann MW, Pons G. Various pharmacogenetic aspects of antiepileptic drug therapy: a review. CNS Drugs. 2007;21(2):143-64.

[63]   Löscher W, Klotz U, Zimprich F, Schmidt D. The clinical impact of pharmacogenetics on the treatment of epilepsy. Epilepsia. 2009;50(1):1-23.

[64]   Perucca E. Age-related changes in pharmacokinetics: predictability and assessment methods. Int Rev Neurobiol. 2007;81:183-199.

[65]   Perucca E. Clinical pharmacokinetics of new-generation antiepileptic drugs at the extremes of age. Clin Pharmacokinet. 2006;45(4):351-63.

[66]   Bartelink IH, Rademaker CM, Schobben AF, van den Anker JN. Guidelines on paediatric dosing on the basis of developmental physiology and pharmacokinetic considerations. Clin Pharmacokinet. 2006;45(11):1077-97.

[67]   Pennell PB. Antiepileptic drug pharmacokinetics during pregnancy and lactation. Neurology. 2003;61(S2):35-42.

[68]   Tomson T, Battino D. Pharmacokinetics and therapeutic drug monitoring of newer antiepileptic drugs during pregnancy and the puerperium. Clin Pharmacokinet. 2007;46(3):209-219.

[69]   Lacerda G, Krummel T, Sabourdy C, Ryvlin P, Hirsch E. Optimizing therapy of seizures in patients with renal or hepatic dysfunction. Neurology. 2006;67(12 Suppl 4):S28-33.

[70]   Diaz A, Deliz B, Benbadis SR. The use of newer antiepileptic drugs in patients with renal failure. Expert Rev Neurother. 2012;12(1):99-105.

[71]   Pisani F. Pharmacokinetics and therapeutic plasma monitoring of antiepileptic drugs". In: Handbook of Clinical Neurology, The Epilepsies, P.J. Vinken and G.W. Bruyn (Eds.), vol 73, cap 27, pp 357- 368;Elseviers Publishers, 2000.

[72]   Johannessen SI, Landmark CJ. Value of therapeutic drug monitoring in epilepsy. Expert Rev Neurother. 2008; 8(6):929-39.

[73]   Striano S, Striano P, Capone D, Pisani F.Limited place for plasma monitoring of new antiepileptic drugs in clinical practice. Med Sci Monit. 2008;14(10):RA173-8.

[74]   Patsalos PN, Berry DJ, Bourgeois BF, Cloyd JC, Glauser TA, Johannessen SI, Leppik IE, Tomson T, Perucca E. Antiepileptic drugs--best practice guidelines for therapeutic drug monitoring: a position paper by the subcommission on therapeutic drug monitoring, ILAE Commission on Therapeutic Strategies. Epilepsia. 2008; 49(7):1239-76.

[75]   Patsalos P.N., Froscher W., Pisani F., and van Rijn C.M. The importance of drug interactions in epilepsy therapy. Epilepsia 43:365-385, 2002.

[76]   Perucca E. Clinically relevant drug interactions with antiepileptic drugs. Br J Clin Pharmacol. 2006;61(3):246- 255.

[77]   Johannessen SI, Landmark CJ. Antiepileptic drug interactions - principles and clinical implications. Curr Neuropharmacol. 2010;8(3):254-267.

[78]   Anderson GD. Pharmacogenetics and enzyme induction/inhibition properties of antiepileptic drugs. Neurology. 2004;63(10 Suppl 4):S3-8.

[79]   Pisani F, Caputo M, Fazio A, Oteri G, Russo M, Spina E, Perucca E, Bertilsson L. Interaction of carbamazepine-10,11-epoxide, an active metabolite of carbamazepine, with valproate: a pharmacokinetic study. Epilepsia. 1990;31(3):339-42.

[80]   Johannessen Landmark C, Patsalos PN. Drug interactions involving the new second- and third-generation antiepileptic drugs. Expert Rev Neurother. 2010;10(1):119-140.

[81]   Panegyres PK, Rischbieth RH. Fatal phenytoin warfarin interaction. Postgrad Med J. 1991;67(783):98.

[82]   Sabers, A.; Öhman, I.; Christensen, J.; Tomson, T. Oral contraceptives reduce lamotrigine plasma levels. Neurology 2003, 61, 570- 571.

[83]   Reimers, A.; Helde, G.; Brodtkorb, E. Ethinyl estradiol, not progestogens, reduces lamotrigine serum concentrations. Epilepsia 2005,46, 1414-1417.

[84]   Wegner I, Edelbroek PM, Bulk S, Lindhout D. Lamotrigine kinetics within the menstrual cycle, after menopause, and with oral contraceptives. Neurology. 2009 Oct 27;73(17):1388-93.

[85]   Cornaggia CM, Beghi M, Beghi E for the REST-1 Group. Psychiatric events in epilepsy. Seizure 2007;16:586- 592.

[86]   Mula M, Jauch R, Cavanna A, Gaus V, Kretz R, Collimedaglia L, Barbagli D, Cantello R, Monaco F, Schmitz B. Interictal dysphoric disorder and periictal dysphoric symptoms in patients with epilepsy. Epilepsia. 2010; 51(7):1139-1145.

[87]   Mula M, Monaco F, Trimble MR. Use of psychotropic drugs in patients with epilepsy: interactions and seizure risk. Expert Rev Neurother 2004;4(6):953-964.

[88]   Spina E, Perucca E. Clinical significance of pharmacokinetic interactions between antiepileptic and psychotropic drugs. Epilepsia. 2002;43 Suppl 2:37-44.

[89]   Spina E, Trifirò G, Caraci F. Clinically significant drug interactions with newer antidepressants. CNS Drugs. 2012 1;26(1):39-67.

[90]   de Leon J, Santoro V, D'Arrigo C, Spina E. Interactions between antiepileptics and second-generation antipsychotics. Expert Opin Drug Metab Toxicol. 2012;8(3):311-334.

[91]   Haddad PM, Dursun SM. Neurological complications of psychiatric drugs: clinical features and management. Hum Psychopharmacol. 2008;23(S1):15-26.

[92]   French JA, Faught E. Rational polytherapy. Epilepsia. 2009;50(S8):63-68.

[93]   Rowan AJ, Meijer JW, de Beer-Pawlikowski N, van der Geest P, Meinardi H. Valproate-ethosuximide combination therapy for refractory absence seizures. Arch Neurol 1983;40(13):797-802.

[94] Pisani F., Russo M., Trio R., Artesi C., Fazio A., Oteri G., Di Perri R. Efficacia dell'associazione lamotrigina- valproato di sodio nelle crisi epilettiche parziali resistenti. Quaderni di Acta Neurologica 1992; 58:71-73.

[95] Brodie MJ, Yuen AW. Lamotrigine substitution study: evidence for synergism with sodium valproate? 105 Study Group. Epilepsy Res. 1997;26(3):423-432.

[96] Pisani F., Oteri G., Russo MF, Di Perri R, Perucca E, Richens A. The efficacy of valproate-lamotrigine comedication in refractory complex partial seizures: evidence for a pharmacodynamic interaction. Epilepsia 1999;40:1141-1146.

[97] Yuen AW, Land G, Weatherley BC, Peck AW. Sodium valproate acutely inhibits lamotrigine metabolism. Br J Clin Pharmacol. 1992;33(5):511-3.

[98] Reutens DC, Duncan JS, Patsalos PN. Disabling tremor after lamotrigine with sodium valproate. Lancet. 1993;17;342(8864):185-6.

[99] Li LM, Russo M, O'Donoghue MF, Duncan JS, Sander JW. Allergic skin rash with lamotrigine and concomitant valproate therapy: evidence for an increased risk. Arq Neuropsiquiatr. 1996;54(1):47-49.

[100] Wong IC, Mawer GE, Sander JW. Factors influencing the incidence of lamotrigine-related skin rash. Ann Pharmacother. 1999;33(10):1037-1042.

[101] Kennedy GM, Lhatoo SD. CNS adverse events associated with antiepileptic drugs. CNS Drugs. 2008;22(9):739-60.

[102] Cramer JA, Mintzer S, Wheless J, Mattson RH. Adverse effects of antiepileptic drugs: a brief overview of important issues. Expert Rev Neurother 2010;10(6):885-91.

[103] Sarco DP, Bourgeois BF. The safety and tolerability of newer antiepileptic drugs in children and adolescents. CNS Drugs. 2010;24(5):399-430.

[104] Anderson M, Choonara I. A systematic review of safety monitoring and drug toxicity in published randomised controlled trials of antiepileptic drugs in children over a 10-year period. Arch Dis Child. 2010;95(9):731-738.

[105] Canevini MP, De Sarro G, Galimberti CA, Gatti G, Licchetta L, Malerba A, Muscas G, La Neve A, Striano P, Perucca E; SOPHIE Study Group Relationship between adverse effects of antiepileptic drugs, number of co-prescribed drugs, and drug load in a large cohort of consecutive patients with drug-refractory epilepsy. Epilepsia. 2010;51(5):797-804.

[106] Landmark CJ, Johannessen SI. Safety aspects of antiepileptic drugs--focus on pharmacovigilance. Pharmacoepidemiol Drug Saf. 2012;21(1):11-20.

[107] Perucca P., Gilliam FG. Adverse effects of antiepileptic drugs. Lancet Neurol 2012 (in press).

[108] Perucca E, Kwan P. Overtreatment in epilepsy: how it occurs and how it can be avoided. CNS Drugs. 2005;19(11):897-908.

[109] Zaccara G, Gangemi PF and Cincotta M. Central nervous system adverse effects of new anti-epileptic drugs. A meta-analysis of placebo- controlled studies. Seizure 2008;17:405-21.

[110] Zaccara G, Franciotta D, Perucca E. Idiosyncratic adverse reactions to antiepileptic drugs. Epilepsia. 2007;48(7):1223-1244.

[111] Beghi E, Shorvon S. Antiepileptic drugs and the immune system. Epilepsia. 2011;52(S3):40-44.

[112] Rzany B, Correia O, Kelly JP, Naldi L, Auquier A, Stern R. Risk of Stevens-Johnson syndrome and toxic epidermal necrolysis during first weeks of antiepileptic therapy: a case-

control study. Study Group of the International Case Control Study on Severe Cutaneous Adverse Reactions. Lancet. 1999;353(9171):2190- 2194.

[113]   Arif H, Buchsbaum R, Weintraub D, Koyfman S, Salas-Humara C, Bazil CW, Resor SR Jr, Hirsch LJ. Comparison and predictors of rash associated with 15 antiepileptic drugs. Neurology. 2007;68(20):1701-1709.

[114]   Hirsch LJ, Arif H, Nahm EA, *et al.* Cross-sensitivity of skin rashes with antiepileptic drug use. Neurology 2008;71:1527.

[115]   Yang CY, Dao RL, Lee TJ, Lu CW, Yang CH, Hung SI, Chung WH. Severe cutaneous adverse reactions to antiepileptic drugs in Asians. Neurology. 2011;77(23):2025-2033.

[116]   Chen P, Lin JJ, Lu CS, *et al.* Carbamazepine-induced toxic effects and HLA-B*1502 screening in Taiwan. N Engl J Med 2011; 364:1126.

[117]   McCormack M, Alfirevic A, Bourgeois S, *et al.* HLA-A*3101 and carbamazepine-induced hypersensitivity reactions in Europeans. N Engl J Med 2011; 364:1134.

[118]   Mula M, Sander JW. Negative effects of antiepileptic drugs on mood in patients with epilepsy. Drug Saf. 2007;30(7):555-567.

[119]   Mula M, Trimble MR. Antiepileptic drug-induced cognitive adverse effects: potential mechanisms and contributing factors. CNS Drugs. 2009;23(2):121-137.

[120]   Eddy CM, Rickards HE, Cavanna AE. Behavioral adverse effects of antiepileptic drugs in epilepsy. J Clin Psychopharmacol. 2012;32(3):362-375.

[121]   Piedad J, Rickards H, Besag FM, Cavanna AE. Beneficial and adverse psychotropic effects of antiepileptic drugs in patients with epilepsy: a summary of prevalence, underlying mechanisms and data limitations. CNS Drugs. 2012 Apr 1;26(4):319-35.

[122]   Kanner AM. Epilepsy, suicidal behaviour, and depression: do they share common pathogenic mechanisms? Lancet Neurol. 2006;5:107–108.

[123]   Reynolds EH. Mental effect of antiepileptic medication: a review. Epilepsia 1983;24(S2):85-95.

[124]   Hesdorffer DC, Kanner AM. The FDA alert on suicidality and antiepileptic drugs: Fire or false alarm? Epilepsia. 2009;50(5):978-86.

[125]   Andersohn F, Schade R, Willich SN, Garbe E. Use of antiepileptic drugs in epilepsy and the risk of self-harm or suicidal behaviour. Neurology. 2010;75:335–340.

[126]   Bell GS, Mula M, Sander JW Suicidality in people taking antiepileptic drugs: What is the evidence? CNS Drugs. 2009;23(4):281-92.

[127]   Arana A, Wentworth CE, Ayuso-Mateos JL, Arellano FM. Suicide-related events in patients treated with antiepileptic drugs. N Engl J Med. 2010;363(6):542-51.

[128]   Harden CL, Meador KJ, Pennell PB, Hauser WA, Gronseth GS, French JA, Wiebe S, Thurman D, Koppel BS, Kaplan PW, Robinson JN, Hopp J, Ting TY, Gidal B, Hovinga CA, Wilner AN, Vazquez B, Holmes L, Krumholz A, Finnell R, Hirtz D, Le Guen C; American Academy of Neurology; American Epilepsy Society. Management issues for women with epilepsy-Focus on pregnancy (an evidence-based review): II. Teratogenesis and perinatal outcomes: Report of the Quality Standards Subcommittee and Therapeutics and Technology Subcommittee of the American Academy of Neurology and the American Epilepsy Society. Epilepsia. 2009;50(5):1237-1246.

[129]   Jentink J, Loane MA, Dolk H, Barisic I, Garne E, Morris JK, de Jong-van den Berg LT; EUROCAT Antiepileptic Study Working Group. Valproic acid monotherapy in pregnancy and major congenital malformations. N Engl J Med 2010; 362: 2185–93.

[130] Tomson T, Battino D. Teratogenic effects of antiepileptic drugs. Lancet Neurol. 2012 (in press).

[131] Harden CL, Pennell PB, Koppel BS, Hovinga CA, Gidal B, Meador KJ, Hopp J, Ting TY, Hauser WA, Thurman D, Kaplan PW, Robinson JN, French JA, Wiebe S, Wilner AN, Vazquez B, Holmes L, Krumholz A, Finnell R, Shafer PO, Le Guen CL; American Academy of Neurology; American Epilepsy Society. Management issues for women with epilepsy--focus on pregnancy (an evidence-based review): III. Vitamin K, folic acid, blood levels, and breast-feeding: Report of the Quality Standards Subcommittee and Therapeutics and Technology Assessment Subcommittee of the American Academy of Neurology and the American Epilepsy Society. Epilepsia. 2009;50(5):1247-1255.

[132] Tomson T, Battino D, Bonizzoni E, Craig J, Lindhout D, Sabers A, Perucca E, Vajda F; EURAP study group. Dose-dependent risk of malformations with antiepileptic drugs: an analysis of data from the EURAP epilepsy and pregnancy registry. Lancet Neurol. 2011;10(7):609-17.

[133] Mølgaard-Nielsen D, Hviid A Newer-generation antiepileptic drugs and the risk of major birth defects. JAMA. 2011;305(19):1996-2002.

[134] Hernández-Díaz S, Smith CR, Shen A, Mittendorf R, Hauser WA, Yerby M, Holmes LB; North American AED Pregnancy Registry. Comparative safety of antiepileptic drugs during pregnancy. Neurology. 2012;78(21):1692- 1699.

[135] Vajda FJ, Graham J, Roten A, Lander CM, O'Brien TJ, Eadie M. Teratogenicity of the newer antiepileptic drugs--the Australian experience. J Clin Neurosci. 2012;19(1):57-59.

[136] Banach R, Boskovic R, Einarson T, Koren G. Long-term developmental outcome of children of women with epilepsy, unexposed or exposed prenatally to antiepileptic drugs: a meta-analysis of cohort studies. Drug Saf. 2010;33(1):73-79.

[137] Forsberg L, Wide K, Källén B. School performance at age 16 in children exposed to antiepileptic drugs in utero- -a population-based study. Epilepsia. 2011;52(2):364-369.

[138] Meador KJ, Baker GA, Browning N, Cohen MJ, Bromley RL, Clayton-Smith J, Kalayjian LA, Kanner A, Liporace JD, Pennell PB, Privitera M, Loring DW; NEAD Study Group. Effects of fetal antiepileptic drug exposure: outcomes at age 4.5 years. Neurology 2012;78(16):1207-1214.

[139] Sukumaran SC, Sarma PS, Thomas SV. Polytherapy increases the risk of infertility in women with epilepsy. Neurology 2010;75(15):1351-1355.

[140] Verrotti A, D'Egidio C, Mohn A, Coppola G, Parisi P, Chiarelli F. Antiepileptic drugs, sex hormones, and PCOS. Epilepsia. 2011;52(2):199-211.

[141] Hu X, Wang J, Dong W, Fang Q, Hu L, Liu C. A meta-analysis of polycystic ovary syndrome in women taking valproate for epilepsy. Epilepsy Res. 2011;97(1-2):73-82.

[142] Belcastro V., Gorgone G., Italiano D., Oteri G., Caccamo D., Pisani L.R., Striano P., Striano S., Ientile R., and Pisani F. Antiepileptic drugs and MTHFR polymorphisms influence hyperhomocysteinemia recurrence in epileptic patients. Epilepsia 48(10):1990-1994, 2007.

[143] Belcastro V, Striano P, Gorgone G, Costa C, Campa C, Caccamo D, Pisani LR, Oteri G, Marciani MG, Aguglia U, Striano S, Ientile R, Calabresi P, Pisani F. Hyperhomocysteinemia in epileptic patients on new antiepileptic drugs. Epilepsia 51: 274-279, 2010.

[144] Gerstner T, Teich M, Bell N, Longin E, Dempfle CE, Brand J, König S. Valproate-associated coagulopathies are frequent and variable in children. Epilepsia. 2006;47(7):1136-1143.

[145]   Acharya S., Bussel J. Hematologic Toxicity of Sodium Valproate. J Ped Hemat Onc 2000;22(1):62-65.

[146]   Dreifuss FE, Langer DH. Hepatic Considerations in the Use of Antiepileptic Drugs. Epilepsia.

[147]   Sonmez FM, Demir E, Orem A, Yildirmis S, Orhan F, Aslan A, Topbas M. Effect of antiepileptic drugs on plasma lipids, lipoprotein (a), and liver enzymes. J Child Neurol. 2006 Jan;21(1):70-74.

[148]   Chuang YC, Chuang HY, Lin TK, Chang CC, Lu CH, Chang WN, Chen SD, Tan TY, Huang CR, Chan SH. Effects of long-term antiepileptic drug monotherapy on vascular risk factors and atherosclerosis. Epilepsia. 2012;53(1):120-128.

[149]   Hamed SA, Hamed EA, Hamdy R, Nabeshima T. Vascular risk factors and oxidative stress as independent predictors of asymptomatic atherosclerosis in adult patients with epilepsy. Epilepsy Res 2007;74:183–192.

[150]   Gorgone G, Ursini F, Altamura C, Bressi F, Tombini Ma, Curcio G, Chiovenda P, Squitti R, Silvestrini M, Ientile R, Pisani F, Rossini PM, Vernieri F. Hyperhomocysteinemia, intima-media thickness and C677T MTHFR gene polymorphism: a correlation study in patients with cognitive impairment. Atherosclerosis 209:309-313, 2009.

[151]   Gorgone G, Caccamo D, Pisani LR, Curro` M, Parisi G, Oteri G, Ientile R, Rossini PM, and Pisani F. Hyperhomocysteinemia in Patients with epilepsy: Does it play a role in the pathogenesis of brain Atrophy ? A preliminary report. Epilepsia 50:33-36, 2009.

[152]   Ben-Menachem E. Weight issues for people with epilepsy – a review. Epilepsia 2007;48: 42–45.

[153]   Cansu A. Antiepileptic drugs and hormones in children. Epilepsy Res. 2010;89(1):89-95.

[154]   Lossius MI, Taubøll E, Mowinckel P, Gjerstad L. Reversible effects of antiepileptic drugs on thyroid hormones in men and women with epilepsy: a prospective randomized double-blind withdrawal study. Epilepsy Behav. 2009;16(1):64-68.

[155]   Arzu Babayigit, Eray Dirik, Ece Bober, Handan Cakmakcı. Adverse Effects of Antiepileptic Drugs on Bone Mineral Density. Pediatric Neurol 2006;35:177-181.

[156]   Beniczky SA, Viken J, Jensen LT, Andersen NB.Bone mineral density in adult patients treated with various antiepileptic drugs. Seizure. 2012;21(6):471-2.

[157]   Shiek Ahmad B, Hill KD, O'Brien TJ, Gorelik A, Habib N, Wark JD. Falls and fractures in patients chronically treated with antiepileptic drugs. Neurology. 2012 Jul 10;79(2):145-151.

[158]   Wang Z, Lin YS, Zheng XE, Senn T, Hashizume T, Scian M, Dickmann LJ, Nelson SD, Baillie TA, Hebert MF, Blough D, Davis CL, Thummel KE. An inducible cytochrome P450 3A4-dependent vitamin D catabolic pathway. Mol Pharmacol. 2012;81(4):498-509.

[159]   Schachter SC. Antiepileptic Drug Therapy: General Treatment Principles and Application for Special Patient Populations. Epilepsia 1999;4O(S9):20-25.

[160]   Sander JW. The use of antiepileptic drugs--principles and practice. Epilepsia. 2004;45(S6):28-34.

[161]   Perucca E, Tomson T. The pharmacological treatment of epilepsy in adults. Lancet Neurol. 2011;10(5):446-56.

[162]   Faught E. Monotherapy in adults and elderly persons. Neurology 2007 11;69(S3):3-9.

[163]   Wilfong AA. Monotherapy in children and infants. Neurology. 2007;69(S3):17-22.

[164]   Perucca P, Jacoby A, Marson AG, Baker GA, Lane S, Benn EK, Thurman DJ, Hauser WA, Gilliam FG, Hesdorffer DC. Adverse antiepileptic drug effects in new-onset seizures: a case-control study. Neurology. 2011;76(3):273-9.

[165] Berg AT, Shinnar S. The risk of seizure recurrence following a first unprovoked seizure: a quantitative review. Neurology 1991;41: 965–972.

[166] Hauser WA, Rich SS, Lee JR, Annegers JF, Anderson VE. Risk of recurrent seizures after two unprovoked seizures. N Engl J Med 1998; 338: 429–34.

[167] Schmidt D, Sillanpää M. Evidence-based review on the natural history of the epilepsies. Curr Opin Neurol. 2012;25(2):159-163.

[168] Berg AT, Shinnar S. The risk of seizure recurrence following a first unprovoked seizure: a quantitative review. Neurology 1991;41: 965–72.

[169] Fisher RS, van Emde Boas W, Blume W, Elger C, Genton P, Lee P, Engel J Jr. Epileptic seizure and epilepsy: definitions proposed by the International League against Epilepsy (ILAE) and the International Bureau for Epilepsy (IBE). Epilepsia 2005; 46: 470–472.

[170] Marson A, Jacoby A, Johnson A, Kim L, Gamble C, Chadwick D. Medical Research Council MESS Study Group. Immediate *vs.* deferred antiepileptic drug treatment for early epilepsy and single seizures: a randomised controlled trial. Lancet 2005;365: 2007–13.

[171] Leone MA, Solari A, Beghi E, FIRST Group. Treatment of the first tonic-clonic seizure does not affect long-term remission of epilepsy. Neurology 2006; 67: 2227–29.

[172] Marson AG, Al-Kharusi AM, Alwaidh M, Appleton R, Baker GA, Chadwick DW, Cramp C, Cockerell OC, Cooper PN, Doughty J, Eaton B, Gamble C, Goulding PJ, Howell SJ, Hughes A, Jackson M, Jacoby A, Kellett M, Lawson GR, Leach JP, Nicolaides P, Roberts R, Shackley P, Shen J, Smith DF, Smith PE, Smith CT, Vanoli A, Williamson PR; SANAD Study group. The SANAD study of effectiveness of carbamazepine, gabapentin, lamotrigine, oxcarbazepine, or topiramate for treatment of partial epilepsy: an unblinded randomised controlled trial. Lancet 2007;369(9566):1000-15.

[173] Saetre E, Perucca E, Isojärvi J, Gjerstad L; LAM 40089 Study Group. An international multicenter randomized double-blind controlled trial of lamotrigine and sustained-release carbamazepine in the treatment of newly diagnosed epilepsy in the elderly. Epilepsia. 2007;48(7):1292-302.

[174] Brodie MJ, Perucca E, Ryvlin P, Ben-Menachem E, Meencke HJ. Levetiracetam Monotherapy Study Group. Comparison of levetiracetam and controlled-release carbamazepine in newly diagnosed epilepsy. Neurology 2007; 68: 402–08.

[175] Baulac M, Brodie MJ, Patten A, Segieth J, Giorgi L. Efficacy and tolerability of zonisamide *vs.* controlled- release carbamazepine for newly diagnosed partial epilepsy: a phase 3, randomised, double-blind, non-inferiority trial. Lancet Neurol. 2012;11(7):579-88.

[176] Kwan P, Brodie MJ, Kälviäinen R, Yurkewicz L, Weaver J, Knapp LE. Efficacy and safety of pregabalin *vs.* lamotrigine in patients with newly diagnosed partial seizures: a phase 3, double-blind, randomised, parallel-group trial. Lancet Neurol. 2011;10(10):881-90. Epub 2011 Aug 31.

[177] Rowan AJ, Ramsay RE, Collins JF, Pryor F, Boardman KD, Uthman BM, Spitz M, Frederick T, Towne A, Carter GS, Marks W, Felicetta J, Tomyanovich ML; VA Cooperative Study 428 Group. New onset geriatric epilepsy: a randomized study of gabapentin, lamotrigine, and carbamazepine. Neurology 2005;64(11):1868-73.

[178] Sankar R. Initial treatment of epilepsy with antiepileptic drugs: pediatric issues. Neurology 2004;63(S4):30-9.

[179] Weijenberg A, Offringa M, Brouwer OF, Callenbach PM. RCTs with new antiepileptic drugs in children: a systematic review of monotherapy studies and their methodology. Epilepsy Res. 2010;91(1):1-9.

[180] Asconapé JJ. The selection of antiepileptic drugs for the treatment of epilepsy in children and adults. Neurol Clin. 2010;28(4):843-52.

[181] Guerrini R. Valproate as a mainstay of therapy for pediatric epilepsy. Paediatr Drugs. 2006;8(2):113-29.

[182] Marson AG, Al-Kharusi AM, Alwaidh M, Appleton R, Baker GA, Chadwick DW, Cramp C, Cockerell OC, Cooper PN, Doughty J, Eaton B, Gamble C, Goulding PJ, Howell SJ, Hughes A, Jackson M, Jacoby A, Kellett M, Lawson GR, Leach JP, Nicolaides P, Roberts R, Shackley P, Shen J, Smith DF, Smith PE, Smith CT, Vanoli A, Williamson PR; SANAD Study group. The SANAD study of effectiveness of valproate, lamotrigine, or topiramate for generalised and unclassifiable epilepsy: an unblinded randomised controlled trial. Lancet. 200;369(9566):1016-26.

[183] Beydoun A, D'Souza J Treatment of idiopathic generalized epilepsy - a review of the evidence. Expert Opin Pharmacother. 2012;13(9):1283-98.

[184] Patsalos PN. Properties of antiepileptic drugs in the treatment of idiopathic generalized epilepsies. Epilepsia. 2005;46(S9):140-148.

[185] Glauser T, Ben-Menachem E, Bourgeois B, Cnaan A, Chadwick D, Guerreiro C, Kalviainen R, Mattson R, Perucca E, Tomson T. ILAE treatment guidelines: evidence-based analysis of antiepileptic drug efficacy and effectiveness as initial monotherapy for epileptic seizures and syndromes. Epilepsia. 2006;47(7):1094-120.

[186] French JA. First-choice drug for newly diagnosed epilepsy. Lancet. 2007;369(9566):970-9711.

[187] Rheims S, Perucca E, Cucherat M, Ryvlin P. Factors determining response to antiepileptic drugs in randomized controlled trials. A systematic review and meta-analysis. Epilepsia. 2011;52(2):219-33.

[188] Costa J, Fareleira F, Ascenção R, Borges M, Sampaio C, Vaz-Carneiro A. Clinical comparability of the new antiepileptic drugs in refractory partial epilepsy: a systematic review and meta-analysis. Epilepsia 2011;52(7):1280-91.

[189] Beyenburg S, Stavem K, Schmidt D. Placebo-corrected efficacy of modern nonenzyme-inducing AEDs for refractory focal epilepsy: systematic review and meta-analysis. Epilepsia. 2012;53(3):512-20.

[190] Guerrini R, Belmonte A, Genton P. Antiepileptic drug-induced worsening of seizures in children. Epilepsia. 1998;39 (S 3):2-10.

[191] Chaves J, Sander JW. Seizure aggravation in idiopathic generalized epilepsies. Epilepsia. 2005;46(S9):133-9.

[192] Gayatri NA, Livingston JH. Aggravation of epilepsy by anti-epileptic drugs. Dev Med Child Neurol. 2006;48(5):394-8.

[193] Pisani F., Spina E., Oteri G. Antidepressant drugs and seizure susceptibility. From *in vitro* data to clinical practice. Epilepsia 1999; 40:48-56.

[194] Pisani F., Oteri G., Costa C., Di Raimondo G. , Di Perri R. Effect of psychotropic drugs on seizure threshold. Drug Safety 2002;25: 91-110.

[195] Guerrini R, Dravet C, Genton P, Belmonte A, Kaminska A, Dulac O. Lamotrigine and seizure aggravation in severe myoclonic epilepsy. Epilepsia. 1998;39(5):508-12.

[196] Carrazana EJ, Wheeler SD. Exacerbation of juvenile myoclonic epilepsy with lamotrigine. Neurology 2001;56(10):1424-5.

[197] Crespel A, Genton P, Berramdane M, Coubes P, Monicard C, Baldy-Moulinier M, Gelisse P. Lamotrigine associated with exacerbation or *de novo* myoclonus in idiopathic generalized epilepsies. Neurology 2005 13;65(5):762-4.

[198] Glauser TA, Cnaan A, Shinnar S, Hirtz DG, Dlugos D, Masur D, Clark PO, Capparelli EV, Adamson PC; Childhood Absence Epilepsy Study Group. Ethosuximide, valproic acid, and lamotrigine in childhood absence epilepsy. N Engl J Med. 2010;362(9):790-9.

[199] Tassinari CA, Dravet C, Roger J, Cano JP, Gastaut H. Tonic status epilepticus precipitated by intravenous benzodiazepine in five patients with Lennox-Gastaut syndrome. Epilepsia. 1972;13(3):421-35.

[200] Bittencourt PR, Richens A. Anticonvulsant-induced status epilepticus in Lennox-Gastaut syndrome. Epilepsia. 1981;22(1):129-34.

[201] Hirsch E, Genton P. Antiepileptic drug-induced pharmacodynamic aggravation of seizures: does valproate have a lower potential? CNS Drugs. 2003;17(9):633-40.

[202] Tsuyusaki Y, Shimbo H, Wada T, Iai M, Tsuji M, Yamashita S, Aida N, Kure S, Osaka H. Paradoxical increase in seizure frequency with valproate in nonketotic hyperglycinemia. Brain Dev. 2012;34(1):72-5.

[203] Kwan P, Schachter SC, Brodie MJ. Drug-resistant epilepsy. N Engl J Med. 2011;365(10):919-26.

[204] Berg AT, Testa FM, Levy SR. Complete remission in nonsyndromic childhood-onset epilepsy. Ann Neurol. 2011;70(4):566- 73.

[205] Brodie MJ, Barry SJ, Bamagous GA, Norrie JD, Kwan P. Patterns of treatment response in newly diagnosed epilepsy Neurology. 2012;78(20):1548-54.

[206] Bonnett L, Smith CT, Smith D, Williamson P, Chadwick D, Marson AG. Prognostic factors for time to treatment failure and time to 12 months of remission for patients with focal epilepsy: post-hoc, subgroup analyses of data from the SANAD trial. Lancet Neurol. 2012;11(4):331-40.

[207] Gilioli I, Vignoli A, Visani E, Casazza M, Canafoglia L, Chiesa V, Gardella E, La Briola F, Panzica F, Avanzini G, Canevini MP, Franceschetti S, Binelli S. Focal epilepsies in adult patients attending two epilepsy centers: classification of drug- resistance, assessment of risk factors, and usefulness of "new" antiepileptic drugs. Epilepsia. 2012;53(4):733-40.

[208] Siddiqui A., Kerb R., Weale M.E., Brinkmann U., Smith A., Goldstein D.B., Wood N.W., Sisodiya S.M. Association of Multidrug Resistance in Epilepsy with a Polymorphism in the Drug-Transporter Gene ABCB1N Engl J Med 2003;348:1442-8.

[209] Qu J, Zhou BT, Yin JY, Xu XJ, Zhao YC, Lei GH, Tang Q, Zhou HH, Liu ZQ. ABCC2 Polymorphisms and Haplotype are Associated with Drug Resistance in Chinese Epileptic Patients. CNS Neurosci Ther. 2012 Aug;18(8):647-651.

[210] Luciano AL, Shorvon SD. Results of treatment changes in patients with apparently drug-resistant chronic epilepsy. Ann Neurol. 2007 ;62(4):375-81.

[211] Schiller Y, Najjar Y. Quantifying the response to antiepileptic drugs: effect of past treatment history. Neurology 2008;70(1):54-65.

[212] Berg AT, Levy SR, Testa FM, D'Souza R. Remission of epilepsy after two drug failures in children: a prospective study. Ann Neurology 2009;65(5):510-9.

[213] Neligan A, Bell GS, Elsayed M, Sander JW, Shorvon SD. Treatment changes in a cohort of people with apparently drug-resistant epilepsy: an extended follow-up. J Neurol Neurosurg Psychiatry. 2012;83(8):810-813.

[214] Sillanpää M, Schmidt D. Is incident drug-resistance of childhood-onset epilepsy reversible? A long-term follow-up study. Brain. 2012;135:2256-62.

[215] Kwan P, Arzimanoglou A, Berg AT, Brodie MJ, Allen Hauser W, Mathern G, Moshé SL, Perucca E, Wiebe S, French J. Definition of drug resistant epilepsy: consensus proposal by the ad hoc Task Force of the ILAE Commission on Therapeutic Strategies. Epilepsia 2010;51(6):1069-77.

[216] Friedman D, French JA. Clinical trials for therapeutic assessment of antiepileptic drugs in the 21st century: obstacles and solutions. Lancet Neurol. 2012;11(9):827-34.

[217] Geerts A, Brouwer O, Stroink H, van Donselaar C, Peters B, Peeters E, Arts WF. Onset of intractability and its course over time: the Dutch study of epilepsy in childhood. Epilepsia 2012;53(4):741-51.

[218] Schmidt D, Löscher W. Drug resistance in epilepsy: putative neurobiologic and clinical mechanisms. Epilepsia. 2005;46(6):858-77.

[219] Remy S, Beck H. Molecular and cellular mechanisms of pharmacoresistance in epilepsy. Brain. 2006;129:18-35.

[220] Liu JY, Thom M, Catarino CB, Martinian L, Figarella-Branger D, Bartolomei F, Koepp M, Sisodiya SM. Neuropathology of the blood-brain barrier and pharmaco-resistance in human epilepsy. Brain. 2012. [Epub ahead of print].

[221] Striano P, Coppola A, Paravidino R, Malacarne M, Gimelli S, Robbiano A, Traverso M, Pezzella M, Belcastro V, Bianchi A, Elia M, Falace A, Gazzerro E, Ferlazzo E, Freri E, Galasso R, Gobbi G, Molinatto C, Cavani S, Zuffardi O, Striano S, Ferrero GB, Silengo M, Cavaliere ML, Benelli M, Magi A, Piccione M, Dagna Bricarelli F, Coviello DA, Fichera M, Minetti C, Zara F. Clinical significance of rare copy number variations in epilepsy: a case-control survey using microarray-based comparative genomic hybridization. Arch Neurol. 2012;69(3):322-30.

[222] Galizia EC, Srikantha M, Palmer R, Waters JJ, Lench N, Ogilvie CM, Kasperavičiūtė D, Nashef L, Sisodiya SM. Array comparative genomic hybridization: results from an adult population with drug-resistant epilepsy and co-morbidities. Eur J Med Genet. 2012;55(5):342-8.

[223] Anovadiya AP, Sanmukhani JJ, Tripathi CB. Epilepsy: Novel therapeutic targets. J Pharmacol Pharmacother. 2012;3(2):112-7.

[224] Löscher W. Critical review of current animal models of seizures and epilepsy used in the discovery and development of new antiepileptic drugs. Seizure 2011;20(5):359-68.

[225] Neligan A, Shorvon SD. The history of status epilepticus and its treatment. Epilepsia. 2009 Mar;50 Suppl 3:56-68.

[226] Neligan A, Shorvon SD. Prognostic factors, morbidity and mortality in tonic-clonic status epilepticus: a review. Epilepsy Res. 2011;93(1):1-10.

[227] Brophy GM, Bell R, Claassen J, Alldredge B, Bleck TP, Glauser T, Laroche SM, Riviello JJ Jr, Shutter L, Sperling MR, Treiman DM, Vespa PM; Neurocritical Care Society Status Epilepticus Guideline Writing Committee. Guidelines for the evaluation and management of status epilepticus. Neurocrit Care 2012;17(1):3-23.

[228] Treiman DM, Meyers PD, Walton NY, Collins JF, Colling C, Rowan AJ, Handforth A, Faught E, Calabrese VP, Uthman BM, Ramsay RE, Mamdani MB. A comparison of four treatments for generalized convulsive status epilepticus. Veterans Affairs Status Epilepticus Cooperative Study Group.N Engl J Med. 1998;339(12):792-8.

[229] Appleton R, Macleod S, Martland T. Drug management for acute tonic-clonic convulsions including convulsive status epilepticus in children. Cochrane Database Syst Rev. 2008;(3):CD001905.

[230] Minicucci F, Muscas G, Perucca E, Capovilla G, Vigevano F, Tinuper P. Treatment of status epilepticus in adults: guidelines of the Italian League against Epilepsy. Epilepsia. 2006;47(S5):9-15.

[231] Shorvon S. The treatment of status epilepticus. Curr Opin Neurol. 2011;24(2):165-70.

[232] Trinka E. What is the evidence to use new intravenous AEDs in status epilepticus? Epilepsia. 2011;52(S8):35-8.

[233] Shorvon S, Ferlisi M. The outcome of therapies in refractory and super-refractory convulsive status epilepticus and recommendations for therapy. Brain. 2012;135:2314-28.

[234] Silbergleit R, Durkalski V, Lowenstein D, Conwit R, Pancioli A, Palesch Y, Barsan W; NETT Investigators. Intramuscular *vs.* intravenous therapy for prehospital status epilepticus. N Engl J Med. 2012;366(7):591-600.

[235] Rossetti AO, Hurwitz S, Logroscino G, Bromfield EB. Prognosis of status epilepticus: role of aetiology, age, and consciousness impairment at presentation. J Neurol Neurosurg Psychiatry. 2006;77(5):611-5.

[236] Drislane FW, Blum AS, Lopez MR, Gautam S, Schomer DL. Duration of refractory status epilepticus and outcome: loss of prognostic utility after several hours. Epilepsia. 2009;50(6):1566-71.

[237] Shearer P, Riviello J. Generalized convulsive status epilepticus in adults and children: treatment guidelines and protocols. Emerg Med Clin North Am. 2011;29(1):51-64.

[238] Rossetti AO, Alvarez V, Januel JM, Burnand B. Treatment deviating from guidelines does not influence status epilepticus prognosis. J Neurol. 2012. [Epub ahead of print].

[239] Mula M, Monaco F. Antiepileptic drugs and psychopathology of epilepsy: an update. Epileptic Disord. 2009;11(1):1-9.

[240] Hesdorffer DC, Ishihara L, Mynepalli L, Webb DJ, Weil J, Hauser WA., Brodtkorb E. Epilepsy, suicidality, and psychiatric disorders: A bidirectional association. Ann Neurol. 2012. [Epub ahead of print].

[241] Adelöw C, Andersson T, Ahlbom A, Tomson T. Hospitalization for psychiatric disorders before and after onset of unprovoked seizures/epilepsy. Neurology. 2012;78(6):396-401.

[242] Kanner AM, Schachter SC, Barry JJ, Hersdorffer DC, Mula M, Trimble M, Hermann B, Ettinger AE, Dunn D, Caplan R, Ryvlin P, Gilliam F. Depression and epilepsy: epidemiologic and neurobiologic perspectives that may explain their high comorbid occurrence. Epilepsy Behav. 2012;24(2):156-68.

[243] Hesdorffer DC, Hauser WA, Olafsson E, Ludvigsson P, Kjartansson O Depression and suicide attempt as risk factors for incident unprovoked seizures. Ann Neurol. 2006;59(1):35-41.

[244] Sanchez-Gistau V, Sugranyes G, Baillés E, Carreño M, Donaire A, Bargalló N, Pintor L. Is major depressive disorder specifically associated with mesial temporal sclerosis? Epilepsia. 2012;53(2):386-92.

[245] Kerr MP, Mensah S, Besag F, de Toffol B, Ettinger A, Kanemoto K, Kanner A, Kemp S, Krishnamoorthy E, LaFrance WC Jr, Mula M, Schmitz B, van Elst LT, Trollor J, Wilson SJ; International League of Epilepsy (ILAE) Commission on the Neuropsychiatric Aspects of Epilepsy. International consensus clinical practice statements for the treatment of neuropsychiatric conditions associated with epilepsy. Epilepsia. 2011;52(11):2133-8.

[246] Kanner AM, Barry JJ, Gilliam F, Hermann B, Meador KJ. Depressive and anxiety disorders in epilepsy: do they differ in their potential to worsen common antiepileptic drug-related adverse events? Epilepsia. 2012;53(6):1104-8.

[247] Henning OJ, Nakken KO. Psychiatric comorbidity and use of psychotropic drugs in epilepsy patients. Acta Neurol Scand. 2010;(190):18-22.

[248] Kanner AM. The use of psychotropic drugs in epilepsy: what every neurologist should know. Semin Neurol. 2008;28(3):379- 88.

[249] Alper K, Schwartz KA, Kolts RL, Khan A. Seizure incidence in psychopharmacological clinical trials: an analysis of Food and Drug Administration (FDA) summary basis of approval reports. Biol Psychiatry. 2007;62(4):345-54.

[250] Brodtkorb E, Mula M. Optimizing therapy of seizures in adult patients with psychiatric comorbidity. Neurology. 2006;67(S4):39-44.

[251] Ettinger AB. Psychotropic effects of antiepileptic drugs. Neurology. 2006 Dec 12;67(11):1916-25.

[252] Schmitz B. Effects of antiepileptic drugs on mood and behavior. Epilepsia. 2006;47 Suppl 2:28-33.

[253] Battino D, Tomson T. Management of epilepsy during pregnancy. Drugs. 2007;67(18):2727-46.

[254] Chiba S. Comprehensive management for women with epilepsy. Brain Nerve 2011;63(4):301-8.

[255] Montouris G, Abou-Khalil B. The first line of therapy in a girl with juvenile myoclonic epilepsy: should it be valproate or a new agent? Epilepsia. 2009;50(S8):16-20.

[256] Auvin S Treatment of juvenile myoclonic epilepsy. CNS Neurosci Ther. 2008;14(3):227-33.

[257] Mantoan L, Walker M.Treatment options in juvenile myoclonic epilepsy. Curr Treat Options Neurol. 2011;13(4):355-70.

[258] Foldvary-Schaefer N, Falcone T. Catamenial epilepsy: pathophysiology, diagnosis, and management. Neurology. 2003;61(S2):2-15.

[259] Verrotti A, Laus M, Coppola G, Parisi P, Mohn A, Chiarelli F. Catamenial epilepsy: hormonal aspects. Gynecol Endocrinol. 2010;26(11):783-90.

[260] Herzog AG. Catamenial epilepsy: definition, prevalence pathophysiology and treatment. Seizure. 2008;17(2):151-9.

[261] Herzog AG, Fowler KM, Smithson SD, Kalayjian LA, Heck CN, Sperling MR, Liporace JD, Harden CL, Dworetzky BA, Pennell PB, Massaro JM; Progesterone Trial Study Group. Progesterone *vs.* placebo therapy for women with epilepsy: A randomized clinical trial. Neurology 2012;78(24):1959-66.

[262] Klein A. The postpartum period in women with epilepsy. Neurol Clin. 2012;30(3):867-75.

[263] Brodie MJ, Elder AT, Kwan P. Epilepsy in later life. Lancet Neurol. 2009;8(11):1019-30.

[264] Bergey GK. Initial treatment of epilepsy: special issues in treating the elderly. Neurology. 2004;63(S4):40-8.

[265] Ramsay RE, Rowan AJ, Pryor FM. Special considerations in treating the elderly patient with epilepsy. Neurology 2004;62(S2):24-9.

[266] Arain AM, Abou-Khalil BW. Management of new-onset epilepsy in the elderly. Nat Rev Neurol. 2009;5(7):363- 71.

[267] Fife TD, Blum D, Fisher RS Measuring the effects of antiepileptic medications on balance in older people. Epilepsy Res. 2006;70(2-3):103-9.

[268] Brodie MJ, Overstall PW, Giorgi L. Multicentre, double-blind, randomised comparison between lamotrigine and carbamazepine in elderly patients with newly diagnosed epilepsy. The UK Lamotrigine Elderly Study Group. Epilepsy Res. 1999;37(1):81-7.

[269] Rowan AJ, Ramsay RE, Collins JF, Pryor F, Boardman KD, Uthman BM, Spitz M, Frederick T, Towne A, Carter GS, Marks W, Felicetta J, Tomyanovich ML; VA Cooperative Study 428 Group. New onset geriatric epilepsy: a randomized study of gabapentin, lamotrigine, and carbamazepine. Neurology 2005;64(11):1868-73.

[270] Ryvlin P, Montavont A, Nighoghossian N. Optimizing therapy of seizures in stroke patients. Neurology 2006;67(S4):3-9.

[271] García-Escrivá A, López-Hernández N.The use of levetiracetam in monotherapy in post-stroke seizures in the elderly population. Rev Neurol. 2007;45(9):523-5.

[272] Belcastro V, Costa C, Galletti F, Autuori A, Pierguidi L, Pisani F, Calabresi P, Parnetti L. Levetiracetam in newly diagnosed late-onset post-stroke seizures: a prospective observational study. Epilepsy Res. 2008; 82(2- 3):223-6.

[273] Kutlu G, Gomceli YB, Unal Y, Inan LE. Levetiracetam monotherapy for late poststroke seizures in the elderly. Epilepsy Behav. 2008;13(3):542-4.

[274] Gilad R. Management of seizures following a stroke: what are the options? Drugs Aging. 2012 1;29(7):533-8.

[275] Belcastro V., Costa C., Galletti F., Pisani F., Calabresi P., Parnetti L. Levetiracetam monotherapy in Alzheimer patients with late-onset seizures: a prospective observational study. Eur J Neurol. 2007; 14(10):1176-8.

[276] Jenssen S, Schere D. Treatment and management of epilepsy in the elderly demented patient. Am J Alzheimers Dis Other Demen. 2010;25(1):18-26.

[277] Pugh MJ, Foreman PJ, Berlowitz DR. Prescribing antiepileptics for the elderly: differences between guideline recommendations and clinical practice.Drugs Aging. 2006;23(11):861-75.

[278] Pugh MJ, Van Cott AC, Cramer JA, Knoefel JE, Amuan ME, Tabares J, Ramsay RE, Berlowitz DR; Treatment In Geriatric Epilepsy Research (TIGER) team. Collaborators (10) Bokhour B, Mortensen E, Cunningham F, Zeber J, Copeland L, Restrepo M, Glickman M, Devereaux M, Kressin N, Hope O. Trends in antiepileptic drug prescribing for older patients with new-onset epilepsy: 2000-2004. Neurology 2008;70:2171-8.

[279] Sander JW. The natural history of epilepsy in the era of new antiepileptic drugs and surgical treatment. Epilepsia. 2003;44 (S1):17-20.

[280] Sirven JI, Sperling M, Wingerchuk DM. Early *vs.* late antiepileptic drug withdrawal for people with epilepsy in remission. Cochrane Database Syst Rev. 2001;(3):CD001902.

[281] Specchio LM, Beghi E. Should antiepileptic drugs be withdrawn in seizure-free patients? CNS Drugs. 2004;18(4):201-12.

[282] Ranganathan LN, Ramaratnam S. Rapid *vs.* slow withdrawal of antiepileptic drugs. Cochrane Database Syst Rev. 2006;(2):CD005003.

[283] Hixson JD. Stopping antiepileptic drugs: when and why? Curr Treat Options Neurol. 2010;12(5):434-42.

[284] Bonnett LJ, Shukralla A, Tudur-Smith C, Williamson PR, Marson AG. Seizure recurrence after antiepileptic drug withdrawal and the implications for driving: further results from the MRC Antiepileptic Drug Withdrawal Study and a systematic review. J Neurol Neurosurg Psychiatry. 2011;82(12):1328-33.

[285] Hamed SA. The aspects and mechanisms of cognitive alterations in epilepsy: the role of antiepileptic medications. CNS Neurosci Ther. 2009;15(2):134-56.

[286] Berg AT. Epilepsy, cognition, and behaviour: The clinical picture. Epilepsia. 2011;52(S1):7-12.

[287] Onuma T. Cognitive dysfunction and antiepileptic drugs. Brain Nerve. 2011;63(4):379-83.

[288] Shelley BP, Trimble MR. All that spikes is not fits", mistaking the woods for the trees: the interictal spikes--an "EEG chameleon" in the interface disorders of brain and mind: a critical review. Clin EEG Neurosci. 2009;40(4):245-61.

[289] Van Bogaert P, Urbain C, Galer S, Ligot N, Peigneux P, De Tiège X. Impact of focal interictal epileptiform discharges on behaviour and cognition in children. Neurophysiol Clin. 2012;42(1-2):53-8.

[290] Avanzini G. Is tolerance to antiepileptic drugs clinically relevant? Epilepsia. 2006;47(8):1285-7.

[291] Löscher W, Schmidt D. Experimental and clinical evidence for loss of effect (tolerance) during prolonged treatment with antiepileptic drugs. Epilepsia. 2006;47(8):1253-84.

[292] Shorvon S, Tomson T. Sudden unexpected death in epilepsy. Lancet. 2011;378(9808):2028-38.

[293] Surges R, Sander JW.Sudden unexpected death in epilepsy: mechanisms, prevalence, and prevention. Curr Opin Neurol. 2012;25(2):201-7.

[294] Lamberts RJ, Thijs RD, Laffan A, Langan Y, Sander JW. Sudden unexpected death in epilepsy: people with nocturnal seizures may be at highest risk. Epilepsia. 2012;53(2):253-7.

[295] Velagapudi P, Turagam M, Laurence T, Kocheril A. Cardiac Arrhythmias and Sudden Unexpected Death in Epilepsy (SUDEP). Pacing Clin Electrophysiol. 2012;35(3):363-70.

[296] Richerson GB, Buchanan GF. The serotonin axis: Shared mechanisms in seizures, depression, and SUDEP. Epilepsia. 2011;52(S1):28-38.

[297] Aurlien D, Larsen JP, Gjerstad L, Taubøll E. Increased risk of sudden unexpected death in epilepsy in females using lamotrigine: a nested, case-control study. Epilepsia. 2012;53(2):258-66.

[298] Hesdorffer DC, Tomson T, Benn E, Sander JW, Nilsson L, Langan Y, Walczak TS, Beghi E, Brodie MJ, Hauser WA; ILAE Commission on Epidemiology (Subcommission on Mortality). Do antiepileptic drugs or generalized tonic-clonic seizure frequency increase SUDEP risk? A combined analysis. Epilepsia. 2012;53(2):249-52.

[299] Ryvlin P, Cucherat M, Rheims S. Risk of sudden unexpected death in epilepsy in patients given adjunctive antiepileptic treatment for refractory seizures: a meta-analysis of placebo-controlled randomised trials. Lancet Neurol. 2011;10(11):961-8.

[300] Armijo JA, Sanchez B, González AB. Evidence based treatment of epilepsy. Rev Neurol. 2002;35(S1):59-73.

[301] Shorvon SD Drug treatment of epilepsy in the century of the ILAE: the first 50 years, 1909-1958. Epilepsia. 2009;50 (S3):69- 92.

[302] Shorvon SD. Drug treatment of epilepsy in the century of the ILAE: the second 50 years, 1959-2009. Epilepsia. 2009;50(S3):93-130.

[303] French JA. Refractory epilepsy: one size does not fit all. Epilepsy Curr. 2006;6(6):177-80.

[304] Rheims S, Perucca E, Cucherat M, Ryvlin P. Factors determining response to antiepileptic drugs in randomized controlled trials. A systematic review and meta-analysis. Epilepsia. 2011;52(2):219-33.

[305] Welty TE, Faught E, Schmidt D, McAuley JW, Ryan M Be sure to read the fine print: the agency for healthcare research and quality comparative effectiveness report on antiepileptic drugs. Epilepsy Curr. 2012;12(3):84-6.

[306] French JA. Trial Design: How Do We Figure Out if an AED Works Epilepsy Curr. 2012;12(1):24-6.

[307] Maguire MJ, Hemming K, Hutton JL, Marson AG. Overwhelming heterogeneity in systematic reviews of observational anti-epileptic studies. Epilepsy Res. 2008;80(2-3):201-12.

[308] Maguire MJ, Hemming K, Hutton JL, Marson AG. Reporting and analysis of open-label extension studies of anti-epileptic drugs. Epilepsy Res. 2008;81(1):24-9.

[309] Temkin NR. Antiepileptogenesis and seizure prevention trials with antiepileptic drugs: meta-analysis of controlled trials. Epilepsia. 2001;42(4):515-24.

[310] Pitkänen A, Lukasiuk K. Mechanisms of epileptogenesis and potential treatment targets. Lancet Neurol. 2011;10(2):173-86.

[311] Hamed SA. The multimodal prospects for neuroprotection and disease modification in epilepsy: relationship to its challenging neurobiology. Restor Neurol Neurosci. 2010;28(3):323-48.

[312] Loeb JA. Identifying targets for preventing epilepsy using systems biology. Neurosci Lett. 2011;497(3):205-12.

[313] Löscher W, Brandt C. Prevention or modification of epileptogenesis after brain insults: experimental approaches and translational research. Pharmacol Rev. 2010;62(4):668-700.

[314] Schierhout G, Roberts I. WITHDRAWN: Antiepileptic drugs for preventing seizures following acute traumatic brain injury. Cochrane Database Syst Rev. 2012;6:CD000173.

[315] Belcastro V. Neuroprotection in post-stroke epilepsy: a realistic goal? Seizure. 2011;20(8):665.

[316] Szoeke CE, Newton M, Wood JM, Goldstein D, Berkovic SF, OBrien TJ, Sheffield LJ. Update on pharmacogenetics in epilepsy: a brief review. Lancet Neurol. 2006;5(2):189-96.

[317] Mirza N, Vasieva O, Marson AG, Pirmohamed M. Exploring the genomic basis of pharmacoresistance in epilepsy: an integrative analysis of large-scale gene expression profiling studies on brain tissue from epilepsy surgery. Hum Mol Genet. 2011;20(22):4381-94.

[318] Chan A, Pirmohamed M, Comabella M. Pharmacogenomics in neurology: current state and future steps. Ann Neurol. 2011;70(5):684-97.

[319] Noe' F, Nissinen J, Pitkänen A, Gobbi M, Sperk G, During M, Vezzani A. Gene therapy in epilepsy: the focus on NPY. Peptides. 2007;28(2):377-83.

[320] Riban V, Fitzsimons HL, During MJ. Gene therapy in epilepsy. Epilepsia. 2009;50(1):24-32.

[321] Delgado-Escueta AV, Bourgeois BF. Debate: Does genetic information in humans help us treat patients? PRO--genetic information in humans helps us treat patients. CON--genetic information does not help at all. Epilepsia. 2008;49(S9):13-24.

[322] Helmstaedter C, Mihov Y, Toliat MR, Thiele H, Nuernberg P, Schoch S, Surges R, Elger CE, Kunz WS, Hurlemann R. Genetic variation in dopaminergic activity is associated with the risk for psychiatric side effects of levetiracetam. Epilepsia. 2012 [Epub ahead of print].

[323] Cavalleri GL, McCormack M, Alhusaini S, Chaila E, Delanty N. Pharmacogenomics and epilepsy: the road ahead. Pharmacogenomics. 2011;12(10):1429-47.

Send Orders for Reprints on reprints@benthamscience.net

CHAPTER 7

Pharmacological Treatment of Cognitive Impairment in Schizophrenia: A Critical Review of the Clinical Effects of Current and Developing Drugs

Seiya Miyamoto[*]

Department of Neuropsychiatry, St. Marianna University School of Medicine, Kawasaki, Kanagawa, Japan

Abstract: Schizophrenia is a lifelong devastating mental disorder characterized by positive, negative, affective, and cognitive symptoms. Cognitive impairment is a core feature of the illness and exerts a great influence on long-term outcomes, including quality of life, social, and occupational functioning. Currently, effective treatments for cognitive deficits are thought to be the greatest unmet needs. Numerous recent clinical trials have suggested only modest benefits on cognitive function in schizophrenia relative to first- and second-generation antipsychotics when dosed properly. Moreover, both classes appear to have largely similar efficacy. Adjunctive cognitive enhancing agents with various molecular targets should be pursued, because they can be viable future treatments for cognitive impairment associated with schizophrenia. In line with this goal, a wide range of compounds have been developed and some of them have been evaluated in randomized clinical trials. This article provides a critical updated review of the effects of currently available antipsychotics and developing cognitive enhancers on cognition in schizophrenia.

Keywords: Schizophrenia, cognition, antipsychotic drug, drug development, clinical trials, cognitive-enhancing agents, cognitive impairment associated with schizophrenia.

INTRODUCTION

Schizophrenia is a lifelong devastating mental illness which affects approximately 1% of the population worldwide [1]. Symptoms of schizophrenia generally consist of positive, negative, affective, and cognitive symptoms. Cognitive impairment is a core feature of schizophrenia consistent with the original terminology for this

*Address correspondence to Seiya Miyamoto:** Director of Schizophrenia Treatment Center, Department of Neuropsychiatry, St. Marianna University School of Medicine, 2-16-1, Sugao, Miyamae-ku, Kawasaki, Kanagawa, 216-8511, Japan; Tel: +81 44 977 8111; Fax: +81 44 976 3341; E-mail: s2miya@marianna-u.ac.jp

Atta-ur-Rahman (Ed)

disorder, dementia praecox. It exerts a great influence on long-term outcomes, including quality of life (QOL), social and occupational functioning [2]. Despite advances in the safety and efficacy of antipsychotic drugs (APDs) for the treatment of schizophrenia, a number of patients continue to be plagued by impairments in social and occupational functioning. At present, cognitive deficits constitute one of the most important unmet medical needs in schizophrenia and there is an urgent need for its effective treatments [3, 4].

Until now, numerous clinical studies have investigated the efficacy of first-generation antipsychotics (FGAs) and second-generation antipsychotics (SGAs) on cognitive impairment associated with schizophrenia (CIAS). However, their effects were modest at best [5, 6]. Accordingly, in the United States, the Food and Drug Administration (FDA), the National Institute of Mental Health, representatives of academia and industry developed many initiatives, including Measurement and Treatment Research to Improve Cognition in Schizophrenia (MATRICS) and Treatment Units for Research on Neurocognition and Schizophrenia (TURNS) to facilitate the development of cognitive-enhancing drugs for CIAS [7, 8].

Cognition is a highly complex central nervous system (CNS) function and its impairment is associated with alterations in a variety of neurotransmitter systems [9]. As there is no consensus on optimal CIAS-relevant drug targets, a number of strategies for drug development have been employed for the enhancement of cognition in schizophrenia [9, 10]. For example, receptor targets that are involved in dopaminergic, serotonergic, cholinergic, glutamatergic, or gamma-aminobutyric acid (GABA) neurotransmission have been identified on the basis of pharmacologic challenges using animal models of schizophrenia, typically in rodents, receptor abnormalities observed in the postmortem brains of schizophrenia, and genetic linkage studies [4, 9, 10]. This intervention strategy is predicated on the idea that neuronal receptor targets are intact and available [11]. Other approaches have focused on the use of agents that enhance learning and memory in animal models, or that are effective in other CNS diseases (*e.g.*, Alzheimer's disease) with cognitive impairments. Furthermore, researchers have recently focused on neuroprotective agents and drugs that may facilitate neuroplasticity or neurogenesis [9, 10]. It should be noted, however, that many of these animal models may not represent the same cognitive or biological mechanisms operating in humans, because of species differences both in behaviors and in neural systems [12].

To date, a wide range of compounds with a variety of molecular targets have been developed and some of them have been evaluated in randomized clinical trials (RCTs) [3, 4]. This article provides a critical updated review of the effects of currently available APDs and developing cognitive-enhancing drugs on cognition in schizophrenia.

COGNITIVE IMPAIRMENT IN SCHIZOPHRENIA

Patients with schizophrenia typically perform 1-3 standard deviations under psychiatrically normal controls on a range of standardized testing batteries of cognitive functions, including verbal memory and learning, attention, working memory, processing speed, executive function, and social cognition [5, 13]. Cognitive deficits are usually present by the time of the first psychotic episode, and tend to remain relatively stable, or to worsen gradually after the onset [14-16]. The pattern of cognitive deficits may vary widely among individuals with schizophrenia and about 15% of patients with schizophrenia test within the normal range in all domains [17]. However, almost all patients may have illness-associated performance deficits, even those who perform in the normal range [18].

The side effects of APDs can cause secondary cognitive deficits as described below. In addition, the long-term concomitant use of anticholinergic agents or benzodiazepines has been shown to induce decreases in a variety of cognitive functions, including attention, verbal learning, memory, processing speed, and visuospatial ability [19-23]. Cognitive deficits are more strongly related to social and vocational functioning than psychotic symptoms, and may substantially impair activities of daily living and QOL [2, 24]. The severity of cognitive deficits predicts poorer medication adherence [25, 26] and increased relapse risk in first-episode patients [27]. Therefore, cognitive deficits have become a major target of treatment for schizophrenia.

COGNITIVE EFFECTS OF ANTIPSYCHOTICS

First-Generation Antipsychotics

As a class, FGAs have only modest effects on CIAS [28]. In some cases, FGAs may exacerbate preexisting cognitive deficits as a result of extrapyramidal symptoms (EPS), anticholinergic effects or sedation [29-31]. As mentioned above, cognitive functioning, particularly attention and memory, may also be worsened by anticholinergics, which are frequently required to treat FGA-induced EPS [32]. Such variables should be taken into account when assessing clinical studies of the impact of APDs on CIAS. Excessive dosing of FGAs can also impair performance on time-sensitive tasks *via* EPS, sedation, and anticholinergic activity.

Second-Generation Antipsychotics

Earlier quantitative reviews reported a significant advantage for SGAs over FGAs in several cognitive domains [33, 34], although these reviews have been criticized for the small number of studies they included [5]. Woodward and colleagues conducted a meta-analysis of 27 open label and 14 double-blind studies (many of which involved a switch from prior treatments) comparing FGAs and SGAs [5].The analysis demonstrated somewhat greater cognitive improvement for SGAs (effect sizes=0.17 to 0.46) compared to FGAs (effect sizes=0.00 to 0.24). It is, however, difficult to draw firm conclusions from "switch" studies, in which patients who have been maintained on FGAs are tested before and after switching to an SGA [35]. It remains also uncertain whether the cognitive enhancement observed with SGAs may imply true improvements in cognitive function or merely a relative decrease in EPS-, anticholinergic-, or sedation-related cognitive impact [34, 36]. This has led to discussions with regard to whether lower dose FGAs treatment could demonstrate equal efficacy to SGAs for CIAS [28].

Recent clinical trials with methodological improvements indicate that low-dose haloperidol may have a less deleterious impact on cognitive function [37-40]. Furthermore, several studies in first-episode patients with schizophrenia reported moderate improvement in the cognitive test performance after long-term (several months to 2 years) treatment with SGAs or low-dose haloperidol, but the magnitude of improvement did not differ between and among them [39, 41, 42]. It also should be noted that practice (retest) effects for many cognitive measures can be confounded with cognitive enhancement associated with APD treatments [43].

In summary, available APDs have a small effect on cognition. The observed differences in cognitive improvement between SGAs and FGAs might not be as large as initially reported if both classes of medication are dosed appropriately [33, 44]. Moreover, SGAs generally do not show differential effects when compared with one another [41, 45, 46].

PHARMACOLOGICAL TARGETS FOR ENHANCING COGNITION IN SCHIZOPHRENIA

Researchers have been searching for mechanisms and agents that can target CIAS. As a result, a large number of cognitive-enhancing drugs have been developed and investigated in clinical trials.

Dopaminergic Agents

Dopamine D_1 receptors are closely associated with working memory and executive function. It is proposed that either low or excessively high levels of D_1 receptor activation in the prefrontal cortex (PFC) is associated with impaired cognitive function, such that an optimal level of D_1 receptor stimulation may be required to obtain full cognitive benefits [47, 48]. In rodents and monkeys, treatment with selective full D_1 receptor agonists, including dihydrexidine (DAR-0100), A77636, and SKF81297, at low doses demonstrated cognitive-enhancing effects [49-52]. In patients with schizophrenia, a single low dose of dihydrexidine was adequately tolerated, but did not improve delayed neuropsychological performance [53]. The half-life of dihydrexidine is 30 min, thus its blood levels were essentially undetectable at the time of cognitive testing. However, dihydrexidine acutely increased both prefrontal and non-prefrontal cortical perfusion [54]. It is notable that chronic treatment with full D_1 agonists may induce down-regulation of D_1 receptors, which could worsen CIAS [55]. These pitfalls may be circumvented by an intermittent pattern of administration to optimally sensitize prefrontal D_1 receptors [56].

Serotonergic Agents

5-HT$_{1A}$ Receptor Agonists

There is evidence from postmortem studies that 5-HT$_{1A}$ receptors may be up-regulated in dorsolateral PFC in schizophrenia, suggesting dysfunction of

5-HT$_{1A}$ receptors in this disorder [57, 58]. In addition, 5-HT$_{1A}$ agonism has been suggested to enhance dopamine and acetylcholine (ACh) efflux in the PFC and hippocampus [59]. The 5-HT$_{1A}$ partial agonists tandospirone and buspirone may have a modest ability to improve some domains of cognition, including attention, executive function, and verbal memory, in patients with schizophrenia receiving FGAs or SGAs [60, 61]. However, these studies provide limited positive evidence of the benefits of 5-HT$_{1A}$ partial agonism on cognition in schizophrenia. Further studies will need to use larger sample sizes and more thorough cognitive assessment procedures [11].

5-HT$_3$ Receptor Antagonists

5-HT$_3$ receptors are located in multiple brain regions and tonically inhibit Ach release [62]. Moreover, 5-HT$_3$ receptors can inhibit GABA$_B$ inhibitory interneurons, thus blockade of 5-HT$_3$ receptors may potentiate the action of GABA$_B$ interneurons. In animal behavioral assays, 5-HT$_3$ antagonists can correct psychotic-like behavior and improve cognitive function [63]. Adjunctive administration of ondansetron, a selective 5-HT$_3$ receptor antagonist, enhanced P50 auditory sensory gating [64] and improved visual memory in patients with chronic schizophrenia on stable APDs [65-67]. However, ondansetron failed to improve other domains of cognition [66]. 5-HT$_3$ antagonists may be a novel therapeutic agent for CIAS, but further studies are warranted with larger sample sizes to determine whether ondansetron has pro-cognitive effects in schizophrenia.

5-HT$_6$ Receptor Antagonists

5-HT$_6$ receptors are exclusively located in the brain. In limbic and learning- and memory-related areas, the subtype is highly expressed in the nucleus accumbens, cerebral cortex, and subfield of the hippocampus [68, 69]. 5-HT$_6$ receptors may be important to the actions of several SGAs (*e.g.*, clozapine, olanzapine, iloperidone) and the pathophysiology of schizophrenia [70]. Indeed, 5-HT$_6$ receptor antagonists such as SB-271046, SB-258510A, and SB-399885 can improve cognition *via* facilitation of the release of cortical dopamine and cortical and hippocampal ACh and glutamate and neurotrophic actions in normal adult and aged rats [68, 69]. 5-HT$_6$ receptor antagonists can also improve cognitive deficits in rodent models of schizophrenia [71-73]. Of particular interest, SB-399885 potentiated haloperidol- and risperidone-induced dopamine release in the medial PFC and hippocampus in rats [74]. Thus, adjunctive 5-HT$_6$ antagonists may enhance cognitive functioning in schizophrenia. Currently, the 5-HT$_6$ antagonists SB-742457 and SGS-518 have entered phase II clinical trials for CIAS [73].

Cholinergic Agents

Alpha-7 Nicotinic Receptor Agonists

Studies in animals and humans have demonstrated an essential role for nicotinic ACh receptors (nAChRs) in the regulation of attention, memory, and sensory processing. The α_7 nAChR subtype is a drug target of considerable interest for treating schizophrenia, because animal, genetic, and postmortem studies have shown altered α_7 nAChR levels and function related to the illness [75]. Researchers have been exploring the development of selective α_7 nAChR agonists, including 3-2,4-dimethoxybenzylidene anabaseine (DMXB-A), R3487/MEM3454, PHA-709829, PH-399733, EVP-6124, and TC-5619, as adjunctive treatments for CIAS [76-80]. However, long-term use of α_7 nAChR agonists may induce the desensitization of nAChRs, leading to a limited duration of efficacy.

The results of a phase II double-blind, randomized crossover trial of DMXB-A, an α_7 partial agonist, showed significant improvements in negative symptoms, but no detectable cognitive effects [81]. However, DMXB-A diminished the activity of the hippocampus during a smooth pursuit eye movement task [82]. Moreover, DMXB-A altered default network activity, including a reduction in posterior cingulate, inferior parietal cortex, and medial frontal gyrus activity and an increase in precuneus activity in non-smoking patients with schizophrenia [83]. In a 12-week phase IIb double-blind, RCT, adjunctive EVP-6124 showed significant improvement in global cognitive functionas measured by the CogState overall cognitive index, clinical functionas measured by the Schizophrenia Cognition Rating Scale (SCoRS), and negative symptoms in patients with schizophrenia taking SGAs (except clozapine) [78]. In a 12-week phase IIa trial, adjunctive TC-5619 improved executive function (primary outcome measure) as well as several other cognitive measures, negative symptoms, and global function in patients with schizophrenia receiving quetiapine or risperidone. It should be noted that these effects were observed predominantly in tobacco users [84]. In an 8-week double-blind RCT of adjunctive tropisetron, a high-affinity partial agonist for α_7 nAChR and a potent 5-HT$_3$ receptor antagonist, auditory sensory gating P50 deficits and sustained visual attention were improved compared with baseline in non-smoking patients with chronic schizophrenia [85]. However, the sample size was small and no difference was observed in cognitive performance between tropisetron group and placebo group. While encouraging, further studies are necessary to determine the therapeutic potential of α_7 nAChR agonists in schizophrenia.

Alpha4-Beta2 Nicotinic Receptor Agonists

Another nAChR with important roles in cognition is the $\alpha_4\beta_2$ subtype. Agonists at $\alpha_4\beta_2$ nAChR such as AZD3480 (TC-1734), varenicline, SIB-1553A, and RJR2403 have been studied as adjunctive drugs to APDs for CIAS [55, 86]. A phase IIb trial of AZD3480 showed no evidence of cognitive improvement [87]. An open-label trial of varenicline, a partial agonist at $\alpha_4\beta_2$ nAChR and a much weaker full agonist at α_7 nAChR, demonstrated some beneficial effects on cognition (verbal learning and memory) and decreased smoking behavior in patients with schizophrenia [88]. In a recent 8-week placebo-controlled RCT of adjunctive varenicline, the primary analyses of cognitive function exhibited no drug-placebo differences, but secondary analyses found advantages over placebo on the Digital Symbol Substitution Test and the Wisconsin Card Sorting Test non-perseverative errors with some differences between smokers and non-smokers [89]. It is notable that while none of the aforementioned studies reported evidence of varenicline-associated clinical decompensation, there have been case reports and an FDA advisory about potential behavioral toxicity of varenicline [90, 91]. Further studies of varenicline are necessary to know whether a risk of psychiatric side effects also appears in clinically stable patients with schizophrenia taking APDs.

Acetylcholinesterase Inhibitors

Acetylcholinesterase (AChE) breaks down ACh in the synaptic cleft. AChE inhibitors can increase both the level and duration of action of ACh, and thus can make more ACh available for both nicotinic and muscarinic receptors in a nonspecific manner [12]. Several AChE inhibitors, including donepezil, rivastigmine, and galantamine, have been studied as adjunctive cognitive enhancers to APDs for CIAS. However, double-blind RCTs of donepezil and rivastigmine in schizophrenia demonstrated no beneficial effects on cognition [12].

Galantamine is a non-selective AChE inhibitor at higher doses and a relatively selective positive allosteric modulator of the $\alpha_4\beta_2$ and α_7 nAChRs at lower doses. The double-blind, placebo-controlled studies with galantamine in patients with schizophrenia have had mixed results. Of 7 RCTs, 4 studies reported some evidence suggestive of benefit for memory and processing speed [92-95]. However, 3 recent studies demonstrated no evidence of cognitive improvement with galantamine [96-98]. Taken together, the promise of adjunctive galantamine appears to be limited, although a specific set of cognitive abilities may be improved with galantamine in some patients with schizophrenia.

Glutamatergic Agents

Glycine Site Allosteric Modulators

N-methyl-D-aspartate receptor (NMDA-R) hypofunction has been hypothesized to contribute to the pathophysiology of schizophrenia and agents that enhance NMDA-R activity have been increasingly considered for their therapeutic potential [99]. Given that excessive activation of the NMDA-R can cause excitotoxicity, much of the research on cognitive enhancing effects has focused on the glycine allosteric modulatory site [100]. To date, there have been a number of RCTs of the glycine site agonists, including glycine, D-cycloserine, D-serine, and D-alanine, in patients with schizophrenia. A meta-analysis of 26 RCTs demonstrated that a full glycine site agonist D-serine as adjuvant therapy showed efficacy for cognitive symptoms [101]. A partial glycine site agonist D-cycloserine did not have any positive benefits on any aspects of cognitive function as assessed by a formal cognitive test. In the largest 16-week trial, the Cognitive and Negative Symptoms in Schizophrenia Trial (CONSIST), neither glycine nor D-cycloserine improved CIAS [102]. However, the CONSIST study reported a significant effect of site and lower levels of glycine than was achieved in previous studies. Although adjunctive glycine and D-alanine may have some benefits on positive symptoms, no consistent benefit was reported on cognition.

Ampakines

Positive modulators of the α-amino-3-hydroxy-5-methy-isoxazole-4-propionic acid (AMPA) receptor enhance NMDA-R channel opening and long-term potentiation (LTP) by initiating rapid cellular depolarization. A class of compounds termed ampakines that binds to an allosteric site of the AMPA receptor can enhance glutamatergic transmission and facilitate LTP, learning and memory in rats without the desensitization often seen with direct AMPA agonists [103, 104]. Thus, ampakines have been studied as potential adjunctive treatments for CIAS. A small double-blind RCT of the short-acting, relatively low-potency ampakine CX516 as monotherapy showed no clear beneficial effects on cognition in patients with schizophrenia who were partially refractory to FGAs [105]. Furthermore, a large 4-week RCT of CX516 added to clozapine, olanzapine, or risperidone failed to demonstrate cognitive benefit [106]. Another AMPA potentiator farampator (CX-691, ORG-24448) is now under evaluation in a phase II trial for the treatment of CIAS [107].

GABA Modulating Agents

Studies in non-human primates demonstrated that normal working memory function depends on appropriate neurotransmission of GABA, the inhibitory neurotransmitter of the CNS, in the dorsolateral PFC [108]. Postmortem studies indicated reduced GABA synthesis, as shown by decreased glutamic acid decarboxylase (GAD_{67}), and by a compensatory increase in postsynaptic density of $GABA_A$ receptors containing α2 subunits in the dorsolateral PFC of subjects with schizophrenia [108]. Thus, it has been suggested that selective $GABA_A$ receptor agonists may improve the cognitive deficits associated with dorsolateral PFC dysfunction in patients with schizophrenia.

In a proof-of-concept 4-week double-blind RCT, adjunctive MK-0777, which is a selective $GABA_A$ α2/α3 partial agonist, improved delayed memory and shortened reaction time on selected measures of PFC function in male patients with chronic schizophrenia [109]. Moreover, the results of EEG were suggestive of increased gamma band power in the frontal area. However, a recent 4-week double-blind RCT of MK-0777 did not demonstrated any significant effects on CIAS [110]. MK-0777 is a relatively weak $GABA_A$ $\alpha2$ partial agonist. Thus, a more potent partial agonist with greater intrinsic activity at the $GABA_A$ $\alpha2$ site may be required for cognitive improvement in schizophrenia.

Miscellaneous Agents

COMT Inhibitors

Catechol-*O*-methyl transferase (COMT) deactivates dopamine in the cortex and COMT-dependent dopamine degradation is especially important in the PFC where expression of the presynaptic dopamine transporter is low. Moreover, variation in COMT activity may be associated with PFC function, particularly during cognitive performance tasks such as working memory and executive function [111, 112]. The [158]Val/Val COMT gene variant, which has been associated with schizophrenia in some studies, leads to greater COMT activity and reduced prefrontal cognitive functions [113].

Tolcapone, a centrally-acting COMT inhibitor, can enhance prefrontal-related cognitive function in rats [114] and in healthy individuals [115]. Adjunctive tolcapone and entacapone, a peripherally acting COMT inhibitor, are currently in phase II clinical trials in patients with schizophrenia with and without the [158]Val/Val COMT gene variant [116]. Tolcapone was, however, removed from the market in Europe and Canada because of potential hepatotoxicity [117], and a black box warning issued in the United States, which seriously limits its clinical use [118].

Minocycline

The second-generation tetracycline antibiotic drug minocycline can attenuate cognitive impairments and prepulse inhibition deficits induced by NMDA-R antagonists and can modulate glutamate-induced excitotoxicity [119-121]. Moreover, minocycline attenuated behavioral sensitization and neurotoxicity induced by methamphetamine [122]. It has antioxidant, anti-inflammatory, and neuroprotective properties, possibly *via* nitric oxide synthase inhibition, inhibition of microglial activation, and anti-apoptotic properties [123]. Given that these mechanisms overlap with several more recently hypothesized pathophysiological pathways of schizophrenia, minocycline has emerged as a potential adjunctive treatment of CIAS.

In a 24-week double-blind RCT, minocycline added to SGAs was effective for several cognitive functions, including executive function, spatial working memory, and spatial recognition memory in young patients with early-phase schizophrenia [124]. Interestingly, minocycline significantly improved negative symptoms and general outcomes. However, this study included only patients at the first years after onset and the dropout rates were high (over 60%). Further studies are necessary to determine whether minocycline has pro-cognitive effects in schizophrenia.

Oxytocin

Oxytocin, a posterior pituitary hormone, has widespread effects beyond its characteristic roles in pregnancy and lactation. Oxytocin strongly regulates social cognition, trust and affiliation, each of which is often disturbed in patients with schizophrenia [125, 126]. Surprisingly, preclinical studies have demonstrated that oxytocin has antipsychotic-like effects in several animal models relevant to schizophrenia [127-129]. In clinical studies, single-dose intranasal oxytocin was associated with increased interpersonal trust, eye contact, performance on tests of face emotion recognition and theory of mind, and social reciprocity in normal [130, 131] and autistic individuals [132]. In patients with schizophrenia, higher levels of oxytocin were associated with increased pro-social behaviors [133].

A 3-week placebo-controlled RCT of intranasal oxytocin, given adjunctively to APDs, showed therapeutic effects of oxytocin on clinical symptoms in schizophrenia patients with residual symptoms [134]. In a 2-week double-blind RCT, daily intranasal oxytocin also improved measures of social cognition and psychotic symptoms, particularly paranoia in patients with schizophrenia [135]. Moreover, a recent double-blind, placebo-controlled cross-over study demonstrated that oxytocin improved emotion recognition in male patients with schizophrenia [136]. Although oxytocin, given chronically, significantly improved verbal memory [137], it is unclear whether oxytocin can improve general cognitive functioning in schizophrenia. Additional trials of oxytocin administered adjunctively or as monotherapy are needed with larger sample sizes and longer durations of treatment.

Erythropoietin

The hematopoietic growth factor erythropoietin (EPO) has multifaceted direct neuroprotective effects, including anti-apoptotic, anti-inflammatory, anti-oxidant, neurotrophic, synaptogenic, and angiogenic activity [138, 139]. EPO crosses the blood-brain barrier and is generally considered safe, although more prolonged administration sometimes requires therapeutic phlebotomy to maintain normal hematocrit [138]. In a 12-week double-blind phase IIb RCT, weekly intravenous administration of adjunctive high-dose recombinant human EPO (rhEPO), induced significantly greater cognitive improvement as compared to placebo in patients with chronic schizophrenia on stable APDs [140]. Furthermore, rhEPO was associated with delayed progression of cortical gray matter loss and evidence for gray matter protection by rhEPO was combined with an improvement in attention and memory functions in schizophrenia [141]. Therefore, EPO may have the potential to prevent progressive structural changes in the brain and to enhance cognitive function in schizophrenia.

DISCUSSION

Neuroscientific discoveries over the past decades have enriched our understanding of the neurobiology underlying CIAS. This research has identified new molecular mechanisms and process as promising pharmacological targets as described in this article [3]. However, there has been no major clinical success in the development of medications to enhance cognition in schizophrenia [142]. As of June 2013, none of the drug development programs has led to approval by the FDA. Although selective α_7 nAChR agonists, minocycline, and oxytocin might be promising targets, the relative lack of success to-date indicates the difficulty that this therapeutic area poses. There appears to be several possible reasons for negative study results.

First, the most studies completed to date have not had sufficient statistical power to detect true treatment effects. Keefe *et al.,* [13] recently reviewed 118 CIAS trials and pointed out that the large majority of completed trials are underpowered to detect moderate effect sizes. Second, the etiology of schizophrenia is complex and CIAS may not result from dysregulation of a single neurotransmitter or class of receptors [10]. Patients with schizophrenia may have long-standing impairments in the wiring or developmental connectivity of various neural systems [12]. Accordingly, it is not surprising that such complexity limits the potential efficacy of compounds which target a single mechanism of action [4]. However, some of receptor antagonists (*e.g.*, histamine 3 receptor antagonists) may be able to modulate additional neurotransmitter systems relevant for cognition in animal models [4, 12]. Third, there is a possibility of considerable variability (heterogeneity) in the degree and/or pathophysiology of cognitive deficits among patients. Also, there appears to be little reason to assume that the pathophysiology of schizophrenia overlaps with that of other CNS disorders with cognitive impairment. Thus, exploring more creative approaches for CIAS and matching of treatments to appropriate subgroups of patients may be necessary to establish efficacy [10]. Fourth, the design of clinical trials utilized in the drug development programs may be methodologically inadequate. There are many issues associated with clinical trials such as subject-selection criteria, phase of illness, length of the trials, high placebo response rates, practice effects, dosing of add-on medications, anticholinergic use, high attrition, poor adherence, and surreptitious substance abuse [10, 11, 18]. Fifth, sensitivity of the cognitive test battery may affect the negative results. Thus, the ongoing CIAS trials are more likely to use a widely accepted standardized cognitive battery (*e.g.*, the MATRICS Consensus Cognitive Battery) and MATRICS guidelines [13]. Finally, pharmacodynamic and/or pharmacokinetic interactions may occur between the antipsychotic drugs and the cognitive-enhancing drugs.

As the focus of this article is on pharmacological approaches, nonpharmacological interventions for cognition such as cognitive remediation are not mentioned. The most recent meta-analysis has shown a small to moderate effect of cognitive remediation therapy on cognitive outcomes in schizophrenia [143]. Although no RCTs comparing the efficacy of cognitive remediation or rehabilitation with that of pharmacological interventions have been conducted, a combination of such different approaches may be necessary to produce more robust and potentially longer-lasting cognitive benefits [12]. Developing new clinical protocols aimed at both the pharmacological approach and the cognitive training strategies is a worthwhile treatment target.

CONCLUSIONS

So far, the results of studies on cognitive-enhancing drugs for schizophrenia are not encouraging. It is possible that heterogeneity of mechanisms responsible for CIAS may in part be responsible for the lack of success of any single approach [10]. It is hoped that future research will identify predictive biomarkers and pharmacogenetic approaches will improve our ability to detect treatment effects by reducing biological heterogeneity and allow for more targeted tailor-made treatment [10, 144]. Pharmacological approaches should be combined with cognitive rehabilitation and/or remediation to explore a possible synergist effect on cognition. Because functional recovery is possible and should be an ultimate goal of patients with schizophrenia, potential cognitive-enhancing drugs should improve both cognitive functions and functional outcomes.

ACKNOWLEDGEMENTS

None Declared.

CONFLICT OF INTEREST

The authors confirm that this article content has no conflicts of interest.

REFERENCES

[1]   Insel TR. Rethinking schizophrenia. Nature. 2010;468(7321):187-93.
[2]   Green MF. What are the functional consequences of neurocognitive deficits in schizophrenia? Am J Psychiatry. 1996;153(3):321-30.
[3]   Tcheremissine OV, Castro MA, Gardner DR. Targeting cognitive deficits in schizophrenia: a review of the development of a new class of medicines from the perspective of community mental health researchers. Expert Opin Investig Drugs. 2012;21(1):7-14.
[4]   Miyamoto S, Miyake N, Jarskog LF, Fleischhacker WW, Lieberman JA. Pharmacological treatment of schizophrenia: a critical review of the pharmacology and clinical effects of current and future therapeutic agents. Mol Psychiatry. 2012;17(12):1206-27.
[5]   Woodward ND, Purdon SE, Meltzer HY, Zald DH. A meta-analysis of neuropsychological change to clozapine, olanzapine, quetiapine, and risperidone in schizophrenia. Int J Neuropsychopharmacol. 2005;8(3):457-72.

[6]   Woodward ND, Purdon SE, Meltzer HY, Zald DH. A meta-analysis of cognitive change with haloperidol in clinical trials of atypical antipsychotics: dose effects and comparison to practice effects. Schizophr Res. 2007;89(1-3):211-24.

[7]   Buchanan RW, Davis M, Goff D, *et al.* A summary of the FDA-NIMH-MATRICS workshop on clinical trial design for neurocognitive drugs for schizophrenia. Schizophr Bull. 2005;31(1):5-19.

[8]   Buchanan RW, Keefe RS, Umbricht D, *et al.* The FDA-NIMH-MATRICS guidelines for clinical trial design of cognitive-enhancing drugs: what do we know 5 years later? Schizophr Bull. 2011;37(6):1209-17.

[9]   Wallace TL, Ballard TM, Pouzet B, Riedel WJ, Wettstein JG. Drug targets for cognitive enhancement in neuropsychiatric disorders. Pharmacol Biochem Behav. 2011;99(2):130-45.

[10]   Goff DC, Hill M, Barch D. The treatment of cognitive impairment in schizophrenia. Pharmacol Biochem Behav. 2011;99(2):245-53.

[11]   Harvey PD. Pharmacological cognitive enhancement in schizophrenia. Neuropsychol Rev. 2009;19(3):324-35.

[12]   Barch DM. Pharmacological strategies for enhancing cognition in schizophrenia. Curr Top Behav Neurosci. 2010;4:43-96.

[13]   Keefe RS, Buchanan RW, Marder SR, *et al.* Clinical trials of potential cognitive-enhancing drugs in schizophrenia: what have we learned so far? Schizophr Bull. 2013;39(2):417-35.

[14]   Aleman A, Hijman R, de Haan EH, Kahn RS. Memory impairment in schizophrenia: a meta-analysis. Am J Psychiatry. 1999;156(9):1358-66.

[15]   Gold S, Arndt S, Nopoulos P, O'Leary DS, Andreasen NC. Longitudinal study of cognitive function in first-episode and recent-onset schizophrenia. Am J Psychiatry. 1999;156(9):1342-8.

[16]   Harvey PD, Silverman JM, Mohs RC, *et al.* Cognitive decline in late-life schizophrenia: a longitudinal study of geriatric chronically hospitalized patients. Biol Psychiatry. 1999;45(1):32-40.

[17]   Meltzer HY, Park S, Kessler R. Cognition, schizophrenia, and the atypical antipsychotic drugs. Proc Natl Acad Sci USA. 1999;96(24):13591-3.

[18]   Green MF. Stimulating the development of drug treatments to improve cognition in schizophrenia. Annu Rev Clin Psychol. 2007;3:159-80.

[19]   Stewart SA. The effects of benzodiazepines on cognition. J Clin Psychiatry. 2005;66 Suppl 2:9-13.

[20]   Barker MJ, Greenwood KM, Jackson M, Crowe SF. Cognitive effects of long-term benzodiazepine use: a meta-analysis. CNS Drugs. 2004;18(1):37-48.

[21]   Ashton H. The diagnosis and management of benzodiazepine dependence. Curr Opin Psychiatry. 2005;18(3):249-55.

[22]   Strauss ME, Reynolds KS, Jayaram G, Tune LE. Effects of anticholinergic medication on memory in schizophrenia. Schizophr Res. 1990;3(2):127-9.

[23]   Minzenberg MJ, Poole JH, Benton C, Vinogradov S. Association of anticholinergic load with impairment of complex attention and memory in schizophrenia. Am J Psychiatry. 2004;161(1):116-24.

[24]   Harvey PD, Howanitz E, Parrella M, *et al.* Symptoms, cognitive functioning, and adaptive skills in geriatric patients with lifelong schizophrenia: a comparison across treatment sites. Am J Psychiatry. 1998;155(8):1080-6.

[25]   Burton SC. Strategies for improving adherence to second-generation antipsychotics in patients with schizophrenia by increasing ease of use. J Psychiatr Pract. 2005;11(6):369-78.

[26]   Prouteau A, Verdoux H, Briand C, *et al.* Cognitive predictors of psychosocial functioning outcome in schizophrenia: a follow-up study of subjects participating in a rehabilitation program. Schizophr Res. 2005;77(2-3):343-53.

[27]   Chen EY, Hui CL, Dunn EL, *et al.* A prospective 3-year longitudinal study of cognitive predictors of relapse in first-episode schizophrenic patients. Schizophr Res. 2005;77(1):99-104.

[28]   Mishara AL, Goldberg TE. A meta-analysis and critical review of the effects of conventional neuroleptic treatment on cognition in schizophrenia: opening a closed book. Biol Psychiatry. 2004;55(10):1013-22.

[29]   Mortimer AM. Cognitive function in schizophrenia--do neuroleptics make a difference? Pharmacol Biochem Behav. 1997;56(4):789-95.

[30]   Velligan DI, Miller AL. Cognitive dysfunction in schizophrenia and its importance to outcome: the place of atypical antipsychotics in treatment. J Clin Psychiatry. 1999;60 Suppl 23:25-8.

[31]   Green MF, Braff DL. Translating the basic and clinical cognitive neuroscience of schizophrenia to drug development and clinical trials of antipsychotic medications. Biol Psychiatry. 2001;49(4):374-84.

[32]   Tollefson GD. Cognitive function in schizophrenic patients. J Clin Psychiatry. 1996;57 Suppl 11:31-9.

[33]   Keefe RSE, Silva SG, Perkins DO, Lieberman JA. The effects of atypical antipsychotic drugs on neurocognitive impairment in schizophrenia: A review and meta-analysis. Schizophr Bull. 1999;25(2):201-22.

[34]   Harvey PD, Keefe RS. Studies of cognitive change in patients with schizophrenia following novel antipsychotic treatment. Am J Psychiatry. 2001;158(2):176-84.

[35]   Hill SK, Bishop JR, Palumbo D, Sweeney JA. Effect of second-generation antipsychotics on cognition: current issues and future challenges. Expert Rev Neurother. 2010;10(1):43-57.

[36]   Carpenter WT, Gold JM. Another view of therapy for cognition in schizophrenia. Biol Psychiatry. 2002;51(12):969-71.

[37]   Green MF, Marder SR, Glynn SM, *et al.* The neurocognitive effects of low-dose haloperidol: a two-year comparison with risperidone. Biol Psychiatry. 2002;51(12):972-8.

[38]   Keefe RS, Seidman LJ, Christensen BK, *et al.* Comparative effect of atypical and conventional antipsychotic drugs on neurocognition in first-episode psychosis: a randomized, double-blind trial of olanzapine *vs.* low doses of haloperidol. Am J Psychiatry. 2004;161(6):985-95.

[39]   Keefe RS, Seidman LJ, Christensen BK, *et al.* Long-term neurocognitive effects of olanzapine or low-dose haloperidol in first-episode psychosis. Biol Psychiatry. 2006;59(2):97-105.

[40]   Keefe RS, Young CA, Rock SL, *et al.* One-year double-blind study of the neurocognitive efficacy of olanzapine, risperidone, and haloperidol in schizophrenia. Schizophr Res. 2006;81(1):1-15.

[41]   Davidson M, Galderisi S, Weiser M, *et al.* Cognitive effects of antipsychotic drugs in first-episode schizophrenia and schizophreniform disorder: a randomized, open-label clinical trial (EUFEST). Am J Psychiatry. 2009;166(6):675-82.

[42]   Crespo-Facorro B, Rodriguez-Sanchez JM, Perez-Iglesias R, *et al.* Neurocognitive effectiveness of haloperidol, risperidone, and olanzapine in first-episode psychosis: a randomized, controlled 1-year follow-up comparison. J Clin Psychiatry. 2009;70(5):717-29.

[43]   Goldberg TE, Keefe RS, Goldman RS, Robinson DG, Harvey PD. Circumstances under which practice does not make perfect: a review of the practice effect literature in schizophrenia and its relevance to clinical treatment studies. Neuropsychopharmacology. 2010;35(5):1053-62.

[44]   Purdon SE, Jones BD, Stip E, *et al.* Neuropsychological change in early phase schizophrenia during 12 months of treatment with olanzapine, risperidone, or haloperidol. The Canadian Collaborative Group for research in schizophrenia. Arch Gen Psychiatry. 2000;57(3):249-58.

[45]   Keefe RS, Bilder RM, Davis SM, *et al.* Neurocognitive effects of antipsychotic medications in patients with chronic schizophrenia in the CATIE Trial. Arch Gen Psychiatry. 2007;64(6):633-47.

[46]   Keefe RS, Sweeney JA, Gu H, *et al.* Effects of olanzapine, quetiapine, and risperidone on neurocognitive function in early psychosis: a randomized, double-blind 52-week comparison. Am J Psychiatry. 2007;164(7):1061-71.

[47]   Goldman-Rakic PS, Muly Iii EC, Williams GV. D1 receptors in prefrontal cells and circuits. Brain ResRev. 2000;31:295-301.

[48]   Williams GV, Goldman-Rakic PS. Modulation of memory fields by dopamine D1 receptors in prefrontal cortex. Nature. 1995;376(6541):572-5.

[49]   Arnsten AF, Cai JX, Murphy BL, Goldman-Rakic PS. Dopamine D1 receptor mechanisms in the cognitive performance of young adult and aged monkeys. Psychopharmacology (Berl). 1994;116(2):143-51.

[50]   Schneider JS, Sun ZQ, Roeltgen DP. Effects of dihydrexidine, a full dopamine D-1 receptor agonist, on delayed response performance in chronic low dose MPTP-treated monkeys. Brain Res. 1994;663(1):140-4.

[51]   Cai JX, Arnsten AF. Dose-dependent effects of the dopamine D1 receptor agonists A77636 or SKF81297 on spatial working memory in aged monkeys. J Pharmacol Exp Ther. 1997;283(1):183-9.

[52]   Bubenikova-Valesova V, Svoboda J, Horacek J, Vales K. The effect of a full agonist/antagonist of the D1 receptor on locomotor activity, sensorimotor gating and cognitive function in dizocilpine-treated rats. Int J Neuropsychopharmacol. 2009;12(7):873-83.

[53]   George MS, Molnar CE, Grenesko EL, *et al.* A single 20 mg dose of dihydrexidine (DAR-0100), a full dopamine D1 agonist, is safe and tolerated in patients with schizophrenia. Schizophr Res. 2007;93(1-3):42-50.

[54]   Mu Q, Johnson K, Morgan PS, *et al.* A single 20 mg dose of the full D1 dopamine agonist dihydrexidine (DAR-0100) increases prefrontal perfusion in schizophrenia. Schizophr Res. 2007;94(1-3):332-41.

[55]   Gray JA, Roth BL. Molecular targets for treating cognitive dysfunction in schizophrenia. Schizophr Bull. 2007;33(5):1100-19.

[56]   Goldman-Rakic PS, Castner SA, Svensson TH, Siever LJ, Williams GV. Targeting the dopamine D1 receptor in schizophrenia: insights for cognitive dysfunction. Psychopharmacology (Berl). 2004;174(1):3-16.

[57] Burnet PW, Eastwood SL, Harrison PJ. 5-HT1A and 5-HT2A receptor mRNAs and binding site densities are differentially altered in schizophrenia. Neuropsychopharmacology. 1996;15(5):442-55.

[58] Simpson MD, Lubman DI, Slater P, Deakin JF. Autoradiography with [3H]8-OH-DPAT reveals increases in 5-HT(1A) receptors in ventral prefrontal cortex in schizophrenia. Biol Psychiatry. 1996;39(11):919-28.

[59] Meltzer HY, Sumiyoshi T. Does stimulation of 5-HT(1A) receptors improve cognition in schizophrenia? Behav Brain Res. 2008;195(1):98-102.

[60] Sumiyoshi T, Matsui M, Nohara S, *et al.* Enhancement of cognitive performance in schizophrenia by addition of tandospirone to neuroleptic treatment. Am J Psychiatry. 2001;158(10):1722-5.

[61] Sumiyoshi T, Park S, Jayathilake K, *et al.* Effect of buspirone, a serotonin1A partial agonist, on cognitive function in schizophrenia: a randomized, double-blind, placebo-controlled study. Schizophr Res. 2007;95(1-3):158-68.

[62] Ramirez MJ, Cenarruzabeitia E, Lasheras B, Del Rio J. Involvement of GABA systems in acetylcholine release induced by 5-HT3 receptor blockade in slices from rat entorhinal cortex. Brain Res. 1996;712(2):274-80.

[63] Costall B, Naylor RJ. 5-HT3 receptors. Curr Drug Targets CNS Neurol Disord. 2004;3(1):27-37.

[64] Adler LE, Cawthra EM, Donovan KA, *et al.* Improved p50 auditory gating with ondansetron in medicated schizophrenia patients. Am J Psychiatry. 2005;162(2):386-8.

[65] Zhang ZJ, Kang WH, Li Q, *et al.* Beneficial effects of ondansetron as an adjunct to haloperidol for chronic, treatment-resistant schizophrenia: a double-blind, randomized, placebo-controlled study. Schizophr Res. 2006;88(1-3):102-10.

[66] Akhondzadeh S, Mohammadi N, Noroozian M, *et al.* Added ondansetron for stable schizophrenia: a double blind, placebo controlled trial. Schizophr Res. 2009;107(2-3):206-12.

[67] Levkovitz Y, Arnest G, Mendlovic S, Treves I, Fennig S. The effect of Ondansetron on memory in schizophrenic patients. Brain Res Bull. 2005;65(4):291-5.

[68] Fone KC. An update on the role of the 5-hydroxytryptamine6 receptor in cognitive function. Neuropharmacology. 2008;55(6):1015-22.

[69] King MV, Marsden CA, Fone KC. A role for the 5-HT(1A), 5-HT4 and 5-HT6 receptors in learning and memory. Trends Pharmacol Sci. 2008;29(9):482-92.

[70] Terry AV, Jr., Buccafusco JJ, Wilson C. Cognitive dysfunction in neuropsychiatric disorders: selected serotonin receptor subtypes as therapeutic targets. Behav Brain Res. 2008;195(1):30-8.

[71] Marcos B, Chuang TT, Gil-Bea FJ, Ramirez MJ. Effects of 5-HT6 receptor antagonism and cholinesterase inhibition in models of cognitive impairment in the rat. Br J Pharmacol. 2008;155(3):434-40.

[72] Hirst WD, Stean TO, Rogers DC, *et al.* SB-399885 is a potent, selective 5-HT6 receptor antagonist with cognitive enhancing properties in aged rat water maze and novel object recognition models. Eur J Pharmacol. 2006;553(1-3):109-19.

[73] Rosse G, Schaffhauser H. 5-HT6 receptor antagonists as potential therapeutics for cognitive impairment. Curr Top Med Chem. 2010;10(2):207-21.

[74] Li Z, Huang M, Prus AJ, Dai J, Meltzer HY. 5-HT6 receptor antagonist SB-399885 potentiates haloperidol and risperidone-induced dopamine efflux in the medial prefrontal cortex or hippocampus. Brain Res. 2007;1134(1):70-8.

[75]   Martin LF, Kem WR, Freedman R. Alpha-7 nicotinic receptor agonists: potential new candidates for the treatment of schizophrenia. Psychopharmacology (Berl). 2004;174(1):54-64.

[76]   Mazurov A, Hauser T, Miller CH. Selective alpha7 nicotinic acetylcholine receptor ligands. Curr Med Chem. 2006;13(13):1567-84.

[77]   Hauser TA, Kucinski A, Jordan KG, *et al.* TC-5619: an alpha7 neuronal nicotinic receptor-selective agonist that demonstrates efficacy in animal models of the positive and negative symptoms and cognitive dysfunction of schizophrenia. Biochem Pharmacol. 2009;78(7):803-12.

[78]   EnVivo Pharmaceuticals. EnVivo reports positive results of its EVP-6124 clinical bio-marker study in schizophrenia patients. Available from URL: http://www.envivopharma.com/news-item.php?id=35. Press release, December 5, 2011

[79]   Rezvani AH, Kholdebarin E, Brucato FH, *et al.* Effect of R3487/MEM3454, a novel nicotinic alpha7 receptor partial agonist and 5-HT3 antagonist on sustained attention in rats. Prog Neuropsychopharmacol Biol Psychiatry. 2009;33(2):269-75.

[80]   Acker BA, Jacobsen EJ, Rogers BN, *et al.* Discovery of N-[(3R,5R)-1-azabicyclo[3.2.1]oct-3-yl]furo[2,3-c]pyridine-5-carboxamide as an agonist of the alpha7 nicotinic acetylcholine receptor: *in vitro* and *in vivo* activity. Bioorg Med Chem Lett. 2008;18(12):3611-5.

[81]   Freedman R, Olincy A, Buchanan RW, *et al.* Initial phase 2 trial of a nicotinic agonist in schizophrenia. Am J Psychiatry. 2008;165(8):1040-7.

[82]   Tregellas JR, Olincy A, Johnson L, *et al.* Functional magnetic resonance imaging of effects of a nicotinic agonist in schizophrenia. Neuropsychopharmacology. 2010;35(4):938-42.

[83]   Tregellas JR, Tanabe J, Rojas DC, *et al.* Effects of an alpha 7-nicotinic agonist on default network activity in schizophrenia. Biol Psychiatry. 2011;69(1):7-11.

[84]   Lieberman JA, Dunbar G, Segreti AC, *et al.* A Randomized Exploratory Trial of an Alpha-7 Nicotinic Receptor Agonist (TC-5619) for Cognitive Enhancement in Schizophrenia. Neuropsychopharmacology. 2013.

[85]   Shiina A, Shirayama Y, Niitsu T, *et al.* A randomised, double-blind, placebo-controlled trial of tropisetron in patients with schizophrenia. Ann Gen Psychiatry. 2010;9:27.

[86]   Arneric SP, Holladay M, Williams M. Neuronal nicotinic receptors: a perspective on two decades of drug discovery research. Biochem Pharmacol. 2007;74(8):1092-101.

[87]   Biedermann F, Fleischhacker WW. Antipsychotics in the early stage of development. Curr Opin Psychiatry. 2009;22(3):326-30.

[88]   Smith RC, Lindenmayer JP, Davis JM, *et al.* Cognitive and antismoking effects of varenicline in patients with schizophrenia or schizoaffective disorder. Schizophr Res. 2009;110(1-3):149-55.

[89]   Shim JC, Jung DU, Jung SS, *et al.* Adjunctive varenicline treatment with antipsychotic medications for cognitive impairments in people with schizophrenia: a randomized double-blind placebo-controlled trial. Neuropsychopharmacology. 2012;37(3):660-8.

[90]   Kohen I, Kremen N. Varenicline-induced manic episode in a patient with bipolar disorder. Am J Psychiatry. 2007;164(8):1269-70.

[91]   Freedman R. Exacerbation of schizophrenia by varenicline. Am J Psychiatry. 2007;164(8):1269.

[92]   Buchanan RW, Conley RR, Dickinson D, *et al.* Galantamine for the treatment of cognitive impairments in people with schizophrenia. Am J Psychiatry. 2008;165(1):82-9.

[93]   Lee SW, Lee JG, Lee BJ, Kim YH. A 12-week, double-blind, placebo-controlled trial of galantamine adjunctive treatment to conventional antipsychotics for the cognitive impairments in chronic schizophrenia. Int Clin Psychopharmacol. 2007;22(2):63-8.

[94]   Noren U, Bjorner A, Sonesson O, Eriksson L. Galantamine added to antipsychotic treatment in chronic schizophrenia: cognitive improvement? Schizophr Res. 2006;85(1-3):302-4.

[95]   Schubert MH, Young KA, Hicks PB. Galantamine improves cognition in schizophrenic patients stabilized on risperidone. Biol Psychiatry. 2006;60(6):530-3.

[96]   Sacco KA, Creeden C, Reutenauer EL, George TP. Effects of galantamine on cognitive deficits in smokers and non-smokers with schizophrenia. Schizophr Res. 2008;103(1-3):326-7.

[97]   Lindenmayer JP, Khan A. Galantamine augmentation of long-acting injectable risperidone for cognitive impairments in chronic schizophrenia. Schizophr Res. 2011;125(2-3):267-77.

[98]   Dyer MA, Freudenreich O, Culhane MA, *et al.* High-dose galantamine augmentation inferior to placebo on attention, inhibitory control and working memory performance in nonsmokers with schizophrenia. Schizophr Res. 2008;102(1-3):88-95.

[99]   Miyamoto S, Duncan GE, Marx CE, Lieberman JA. Treatments for schizophrenia: a critical review of pharmacology and mechanisms of action of antipsychotic drugs. Mol Psychiatry. 2005;10(1):79-104.

[100] Leeson PD, Iversen LL. The glycine site on the NMDA receptor: structure-activity relationships and therapeutic potential. J Med Chem. 1994;37(24):4053-67.

[101] Tsai GE, Lin PY. Strategies to enhance N-methyl-D-aspartate receptor-mediated neurotransmission in schizophrenia, a critical review and meta-analysis. Curr Pharm Des. 2010;16(5):522-37.

[102] Buchanan RW, Javitt DC, Marder SR, *et al.* The Cognitive and Negative Symptoms in Schizophrenia Trial (CONSIST): the efficacy of glutamatergic agents for negative symptoms and cognitive impairments. Am J Psychiatry. 2007;164(10):1593-602.

[103] Hampson RE, Rogers G, Lynch G, Deadwyler SA. Facilitative effects of the ampakine CX516 on short-term memory in rats: correlations with hippocampal neuronal activity. J Neurosci. 1998;18(7):2748-63.

[104] Hampson RE, Rogers G, Lynch G, Deadwyler SA. Facilitative effects of the ampakine CX516 on short-term memory in rats: enhancement of delayed-nonmatch-to-sample performance. J Neurosci. 1998;18(7):2740-7.

[105] Marenco S, Egan MF, Goldberg TE, *et al.* Preliminary experience with an ampakine (CX516) as a single agent for the treatment of schizophrenia: a case series. Schizophr Res. 2002;57(2-3):221-6.

[106] Goff DC, Lamberti JS, Leon AC, *et al.* A placebo-controlled add-on trial of the Ampakine, CX516, for cognitive deficits in schizophrenia. Neuropsychopharmacology. 2008;33(3):465-72.

[107] Ward SE, Bax BD, Harries M. Challenges for and current status of research into positive modulators of AMPA receptors. Br J Pharmacol. 2010;160(2):181-90.

[108] Lewis DA, Hashimoto T, Volk DW. Cortical inhibitory neurons and schizophrenia. Nat Rev Neurosci. 2005;6(4):312-24.

[109] Lewis DA, Cho RY, Carter CS, *et al.* Subunit-selective modulation of GABA type A receptor neurotransmission and cognition in schizophrenia. Am J Psychiatry. 2008;165(12):1585-93.

[110] Buchanan RW, Keefe RS, Lieberman JA, *et al.* A randomized clinical trial of MK-0777 for the treatment of cognitive impairments in people with schizophrenia. Biol Psychiatry. 2011;69(5):442-9.

[111] Weinberger DR, Egan MF, Bertolino A, *et al.* Prefrontal neurons and the genetics of schizophrenia. Biol Psychiatry. 2001;50(11):825-44.

[112] Egan MF, Goldberg TE, Kolachana BS, *et al.* Effect of COMT Val108/158 Met genotype on frontal lobe function and risk for schizophrenia. Proc Natl Acad Sci USA. 2001;98(12):6917-22.

[113] Goldberg TE, Egan MF, Gscheidle T, *et al.* Executive subprocesses in working memory: relationship to catechol-O-methyltransferase Val158Met genotype and schizophrenia. Arch Gen Psychiatry. 2003;60(9):889-96.

[114] Liljequist R, Haapalinna A, Ahlander M, Li YH, Mannisto PT. Catechol O-methyltransferase inhibitor tolcapone has minor influence on performance in experimental memory models in rats. Behav Brain Res. 1997;82(2):195-202.

[115] Apud JA, Mattay V, Chen J, *et al.* Tolcapone improves cognition and cortical information processing in normal human subjects. Neuropsychopharmacology. 2007;32(5):1011-20.

[116] Holden C. Neuroscience. Deconstructing schizophrenia. Science. 2003;299(5605):333-5.

[117] Watkins P. COMT inhibitors and liver toxicity. Neurology. 2000;55(11 Suppl 4):S51-S2.

[118] Borges N. Tolcapone-related liver dysfunction: implications for use in Parkinson's disease therapy. Drug Saf. 2003;26(11):743-7.

[119] Fujita Y, Ishima T, Kunitachi S, *et al.* Phencyclidine-induced cognitive deficits in mice are improved by subsequent subchronic administration of the antibiotic drug minocycline. Prog Neuropsychopharmacol Biol Psychiatry. 2008;32(2):336-9.

[120] Levkovitz Y, Levi U, Braw Y, Cohen H. Minocycline, a second-generation tetracycline, as a neuroprotective agent in an animal model of schizophrenia. Brain Res. 2007;1154:154-62.

[121] Zhang L, Shirayama Y, Iyo M, Hashimoto K. Minocycline attenuates hyperlocomotion and prepulse inhibition deficits in mice after administration of the NMDA receptor antagonist dizocilpine. Neuropsychopharmacology. 2007;32(9):2004-10.

[122] Zhang L, Kitaichi K, Fujimoto Y, *et al.* Protective effects of minocycline on behavioral changes and neurotoxicity in mice after administration of methamphetamine. Prog Neuropsychopharmacol Biol Psychiatry. 2006;30(8):1381-93.

[123] Plane JM, Shen Y, Pleasure DE, Deng W. Prospects for minocycline neuroprotection. Arch Neurol. 2010;67(12):1442-8.

[124] Levkovitz Y, Mendlovich S, Riwkes S, *et al.* A double-blind, randomized study of minocycline for the treatment of negative and cognitive symptoms in early-phase schizophrenia. J Clin Psychiatry. 2010;71(2):138-49.

[125] Uvnas-Moberg K. Oxytocin may mediate the benefits of positive social interaction and emotions. Psychoneuroendocrinology. 1998;23(8):819-35.

[126] Carter CS. Neuroendocrine perspectives on social attachment and love. Psychoneuroendocrinology. 1998;23(8):779-818.

[127] Feifel D, Reza T. Oxytocin modulates psychotomimetic-induced deficits in sensorimotor gating. Psychopharmacology (Berl). 1999;141(1):93-8.

[128] Lee PR, Brady DL, Shapiro RA, Dorsa DM, Koenig JI. Social interaction deficits caused by chronic phencyclidine administration are reversed by oxytocin. Neuropsychopharmacology. 2005;30(10):1883-94.

[129]   Caldwell HK, Stephens SL, Young WS, 3rd. Oxytocin as a natural antipsychotic: a study using oxytocin knockout mice. Mol Psychiatry. 2009;14(2):190-6.

[130]   Kosfeld M, Heinrichs M, Zak PJ, Fischbacher U, Fehr E. Oxytocin increases trust in humans. Nature. 2005;435(7042):673-6.

[131]   Domes G, Heinrichs M, Michel A, Berger C, Herpertz SC. Oxytocin improves "mind-reading" in humans. Biol Psychiatry. 2007;61(6):731-3.

[132]   Guastella AJ, Einfeld SL, Gray KM, *et al.* Intranasal oxytocin improves emotion recognition for youth with autism spectrum disorders. Biol Psychiatry. 2010;67(7):692-4.

[133]   Rubin LH, Carter CS, Drogos L, *et al.* Peripheral oxytocin is associated with reduced symptom severity in schizophrenia. Schizophr Res. 2010;124(1-3):13-21.

[134]   Feifel D, Macdonald K, Nguyen A, *et al.* Adjunctive intranasal oxytocin reduces symptoms in schizophrenia patients. Biol Psychiatry. 2010;68(7):678-80.

[135]   Pedersen CA, Gibson CM, Rau SW, *et al.* Intranasal oxytocin reduces psychotic symptoms and improves Theory of Mind and social perception in schizophrenia. Schizophr Res. 2011;132(1):50-3.

[136]   Averbeck BB, Bobin T, Evans S, Shergill SS. Emotion recognition and oxytocin in patients with schizophrenia. Psychol Med. 2012;42:259-66.

[137]   Feifel D. Is oxytocin a promising treatment for schizophrenia? Expert Rev Neurother. 2011;11(2):157-9.

[138]   Ehrenreich H, Degner D, Meller J, *et al.* Erythropoietin: a candidate compound for neuroprotection in schizophrenia. Mol Psychiatry. 2004;9(1):42-54.

[139]   Siren AL, Fasshauer T, Bartels C, Ehrenreich H. Therapeutic potential of erythropoietin and its structural or functional variants in the nervous system. Neurotherapeutics. 2009;6(1):108-27.

[140]   Ehrenreich H, Hinze-Selch D, Stawicki S, *et al.* Improvement of cognitive functions in chronic schizophrenic patients by recombinant human erythropoietin. Mol Psychiatry. 2007;12(2):206-20.

[141]   Wustenberg T, Begemann M, Bartels C, *et al.* Recombinant human erythropoietin delays loss of gray matter in chronic schizophrenia. Mol Psychiatry. 2011;16(1):26-36.

[142]   Buchanan RW, Kreyenbuhl J, Kelly DL, *et al.* The 2009 schizophrenia PORT psychopharmacological treatment recommendations and summary statements. Schizophr Bull. 2010;36(1):71-93.

[143]   Wykes T, Huddy V, Cellard C, McGurk SR, Czobor P. A meta-analysis of cognitive remediation for schizophrenia: methodology and effect sizes. Am J Psychiatry. 2011;168(5):472-85.

[144]   Correll CU. What are we looking for in new antipsychotics? J Clin Psychiatry. 2011;72 Suppl 1:9-13.

CHAPTER 8

Current Pharmacologic Treatments in Impaired Social Interaction in Autism Spectrum Disorders

Kunio Yui[*]

Research Institute of Pervasive Developmental Disorders, Ashiya University Graduate School of Education, Ashiya, Japan

Abstract: Autism spectrum disorders (ASD) are neurodevelopmental disorders with reduced cortical functional connectivity relating to social cognition. The evaluation of pharmacological treatment in ASD has been directed at abnormal developmental trajectories or toward enhancing plasticity during the development of brain. Accumulating evidence indicates that the gross abnormalities in these neurotransmitter systems such as serotonin and dopamine systems may underpin the neurophysiologic mechanism of ASD. Particularly, the serotonergic system may be especially implicated in pathophysiology of social impairment of ASD. Abnormal functional connectivity, which affects the delivery of afferent signals, may be involved in the pathophysiology of autism spectrum disorders (ASD). Arachidonic acid in the nervous system is important in signal transduction related to neuronal maturation. Risperidone solution, a novel antipsychotic which combined dopaminergic and serotonergic action, has shown to be effective in impaired social interaction. Oxytocin may mediate the benefits of positive social interaction and emotions. It is therefore worth noticing that risperidone solution, intranasal administration of oxytocin, and dietary supplementation with arachidonic acid have been found promising to maximize social interaction. Atypical antipsychotics ariprazole and SSRI fluoxetine exhibited their efficacy in treating some aspects of social relatedness or the core deficits of communication and socialization. There is evidence that abnormalities exist in peptide systems such as neuropeptides. $GABA_B$ antagonist STX209 has proved its efficacy in improving the ABC-irritability and Social withdrawal subscales. D-cycloserine exhibited significant improvement on social withdrawal subscale of the ABC in some subjects with ASD. It is hoped that improved strategies for early identification with phenotypic characteristics and biological markers (*e.g.*, brain physiological and biochemical changes) would remarkably improve the effectiveness of treatment.

The evaluation of treatments for ASD needs to be directed towards neurobiological targets known to be important in the brain's response to abnormal developmental trajectories or toward enhancing plasticity during the high sensitive period in gene-environment interaction (epigenetic mechanism). Further research towards neurobiology and effective treatments for ASD is required.

*Address correspondence to Kunio Yui: Research Institute of Pervasive Developmental Disorders, Ashiya University Graduate School of Education, Hyogo 659-8511, Japan; Tel: 81-797-23-0661; Fax: 81-797-23-1901; Email: yui16@bell.ocn.ne.jp

Keywords: Autism   spectrum   disorders,   impaired   social   interaction, neurobiological bases, pharmacologic treatment, risperidone solution, arachidonic acid

INTRODUCTION

ASD are characterized by impairment in social reciprocity and communication, coupled with restricted and stereotyped patterns of interests and behavior) [1]. The discovery of novel treatments for ASD has been a challenge. The stigma behind the lack of existing biomarkers is undoubtedly the complexity of the Autism Spectrum Disorders (ASD) condition itself, that is, the comorbidity, and variations in the type and severity of symptoms expressed by different individuals [2]. Since the putative effects of pharmacologic treatment on the developing brain are unclear neurobiological abnormalities need to be investigated. Moreover, the evaluation of treatments for social impairment in ASD may be directed at neurobiological targets known to be important in the brain's response to abnormal developmental trajectories [2]. The recent development of powerful multivariate analytical techniques have now encouraged the use of multi-modal information for developing complex 'biomarker systems', which may in the future be promising to assist the behavioral diagnosis, aid patient stratification and predict response to treatment/intervention [2]. This review discusses neurobiological characteristics in impaired social interaction in animal models and children with ASD, and highlights the development in useful pharmacotherapy reported over the past several years.

At present, psychopharmacology in ASD not primarily directed at the social and communication components is therefore considered relatively specific [3]. Recent trials with serotonin reuptake inhibitors (SSRIs) did not yield any satisfactory results despite their promising potential role. Although the present pharmacotherapeutic agents lack efficiency for the treatment of ASD core social impairment, research has provided the impetus to study potential drug effects on core social and language impairment [4]. The present article reviews the current results obtained from human clinical trials of drugs aimed at novel pharmacotherapies to improve social impairment of ASD.

EFFICACY OF RISPERIDONE

In contrast to haloperidol, atypical antipsychotic agents are found to be efficient in blocking postsynaptic serotonin receptors. The affinity of these agents for serotonin receptors is likely to maximize the efficacy for social activity, providing protection against extrapyramidal symptoms [5]. Risperidone (0.5 to 3.5 mg/day, mean ± SD=1.8±0.8 mg/day) reported significant improvements in the restricted, repetitive, and stereotyped patterns of behavior, interests, and activities of ASD children but their deficit in social interaction and communication remained significantly unchanged. Further research is required for introducing effective treatments for the core social and communicative impairments of autism [6].

Efficacy of Risperidone Solution

The delivery of risperidone solution *via* the buccal mucosa has been reported to exhibit the potential for producing therapeutically relevant plasma concentrations for the treatment of schizophrenia [7]. To date, only two placebo-controlled study of risperidone solution in children with ASD and pervasive developmental disorders have been reported. In a previous randomized, double-blind, placebo-controlled study, 40 consecutive children with ASD, whose ages ranged from 2 to 9 years received either risperidone solution (n=19) or placebo (n=20) given orally at a dose of 1 mg/day for 6 months [8]. The results were based on the total scores on the Childhood Autism Rating Scale (CARS) and the Children's Global Assessment Scale (CGAS). In the risperidone solution group, 12 out of 19 children showed significant improvement in the total CARS score and 17 out of 19 children in the total CGAS score compared with the placebo group (Mann-whitney U-test, P<0.05) [8]. Risperidone solution also exhibited an improvement in the social responsiveness and nonverbal communication and reduced the symptoms of hyperactivity and aggression. Risperidone solution was associated with increased appetite and a mild weight gain, mild sedation in 20%, and transient dyskinesias in 3 of 19 children. Another study conducted an 8-week, randomized, double-blind, placebo-controlled trial, in which risperidone/placebo solution (0.01-0.06 mg/kg/day) was administered to 79 children aged between 5 to 13 years with ASD and other pervasive developmental disorders (PDD) [9]. Outcome measures were resulted as the Aberrant Behavior Checklist (ABC) and

Clinical Global Impression-Change. Subjects who were taking risperidone solution (mean dosage: 0.04 mg/kg/day) exhibited significantly greater decreases on the other 4 subscales (irritability, hyperactivity, inappropriate speech, stereotyped behavior, and social withdrawal) of the ABC [9]. More risperidone solution-treated subjects (87%) showed global improvement in their condition compared with the placebo group (40%) [9]. Somnolence, the most frequently reported adverse event, was noted in 72.5% *vs.* 7.7% of subjects (risperidone solution *vs.* placebo) and was likely to manage with dose/dose-schedule modification [9]. Risperidone solution-treated subjects showed significantly greater increases in weight (2.7 *vs.* 1.0 kg), pulse rate, and systolic blood pressure [9]. Extrapyramidal symptoms scores were comparable between groups [9]. Risperidone solution was well tolerated and efficacious in treating behavioral symptoms and social withdrawal in children with ASD and other PDD [9]. Insofar as the assessment challenges of these placebo-controlled study of risperidone solution in ASD are associated with the larger field of child/adolescent psychopharmacology in general.

Delivery Systems of Risperidone Solution

Risperidone solution is absorbed through buccal mucosa [7]. The buccal mucosa is considered highly suitable for the development of sustained-delivery systems [10]. Delivery of drugs through the buccal mucosa is found to be more efficient as compared to the traditional oral route. Since the flow of blood from the oral mucosa drains directly into the jugular vein, drugs administered through the buccal mucosa directly enter into the systemic circulation thereby avoiding hepatic first pass metabolism. These considerations taken together validate the efficacy of risperidone solution in social impairment in children with ASD not exhibiting any adverse effects such as weight gain, fatigue, and drowsiness as was observed earlier in risperidone tablet in a previous clinical trial [5].

OXYTOCIN

Neurobiological Bases of Oxytocin

The neuromodulatory role of neuropeptide oxytocin (OXT) depends on the formation of mother–infant and adult–adult pair bonds and social memory and recognition [11]. OXT is a nonapeptide that is synthesized in magnocellular neurons in the paraventricular nucleus and the supraoptic nucleus of the hypothalamus and then it is released into the bloodstream through axon terminals in the posterior pituitary [12]. Imaging studies have suggested that oxytocin potently reduce activation of the amygdala and its coupling to brainstem regions implicated in autonomic and behavioral manifestations of fear [13], and that OXT appears to reduce activity in the amygdala, regardless of whether participants are shown happy, angry, or fearful faces [14]. In all the observations, OXT has been reported to affect only certain percepts or components of various social deficits, it also facilitates certain types of social behavior by halting arousal to social anxiety-causing stimuli *via* reducing the amygdale activity. Indeed, other studies confirmed that OXT was associated with a reduction in amygdale activity [15]. While in another MRI study, OXT has been reported to maximize the activity of amygdale in response to happy or angry faces only in the inferior frontal gyrus. These new findings suggest that the relationship between amygdala activity and OXT now appears to be more complex than previously thought [12].

A previous study on association between the oxyticin receptor gene and ASD, reported that single-nucleotide polymorphisms and haplotypes in the oxyticin receptor gene confer risk for ASD, and that this gene shapes both cognition and daily living skills that may cross diagnostic boundaries [16]. A previous haplotype analysis exhibited a significant association between a five-SNP haplotype and ASD, suggesting that the oxyticin receptor has a significant role in conferring the risk of ASD [17]. Recent studies have reported that oxytocin receptor knockout mice display deficits in social recognition and social communication [18]. In addition, functional polymorphisms of oxytocin receptor may contribute to ASD risk in a subset of families [19].

Efficacy of Oxytocin in Impaired Social Interaction in ASD

The OXT system can nevertheless be manipulated at various stages in development, which can result in the rescue of certain core symptom domains that characterize ASD [11]. In a recent research on humans, OXT has been reported to improve trust and stress-protective effects of positive social interaction [20]. Continuous infusion of synthetic OXT (pitocin) over a period of 4-hour reported an improvement in the effective speech comprehension (happy, indifferent, angry, and sad) in neutral content sentences, suggesting facilitation of social information processing in those with autism [21]. In another previous double-blind, randomized, placebo-controlled study, OXT nasal spray (18 or 24 IU) or a placebo in 16 male youth with ASD, whose ages ranged from 12 to 19 years, OXT administration was found to enhance the performance on the "Reading the Mind in the Eyes Task", indicating improvement of social communication and interaction [22]. The capability of OXT to ease inference of the effective mental state of others might reduce ambiguity in social situations and in this way encourage social approach, affiliation, and trusting behavior. OXT selectively increased patients' gazing time on the socially informative region of the face, namely the eyes during free viewing of pictures of faces [23]. Thus, under OXT, patients were found to respond more strongly to others and exhibited more appropriate social behavior and effect, validating a therapeutic potential of OXT through its action on a core dimension of autism [24], by attenuating amygdale response to emotional faces [25]. According to recent randomized, double-blind, placebo-controlled trial, intranasal OXT 24 IU in 19 adults with ASD (mean age: 33.20 ± 13.29 years) for 6 weeks reported significantly improved social cognition and quality of life without serious adverse effects [26]. Another double-blind, placebo-controlled trial reported that intranasal OXT administration (24 IU) for one week to fathers of children with ASD enhanced the quality of playful interactions between fathers and their child with ASD [27]. These two placebo-controlled trials indicated enhancement in social interaction social interaction of ASD. Further research is needed to clarify which symptom domains are most reliably influenced, which patients may be most responsive, and the neurobiological mechanisms through which OXT exerts its effects on social interaction [12].

THE CLINICAL EFFECTS OF ARACHIDONIC ACID ON IMPAIRED SOCIAL INTERACTION IN AUTISM SPECTRUM DISORDERS

Pharmacological Bases of Arachidonic Acid

The brain phopspholipids are exceptionally rich in highly polyunsaturated fatty acids (HUFAs). In terms of weight, Dry human brain is composed of approximately 60% lipids with over 20% PUFAs containing high amount of metabolites: docosahexaenoic acid (DHA) and arachidonic acid (ARA) [28, 29].

The plasma membrane phospholipids are supplied as second messenger molecules important for normal functioning of the brain [30]. ARA plays an important neurotransmitter role in addition to influencing the activity of other signaling cascade components, and thus seems to act as a neurotransmitter. While, DHA plays a key role in the structural development of neural and synaptic networks by imparting particular fluidity properties on neural membranes and is found to be crucial to their restoration [31, 32]. ARA modulates neurotransmitter systems *via* PLA_2 [30, 32].

Maternal ARA supplementation (0.5% of fat) reported an improvement in the wire hanging endurance and water maze latency, and exploratory behavior that validates the efficacy of ARA in positively improving the neurodevelopment [33].

There is increasing evidence about the positive effects of ARA on neurodevelopment. Previous clinical studies reported that children of 4–7 years of age in Canada that were not living near a marine environment exhibited relatively low dietary intakes of AA and DHA [34]. Retinal and neuronal development continues throughout childhood, therefore low intake of AA and DHA may be associated to have a negative impact [34]. Infant formula containing 0.36% DHA and 0.72% ARA exhibited only a slight improvement of seven points on the Mental Development Index such as problem solving, discrimination, language, and social skills, promoting mental development such as cognition and performance [35].

Effects of ARA on Impaired Social Interaction in ASD

There are only few reports justifying the clinical effects of ARA supplementation in ASD. Up to now, only two studies have been reported to significantly improve behavioral problems in AD/HD or learning disability. According to a previous double-blind placebo-controlled studies in 50 children with attention-deficiency/hyperactivity disorders (AD/HD) aged 6-13 years, supplementation with daily doses of 480 mg DHA, 80 mg EPA and 40mg ARA showed significant improvement in 2 out of 16 scales of behavioral problems in 25 children treated with supplementation, compared to 22 children treated with placebo [36]. In another study, administration of daily doses of 480 mg DHA, 186 mg EPA and 42 mg ARA reported significant improvement from baseline on 7 out of 14 scales of behavioral problems among children with learning disability [37]. Previous PUFAs studies were based on the major involvements of DHA with a small amount of ARA, and reported improvement of ADHD-related behavioral problems [36, 37]. There are still fairly few studies on the effects of larger doses of ARA added to DHA on social impairment in ASD. According to another study, supplementation of omega-3 and omega-6 fatty acids (60mg DHA, 12mg gamma-linoenic acid, 13mg eicosapentaenoic acid and 5mg ARA) and vitamin E.66% over a period of three months reported remarkable improvements in the scores of the Childhood Autism Rating Scale (CARS) [38].

At present, there is a paucity of randomized controlled studies investigating the effects of large doses of ARA added to DHA on the impaired social interaction characteristics of ASD. Yui *et al.* [39-41] conducted a double-blind, randomized placebo- controlled trial to calculate the effects of supplementation including larger doses of ARA in addition to DHA on social and communication impairment in ASD. Autism spectrum disorders (ASD) are neurodevelopmental disorders with defect in neural maturation relating to social interaction. There is little information available about medical cures for core social impairment in ASD. The PUFAS ARA and DHA are likely to play significant roles in brain network maturation. Particularly, ARA actively participates in signal transduction related to neuronal maturation. Supplementing DHA with larger ARA doses may therefore mitigate social impairment. This study is set out with the aim to evaluate the efficacy of supplementing DHA with larger doses of ARA in a double-blind, placebo-controlled 16-week trial, followed by an additional 16-weeks open-label study involving 13 participants with ASD aged 6 to 28 years old (mean age ± SD = 14.6 ± 5.9 years). For examining the relationship between the efficacy of the supplementation regimen and alterations in PUFA levels, plasma levels of PUFAs were calculated. The outcome measures were the Autism Diagnostic Interview-Revised (ADI-R) [39], Social Responsiveness Scale (SRS) and the ABC [39-41]. Repeated measures analysis of variance confirmed the efficacy of this supplementation regimen to have significantly improved SRS-measured communication as well as ABC-measured social withdrawal during the placebo-controlled trial. The treatment effect sizes were more favorable for the treatment group compared with the placebo group (communication: treatment groups, 0.87 *vs.* placebo: 0.44; social withdrawal: treatment groups, 0.88 *vs.* placebo: 0.54) [41]. At the end of the placebo-controlled trial, a significant difference was noticed in the change in plasma ARA levels from the baseline and a trend towards a significant difference in plasma ARA levels between the two groups. As to the mechanism of action of ARA, it is well known that ARA preferentially modulates signal transduction, whereas DHA is an important structural component of neural membranes [32]. By studying the mechanisms underlying the effect of our supplementation a significant difference was reported in the change in transferrin levels as a marker of signal transduction between the two groups. Yui *et al.* reported that supplementation with larger doses of ARA added to DHA helps in reducing social impairment by causing the upregulation of signal transduction [39-41].

Aversive Effects of Arachidonic Acid

In a previous study, the administration of high doses of arachidonate-enriched triglycerides (SUNTAGA40S) as a source of ARA did not exhibit any aversive effects on the health or growth of rats [42, 43]. Infant formula containing ARA presents a significantly lower frequency of adverse events than formula without ARA (Gibson *et al.,* 2009) [44]. Supplementation with 240 mg ARA and 240 mg DHA for a period of three months yielded beneficial effects on the coronary microcirculation without any aversive events in 28 elderly individuals [45]. These previously reported findings indicated that supplementation of same doses of ARA and DHA (240 mg/day ARA and 240 mg/day DHA) did not induce any aversive events.

mGluR ANTAGONISTS AND GABA AGONISTS

Alteration of Synapse Development in ASD

A recent article reviews the implicated synaptic maturation and plasticity (strengthening and weakening of synaptic connections) in the pathogenesis of ASD. The specific pathology of synapse maturation and plasticity during development seen in ASD is likely to result into an imbalance of excitation and inhibition, and specifically a disproportionately high level of excitation. It needs to be noted that the timing of alterations in such synapse development exhibits differential effects on various cortical regions as the timeline for synaptogenesis is different across the cortex [46]. Consistent with the role of altered synapse development in ASD, regions related to language production and social skills in the frontal and prefrontal cortex have a spike in synaptogenesis and plasticity between years 1 and 3 when autistic symptoms related to these processes usually become apparent. The compensatory elevation of inhibitory cell excitability partially rescues social deficits caused by excitation/inhibitory balance elevation [47].

Clinical Effect of GABA$_B$ Antagonist STX209

An open-label study has recently been conducted in 25 boys with ASD aged between 6 -17 years. Their ABC scores were found to be relatively high (>16). Participants were treated with GABA$_B$ antagonist STX209 for a period of 8 weeks. Overall, STX209 was well tolerated. A significant improvement was noted on the ABC-irritability and Social withdrawal subscales. Sixty-three percent of subjects were found to be 'very much improved' or 'much improved' on the Clinical Global Impression-Improvement Scale. Improvements on other measures of social and communicative function were also observed. Thus, STX209 has been found to exhibit broad beneficial effects on core and associated symptoms in ASD [46]. Double-blind, placebo-controlled trials will provide the necessary data to validate these initial findings.

Clinical Effect of Acamprosate: mGluR antagonist and GABA A Agonist

Acamprosate is a mGluR antagonist and GABA A agonist considered likely to impact both gamma-aminobutyric acid and glutamate neurotransmission. According to a preliminary open-label treatment with acamprosate (mean dose is 1,110 mg/day) targeting social impairment in youth with six autism, five of six youth (mean age, 9.5 years) were judged treatment responders to acamprosate over 10 to 30 weeks (mean duration, 20 weeks) of treatment. Acamprosate was well tolerated with only mild gastrointestinal adverse effects noted in three (50%) subjects [52].

OTHER PHSRMACOLOGICAL TREATMENT

Recent pharmacologic approach has mostly been focused at behavioral problems but not directly at the social impairment in ASD. There lies a paucity of randomized controlled studies investigating the effects of medications on social impairment in ASD. Most atypical antipsychotic agents exhibit their efficacy in the treatment of disruptive, impulsive and maladaptive behaviors but they are not found to be effective for treating the core impaired social interaction and communication in ASD [5]. Glutamate antagonists block the metabotropic glutamate receptor (mGluR), thereby reducing excessive excitation, and GAB (γ-aminobutyric acid) agonists that not only maximize the inhibition but also negatively modulate the mGluR (excitatory) pathway [46]. ASD may be caused due to an imbalance in excitation and inhibition. These pharmacological agents that either directly decrease excitation or increase inhibition may exhibit significant therapeutic benefit for behavioral problems as well as social deficits in patients with ASD [46].

Selective Serotonin Reuptake Inhibitors (SSRI)

Selective Serotonin Reuptake Inhibitors (SSRI) are found to be effective to a lesser extent in treating the core deficits of communication and socialization as well as repetitive, irritable, and anxiety symptoms of ASD, but treatment with SSRI is often accompanied with the dilemma of side effects. The most frequently cited side effects of SSRI treatment include behavioral activation (hyperactivity and agitation), aggression, and suicidal ideation [51].

Fluoxetine

Serotonin, a neurotransmitter found throughout the brain and body, has long been of interest in autism. Repeated findings of elevated platelet serotonin levels in approximately one third of children with autism has led some to believe that dysfunctional serotonin signaling may be a causal mechanism for the disorder. The cause of ASD is poorly understood and is most likely multifactorial; however, the serotonergic system may be especially implicated. Subjects with ASD are reported to have a significant decrease in platelet 5-HT_{2A} receptor binding and the 5-HT_{2A} receptor gene is a functional candidate gene in autism [53]. In addition, serotonin dysfunction may be implicated in altered brain activity during facial emotion in relation to social interaction of ASD [54]. Serotonin reuptake inhibitor (SSRI) fluoxetine shows its efficacy in the treatment of the core deficits of communication and socialization in some individuals with ASD [55,

56], however, aversive effects such as hyperactivity and aggression are also reported [55, 56]. An open-label trial of fluoxetine treatment for a period of 13 to 33 months (mean 21 months) in 37 children with ASD (aged between 2 and 7 years), revealed that twenty-two of the 37 children had a beneficial treatment response sustained during continuing treatment, and that eleven had an excellent response and were able to attend mainstream classrooms [55]. Responders showed behavioral, language, cognitive, affective, and social improvements. Responders with adequate testing showed marked increases in language acquisition at every stage of development as compared with (1) pretreatment status, (2) responses to other treatments, (3) ability in non-language (matching) tasks, and (4) historical controls from the literature [55]. The response to fluoxetine strongly correlated with a family history of major affective disorder. Further studies were conducted on 129 children with ASD, aged 2 to 8 years old who were treated with fluoxetine (0.15 to 0.5mg/kg) for 5 to 76 months (mean 32 to 36 months), with discontinuation trials. Out of the total 129 children, 22 (17%) of the children showed an excellent response, and 67 (52%) good response in Childhood Autism Ratings Scale and Autism Diagnostic Observation Schedule [56]. Family history of bipolar disorder and of unusual intellectual achievement was found to be highly correlated. Fluoxetine response, family history of major affective disorder (especially bipolar), unusual achievement, and hyperlexia in the children appear to define a homogeneous autistic subgroup [56].

EMERGING TREATMENT

D-Cycliserine

Important and useful findings on the other pharmacologic treatment in impaired social interaction in ASD have also been reported. A single-blind placebo lead-in phase, three different doses of D-cycloserine (30 mg/day, 50 mg/day, and 85 mg/day) during each of three 2-week periods exhibited significant improvement on social withdrawal subscale of the ABC in 10 subjects with ASD aged between 5 and older [58].

Secretin

Secretin is a gastrointestinal polypeptide used to treat peptic ulcers and helps in the evaluation of pancreatic function (Krishnaswami *et al.*, 2011) [59]. Secretin affects the central nervous system and may function as a neurotransmitter (Krishnaswami *et al.*, 2011) [59]. In an earlier study, secret in infusion (7.5-10 ml/min) for a period of 5 weeks in three children with ASD exhibited a miraculous improvement in their behavior, manifested by improved eye contact, alertness, and expansion of expressive language. These clinical observations suggest an association between gastrointestinal and brain function in patients with autistic behavior. However, the findings in seven randomized controlled trials showed a lack of effectiveness of secretin for the treatment of ASD symptoms such as the language and communication impairment, symptom severity, and cognitive and social skill deficits [59].

Aripiprazole

Aripiprazole is a novel antipsychotic with potent partial agonism at dopamine D2 receptors in addition to properties as a 5HT1A agonist and 5HT2 antagonist [48]. A previous retrospective study reported that aripiprazole (mean doses 8.1 ± 4.9 mg/day) was associated with a significant improvement in the Children's Global Assessment Scale and the Autism Rating Scale in 34 children and young adolescents aged 10.2 ± 3.3 years [49]. Another study found that during an 8-week period in a double-blind, randomized, placebo-controlled, parallel-group study aripiprazole (target dosage: 5, 10, or 15 mg/day) in children and adolescents with ASD (n=98) (aged 6–17 years), aripiprazole treatment group (n=47) demonstrated significantly greater global improvement in the Clinical Global Impression-Improvement score in children and adolescents with ASD compared to placebo treated subjects (n=51). Extrapyramidal symptom rates were reported to be 14.9% for ariprazole and 8.0% for placebo. No serious aversive events were observed.

FUTURE DRUG TARGETS

Neuropeptides play a crucial role in the normal function of the central nervous system and peptide receptors are considered highly promising as therapeutic targets for the treatment of several CNS disorders [57].

Neurotensin

Neurotensin is known to intensify neuronal NMDA-mediated glutamate signaling, which may cause apoptosis in autism. Furthermore, an imbalance of glutamate/GABAergic system in ASD has been described. These observations lead to a postulate that neurotensin may accentuate the hyperglutaminergic state in ASD. Targeting neurotensin might be a possible novel approach for the treatment of autism [60]. Since peptides lack drug like properties, the development of their therapeutics has been limited. Some of these challenges include poor *in vivo* stability, poor solubility and incompatibility with oral administration, shelf stability, and cost of manufacturing [58].

CONCLUSIONS

ASD is characterized by impaired social interaction, socialization, and restrictive and repetitive behaviors. Improvements brought over time in the frequency of these disorders (to present rates of about 1 case per 100 children) may be attributed to the factors such as new administrative classifications, increased awareness, and early identification. However a number of treatments might be available for the core social impairment, only a few medical remedies are available for the impaired social interaction. It is worth noticing that risperidone solution, intranasal administration of oxytocin, and dietary supplementation with arachidonic acid have validated their efficacy in improving the impaired social interaction. The pharmacological agents that either directly decrease excitation or increase inhibition (*e.g.,* mGluR antagonists and GABA and GABA agonists) may have major therapeutic benefit for behavioral problems as well as social deficits in patients with ASD. Indeed, GABA$_B$ antagonist STX209 has been found to improve the ABC-irritability and Social withdrawal subscales. In addition, atypical antipsychotics ariprazole and SSRI fluoxetine were found to be useful in treating some aspects of social relatedness or the core deficits of communication and socialization. Anti-androgenic medication leuprolide acetate administration significantly helped to ameliorate clinical symptoms/behaviors of hyperandrogenemia in some subjects with ASD. D-cycloserine exhibited significant improvement on social withdrawal subscale of the ABC in some subjects with ASD. Previous review articles supported a lack of effectiveness of secretin for the treatment of ASD symptoms including language and communication impairment, symptom severity, and cognitive and social skill deficits It is hoped that introduction of improved strategies for early identification with phenotypic characteristics and biological markers (*e.g.,* brain physiological and biochemical changes) would remarkably maximize the effectiveness of treatment. The evaluation of treatments for ASD should be directed at neurobiological targets known to be important in the brain's response to abnormal developmental trajectories or toward enhancing plasticity during the periods of opportunity. Further research about neurobiology and effective treatments of ASD is required.

ACKNOWLEDGEMENTS

None Declared.

CONFLICT OF INTEREST

The authors confirm that this article content has no conflicts of interest.

REFERENCES

[1]   American Psychiatric Association. Diagnostic and Statistical Manual of Mental Disorders, 4th ed, Text Revision, Washington, DC: American Psychiatric Association 2000.

[2]   Ecker C. Translastional approached to the biology of autism: false dawn or a new era. Mol Psychiatry 2012: doi:10.1038/mp.2012.102.

[3]   West L, Waldrop J, Brunssen S. Pharmacologic treatment for the core deficits and associated symptoms of autism in children. J Pediatr health Care 2009; 23:75-89.

[4]   McPheeters ML, Warren Z, Sathe N, Bruzek JL, Krishnaswami S, Jerome RN *et al.* A systematic review of medical treatments for children with autism spectrum disorders. Pediatrics 2011; 127: e1312-e21.

[5]   Research Units on Pediatric Psychopharmacology Autism Network. Effects of risperidone and parent training on adaptive functioning in children with pervasive developmental disorders and serious behavioral problems. J Am Acad Child Adolesc 2012 51:136-46.

[6]   McDougle CJ, Scahill L, Aman MG, *et al*, Risperidone for the core symptom domains of autism: results from the study by the autism network of the research units on pediatric psychopharmacology. Am J Psychiatry 2005; 162: 1142-48.

[7]   Heemastra LB, Finnin BC, Nicolazzo, JA. The buccal mucosa as an alternative route for the systemic delivery of risperidone. J Pharmacoeutical Sci 2010; 99: 4585-92.

[8]   Nagaraj R, Singhi P, Malhi P. Risperidone in children with autism: randomized, placebo-controlled, double-blind study. J Child Neurol 2006; 21: 450-55.

[9]   Shea S, Turgay A, Carroll A, *et al.* Risperidone in the treatment of disruptive behavioral symptoms in children with autistic and other pervasive developmental disorders. Pediatrics 2004; 114: e634-e41.

[10]   Harris D, Robinson JR. Drug delivery *via* the mucous membranes of the oral cavity. J Pharm Sci, 81: 1-10, 1992.

[11]   Lim MM, Bielsky IF, Young LJ. Neuropeptides and the social brain: potential rodent models of autism. Int J Devl Neuroscience 2005; 23: 235-43.

[12]   Green JJ, Hollander E. Autism and oxytocin: new developments in translational approaches to therapeutics. Neurotherapeutics 2010; 7: 250-57.

[13]   Kirsch P, Esslinger C, Chen Q, et al. Oxytocin modulates neuronal circuitry for social cognition and fear in human. J Neurosci 2005; 11489-93.

[14]   Domes G, Heinrichs M, Michel A, Berger C, Herpertz SC. Oxytocin improves "Mind-Reading" in Humans. Biol Psychiatry 2007, 61: 731-33.

[15]   Petrovic P, Kalisch R, Singer T, Dolan R. Oxytocin attenuates affective evaluations of conditioned faces and amygdale activity. J Neurosci 2008; 28: 6607-15.

[16]   Lerer E, Levi S, Salomon S, Darvasi A, N Yirmiya N, Ebstein RP. Association between the oxytocin receptor (*OXTR*) gene and autism: relationship to Vineland Adaptive Behavior Scales and cognition. Mol Psychiatry 2008; 13: 980-988.

[17]   Liu X, Kawamura Y, Shimada T, Otowa T, Koishi S, Sugiyama T. et al. Association of the oxytocin receptor (OXTR) gene polymorphisms with autism spectrum disorder (ASD) in the Japanese population. Human Genet 2010; 55:137-141.

[18]   Pobbe RLH, Pearson BL, Defensor EB, Bolivar VJ, W. Scott Young III WS, Heon-Jin Lee HJ et al. Oxytocin receptor knockout mice display deficits in the expression of autism-related behaviors. Horm Behav 2012; 61: 436-444.

[19]   Campbell DB, Datta D, Jones ST, Lee EB, Sutcliffe JS, Hammock EAD Pat Levitt P. Association of oxytocin receptor (OXTR) gene variants with multiple phenotype domains of autism spectrum disorder. J Neurodev Disord. 2011; 3: 101–112.

[20]   Kosfeld M, Heinrichc M, Zak PJ, Fischbacher U Fehr E: Oxytocin increase trusts in humans. Nature 2005, 435:673-76.

[21]   Hollander E, Bartz J, Chaplin W, et al. Oxytocin increases retention of social cognition in autism. Biol Psychiatry2007; 61: 498-503.

[22]   Guastella AJ, Einfeld SL, Gray KM, et al. Intranasal oxytocin improves emotion recognition or youth with autism spectrum disorders. Biol Psychiatry 2010; 67: 692-94.

[23]   Petrovic P, Kalisch R, Singer T, Dolan R. Oxytocin attenuates affective evaluations of conditioned faces and amygdale activity. J Neurosci 2008; 25: 6607-15.

[24]   Andari E, Duhamel J-R, Zalla T, Herbrecht E, Leboyer M, Sirigu A. Promoting social behavior with oxytocin in high-functioning autism spectrum disorders. PANS 2010; 107: 4389-94.

[25]   Domes G, Heinrichs M, Glascher J, Buchel C, Braus DF, Herpertz SC. Oxytocin attenuates amygdale responses to emotional faces regardless of valence. Biol Psychiatry 2007a, 62: 1187-90.

[26]   Anagnostou E, Sorya L, Chaplin W, Bartz J, Halpern D, Wasserman S. et al Intranasal oxytocin *versus* placebo in the treatment of adults with autism spectrum disorders: a randomized controlled trial. Mol Autism 2012; 3: 16. doi: 10.1186/2040-2392-3-16.

[27]   Naber FBA, Poslawsky IE, Jzendoorn NH, van Engeland H, and Marian J. Bakermans-Kranenburg MJ. Oxytocin enhances paternal sensitivity to a child with autism: a double-blind within-subject experiment with intranasally administered oxytocin. J Autism Dev Disord. 2013; 43: 224–229.

[28]   Vancassel S, Durand G, Barthelemy C,Martineau J, Guilloteau D, et al. Plasma fatty acid levels in autistic children. Prostagrandins Leukot Essent Fatty Acids 2001; 65:1-7.

[29]   Schuchardt JP, Huss M, Stauss-Grado M, Hahn A. Significance of long-chain polyunsaturated fatty acids (PUFAs) for the development and behaviour of children. Eur J Pediatr 2010; 169: 149-164.

[30]   Tamiji J, Crawford DA. The neurobiology of lipid metabolism in autism spectrum disorders. Neurosignals 2010; 18: 98-112.

[31]   Calder PC. Mechanisms of Action of (n-3) Fatty Acids. J Nutr 2012; 142: 592S-599S.

[32]   Kurlak LO and Stephenson TI, Arch Dis Fetal Neonatal Ed 1999; 80: F148-54.

[33]   Zhao J, Del Bigio MR, Weiler HA. Maternal arachidonic acid supplementation improves neurodevelopment in youth adult offspring from rat dams with and without diabetes. Prostaglandins Leukot Essent Fatty Acids 2010, 84: 63-70.

[34]   Lien VW, Clandinin MT. Dietary assessment of arahidonic acid and docosahexaenoic acid intake in 4-7 year-old children. J Am Cell Nutr 2009; 28: 7-15.

[35]   Birch EE, Garfield S, Hoffman DR, Uauy R, Birch DG. A randomized controlled trial of early dietary supply of long-chain polyunsaturated fatty acids and mental development in term infants. Developmental Medicine & Child Neurology 2000; 42: 174-81.

[36]   Stevens L, Zhang W, Peck L, Kuczek T, Grevstand N, Mahon A, Zentall SS, Arnold LE, Burgess JR. EFA supplementation in children with inattention, hyperactivity, and other disruptive behaviors. Lipids 2002; 38, 1007-21.

[37]   Richardson AJ, Puri BK. A randomized double-blind, placebo- controlled study of the effects of supplementation with highly unsaturated fatty acids on ADHD-related symptoms in children with specific leaning difficulties. Prog Neuro-Psychopharmacol Biol Psychiatry 2002; 26: 233-39.

[38]   Meguid NA, Atta JM, Gouda AS, Khalil RO. Role of polyunsaturated fatty acids in the management of Ezyptian children with autism. Clinical Biochem 2008, 41: 1044-48.

[39]   Yui k, Koshiba M, Nakamura S, Ohnishi M. Efficacy of adding large doses of arachidonic acid to docosahexaenoic acid against restricted repetitive behaviors in individuals with autism spectrum disorders: a placebo-controlled trial. J Addict Res Ther 2011, S4. http://dx.doi.org/10.4172/2155- 6105.S4-006.

[40]   Yui k, Koshiba M, Nakamura S, Kobayashi Y. Effects of large doses of arachidonic acid added to docosahexaenoicacid on social impairment and its underlying mechanism in youth with autism spectrum disorders: a double-blind, placebo-controlled randomized trial. J Clin Psychopharmacol 2012; 32: 200-06.

[41]   Yui k, Koshiba M, Nakamura S. Neurobiological bases and pharmacologic treatment of social impairment in autism spectrum disorders. Curr Psychopharmacol 2012; 1: 233-46.

[42]   Lina BAR, Wolterbeek APM, Suwa Y, Fujikaea S, Ishikura Y, Tsuda S, Dohnalek M. Subchronic (13-week) oral toxicity study, preceded by an in utero exposure phase, with arachidonate-enriched triglyceride oil (SUNTAGA40A) in rats. Food and Chemical Toxicology 2006; 44: 326-35.

[43]   Nisha A, Muthukumar SP, Venkateswaran G. Safety evaluation of arachidonic acid rich Mortierella alpine biomass in albino rats: a subchronic study. Regulatory Toxicology and Pharmacology 2006; 53: 186-94.

[44]   Gibson RA, Barclay D, Marshall H, Moulin J, Maire J-C, Makrides M. Safety of supplementating infant formula with long-chain polyunsatured fatty acids and BIfidobacterium lactis in term infants: a randomized controlled trial. Br J Nutr 2009; 101: 1706-13.

[45]   Oe H, Hozumi T, Murata E, *et al.* Arachidonic acid and docosahexaenoic acid supplementation increases coronary velocity reserve in Japanese elderly individuals. Heart 2008; 94: 316-21.

[46]   Oberman N. nGluR antagonists and GABA agonists as novel pharmacological agents for the treatment of autism spectrum disorders. Expert Opin Investig Drugs 2012; 26 [Epub ahead of print].

[47]   Yizhar O, Fenno LE, Prigge M, Schneider F, Davidson TJ, O'Shea DJ, *et al.* Neocortical excitation/inhibition balance in informing processing and social dysfunction. Nature 2011; 477: 171-8.

[48]   Burris KD, Molski TF, Xu C, Ryan E, Tottori K, Kikuchi T, Yocca FD, Molinoff PB. Aripiprazole, a novel antipsychotic, is a high-affinity partial agonist at human dopamine D2 receptors. J Pharmacol Exp Ther. 2002; 302:381-9.

[49]   Masi G, Cosenza A, Millepiedi S, Muratori F, Pari C, Salvadori F. Aripiprazole monotherapy in children and young adolescents with pervasive developmental disorders: a retrospective study. CNS Drugs 2009; 23511-23521.

[50]   Owem R, Sikich L, Marcus RN, Corey-Lisle P, Manos G, McQuade RD, *et al.* Aripirazole in the treatment of irritability in children and adolescents with autistic disorder. Pediatrics 2009; 124: 1533-41.

[51]   Wink LK, Erikson CA, McDougle CJ. Pharmacologic treatment of behavioral symptoms associated with autism and other perspective developmental disorders. Curr Treat Potions Neurol 2010; 12: 529-538.

[52]   Erickson CA, Early M, Stigler KA, Wink LK, Mullett JE, McDougle CJ. An open-label naturalistic pilot study of acamprosate in youth with autistic disorder. J Child Adolesc Psychopharmacol. 2011 Dec;21(6):565-9. Epub 2011 Dec 2.

[53]   Murphy D, Daly EM, Schmitz, N, Toal F, Murphy K, Curran S. *et al.* Cortical Serotonin 5-HT$_{2A}$ Receptor Binding and Social Communication in Adults With Asperger's Syndrome: An *in Vivo* SPECT Study. Am J Psychiatry 2006; 163: 934-936.

[54]   Daly EM, Deeley Q, Ecker C, Craig M, Hallahan B, Murphy C. *et al.* Serotonin and the neural processing of facial emotions in adults with autism. Arch Gen Psychiatry 2012; 69: 1003-1013.

[55]   DeLong GR, Teague LA, McSwain Kamran M. Effects of fluoxetine treatment in young children with idiopathic autism. Dev Med Child Neurol. 1998; 40: 551-62.

[56]   DeLong GR, Ritch CR, Burch S. Fluoxetine response in children with autistic spectrum disorders: correlation with familial major affective disorder and intellectual achievement. Dev Med Child Neurol 2002; 44:652-59.

[57]   McGonigle P. Peptide therapeutics for CNS indications. Biochem Pharmacol 2012; 83: 559-66.

[58]   Posey D, Kem DL, Swiezy N, Sweeten TL, Wiegand RE, McDougle CJ. A pilot study of D-cycloserine in subjects with autistic disorders. Am J Psychiatry 2001; 161: 2115-17.

[59]   Krishnaswami S, McPheeters ML, Veenstra-VanderWeele J. A systemic review of secretin for children with autism spectrum disorders. Pediatrics, 2011; 127: e1311-25.

[60]   Ghanizadeh A. Targeting neurotensin as a potential novel approach for the treatment of autism. J Neuroinflammation 2010, 7: 58-9.

CHAPTER 9

Pharmacological Treatments, Related Clinical Characteristics and Brain Function in Adolescent Depression

Kunio Yui[*]

Research Institute of Pervasive Developmental Disorders, Ashiya University, Graduate School of Education, Hyogo 659-8511, Japan

Abstract: Major depressive disorder (MDD) in adolescents, which tends to be a particularly malignant and intractable condition, somatic condition, increases the likelihood of recurrence and is a major cause of suicide attempt and death, and chronicity in adulthood. Increases in alpha- or theta- band activity, or asymmetries in the alpha band may predict successful responses to treatment with selective serotonin uptake inhibitors. Reward-related brain function, such as greater reactivity in the lower medial prefrontal cortex, and greater right-side frontal brain activity predict MDD symptoms. Greater activation of both the amygdale, related to functional connections, has been reported as the neurobiological bases of adolescent MDD. An increased imbalance of resting-state brain activity between the frontal cognitive control system and the limbic-striatal emotional processing system was recognized. Inherited risks, such as developmental factors   and psychosocial adversity, interact to increase the risk of depression through hormonal factors and the associated perturbation of the relevant neural pathway. These multiple complex pathophysiological factors might contribute to the development of a particularly malignant and intractable adolescent depression, and cause increased likelihood of recurrence and chronicity in adulthood,

Daily doses of escitalopram and fluoxethine (10-20 mg) have been demonstrated as effective in treating adolescents with MDD. However, a recent review article has proposed that escitalopram should be considered as a second-line treatment option for adolescents with MDD. It should be noted that escitalopram treatment has been recommended on the basis of a single study of positive behavioral cognitive therapy because of lack of evidence.

Keywords: Adolescent depression, somatic symptoms, suicide, recurrent depression, emotional processing system, Amygdale function, escitaoroprum, fluoxethine, behavioral cognitive therapy.

INTRODCTION

During adolescence emotional responses are very strong, while cognitive self-control

Address correspondence to Kunio Yui: Research Institute of Pervasive Developmental Disorders, Ashiya University Graduate School of Education, Hyogo 659-8511, Japan; Tel: 81-797-23-0661; Fax: 81-797-23-1901; E-mail: yui16@bell.ocn.ne.jp

Atta-ur-Rahman (Ed)

and decision-making strategies still are developing, resulting in uncontrollable mood swing, impulsivity, and emotional conflict in social interaction [1]. This predisposition may be related to highly prevalence of suicide attempt in adolescents with depression [1]. Adolescence is referred to by both the terms "teenage years" and "puberty". Puberty refers to the hormonal changes that occur in early youth, yet the period of adolescence can extend well beyond the teenage years [1, 2]. Many researchers and developmental specialists use the age span of 10-24 years as a working definition of adolescence [2]. In this section, an update of currently available pharmacologic treatments and their relevant to clinical characteristics, and brain function in major depressive disorder (MDD) of adolescence aged 10-24 years will be discussed.

MDD is a serious public health concern for children and adolescents because of its frequency and severity [3, 4]. MDD tends to be a particularly malignant and intractable condition and increases the likelihood of recurrence and chronicity in adulthood [4]. The point prevalence of MDD in adolescents ranges from 2-8 % of children and adolescents [5]. One-third of affected children will make a suicide attempt, and 3%-4% will die from suicide [5]; thus, depression is a major cause of death in adolescents [6] (Breton 20). Several somatic symptoms concurrent with adolescent depression are strongly linked to later high rates of recurrent depression, suicide attempts, bipolar disorder, psychotic disorders, post-traumatic stress disorder, recurrent depression, and chronic depression [7, 8]. Thus, effective treatment guidelines are needed for patients with somatic symptoms. Because adolescent MDD appears to be highly prevalent, it is essential that efficacious pharmacologic treatments are identified and that effective pharmacologic treatment strategies are established that best alleviate depressive symptoms in adolescents. Therefore, this article presents an overview of pharmacological treatment and related clinical characteristics and brain function of MDD in adolescents.

CLINICAL CHARACTERRISTICS OF ADOLESCENT DEPRESDION

In general, irritability appears to be more prominent than sadness in adolescents. Adolescent depression occurs more often in combination with other psychopathologies, such as anxiety, conduct problems and learning disabilities

[8]. Furthermore, adolescents suffering from depression have a 40% chance of experiencing a recurrent episode later in life [8]. A previous clinical study on the outcome of adolescent MDD reported that characteristics of depression generally predicted the outcome better than co-morbidity [9]. Longer time to recovery was predicted by earlier lifetime age at onset for depression, poor psychosocial functioning, depressive disorder diagnosis, and longer episode duration by study entry [9]. The negative impacts of parental mental health problems on adolescents are well known. For example, a previous study has reported a significant association between parental mental health and adolescent depressive symptoms; it is thus important to considering the mental health of the parents when treating depressed adolescents [4].

PROGNOSTIC SIGNIFICANCE

To assess the prognostic value of clinical symptoms, a long-term longitudinal study screened a total of 609 populations 16-17 year olds for depression over a 15 years period [7]. The study found that the number of concurrent somatic symptoms predicted adverse adult mental health outcomes in a stepwise manner [7]. The quarter of the depressed adolescents that had the most somatic symptoms subsequently developed recurrent depression (68%), panic disorder (44%), chronic depression (30%), somatoform disorders (26%), bipolar disorder (22%), suicide attempts (16%), and psychotic disorders (8%) [7]. Abdominal pain was a strong independent predictor for depression and anxiety [7]. Adolescents who reported more intense and labile emotions and less effective regulation of these emotions also reported more depressive symptoms and problem behaviors in comparison with adolescents who reported fewer symptoms of MDD [10].

Other sociological or biological factors may affect prognosis of adolescent depression have been reported. The sample, which included 3,732 members of a prospective Canadian cohort study, found that low birth weight may cause an increased risk of depression in the face of chronic stress [11]. In a recent review, the use of electroencephalographic (EEG) techniques allowed the detection of cortical activity [12]. This activity directly reflects subcortical neurotransmitter systems, and antidepressant medications generally target these systems [12]. Increases of alpha or theta band activity, or asymmetry in the alpha band characterized by left lateralization, may predict successful response to treatment with serotonin reuptake inhibitors (SSRI) [12]. EEG derived biomarkers are predictive for treatment response to serotonergic agents [12]. A previous cross-sectional and longitudinal study using the Amsterdam Neurological Tasks Program, which included the Baseline Speed task, the Sustained Attention dot patterns task, the Visual Attention Set Shifting task and the memory Search Letter task, examined the relationship between neurocognitive functioning and affective problems during adolescence in a cohort of 2,179 adolescents (age 10-12 years) [13]. Longitudinally, decreased response time inhibition predicted the presence of affective problems after 5 years follow-up in girls, but not boys [13]. Interestingly, reward-related brain function, such as greater reactivity and lower medial prefrontal cortex reactivity to a monetary reward task before treatment with cognitive behavioral therapy or cognitive behavioral therapy plus SSRIs, predicted both the severity of symptoms at post-treatment and the rate of anxiety symptom reduction during treatment [14]. In addition, reduction in the response to monetary gain was associated with increased depressive symptoms [15]. Greater right-side frontal brain activity has been reported to predict depressive symptoms in 41 adolescent boys one year before the onset of the symptoms [16].

BRAIN FUNCTION IN ADOLESCENT DEPRESION

Threat-related emotional expression evokes abnormal neural activity in young patients with depression [17]. Adolescents with early−onset depressive disorders showed significantly greater left cerebral activity (including amygdala, thalamus, prefrontal and temporal cortex), but significantly reduced activity in the right prefrontal cortex in response to threatening facial expressions; these suggest that there is abnormal activation in the nodes along fronto-limbic pathways in threat-related faces processing [17].

In a recent functional MRI study, 19 depressed and 21 healthy adolescents between the ages of 11-18 years, underwent functional MRI assessment in response to facial expressions [3]. Compared to healthy controls, depressed youths have exaggerated brain activation in multiple regions, including the frontal, temporal, and limbic cortices [3]. Depressed youths had significantly greater bilateral activations of the amygdale, orbitofrontal cortex, and subgenual anterior cingulate cortex in response to fearful facial expression relative to neutral facial expressions when compared to healthy subjects [3]. The MRI scan at the 8 week time point revealed that fluoxetine treatment decreased activations of all three regions of depressed subjects compared to healthy subjects [3]. These findings suggested that increases in brain activation in untreated depressed adolescents, and demonstrated that these aberrant activations were reduced with treatment [3]. This finding represents an initial step in support of using the functional MRI approach to illustrate the relevant circuitry of adolescent depression and uncover potential mechanisms of treatment [18]. The next step will involve further study to determine the specificity of these changes resulting from treatment. In particular, it will be necessary to characterize the overlapping and complementary mechanisms across pharmacologic treatment [18]. Moreover, the prefrontal region and the amygdala may be dysregulated in the resting networks of depressed adolescents, which may mediate emotion processing in their daily lives [19].

In reference to social reward and its relation to brain function, a developmental approach to investigating social reward function in adolescent depression can elucidate the etiology, pathophysiology and course of depression [20]. A pattern of low striatal response and high medial prefrontal response to reward is evident in adolescents [20]. Low social reward function may also be a stable characteristic of people who experience depression [20]. One possible mechanisms underlying altered social reward function in adolescent MDD could involve a disrupted balance in corticostriatal circuit function, with the disruption occurring as the aberrant adolescent brain develops [20]. An increased imbalance of resting-state brain activity between the frontal cognitive control system and the (para) limbic-striatal emotional processing system was recognized in adolescents with MDD [21].

Hypothalamo-pituitary-adrenal axis (HPA) dysregulation is evident in acute stress responses and might result from exposure to chronic early stressors during early childhood [22]. Given the importance of adolescence in the development and recurrence of depressive phenomena over the lifespan, an integrative perspective is important in research investigating the various components of HPA axis functioning among depressed young people [23]. During adolescence, the changes in sex steroid levels induce alterations in neurotransmitter systems, such as the 5-HT system, and alterations in neurotransmitters systems can potently affect mood and behavior. Therefore, changes in sex steroid levels occurring during adolescence may increase vulnerability to depression [8]. In particular, the effect of sex steroids on the maturing HPA makes girls more sensitive to the effects of stress, whereas androgens appear to play a protective role in boys [8]. The combination of genetic predisposition and / or psychosocial factors may more easily trigger depression in girls [8]. Thus, the greater prevalence of depression in adolescent girls likely results from a combination of profound hormonal changes, and psychosocial factors [8]. Whether similar underlying mechanisms are involved in the emergence of sex differences in human psychopathology remains to be studied, and longitudinal human studies in both sexes are required to elucidate a relationship between biological changes during adolescence and the emergence of psychopathology [8].

To summarize, inherited risks, such as developmental factors, sex hormones, and psychosocial adversity, interact to increase the risk of depression through hormonal factors and the associated perturbation of the relevant neural pathways. These multiple complex pathophysiological factors might contribute to the development of a particularly malignant and intractable adolescent depression, and cause increased likelihood of recurrence and chronicity in adulthood, increased DSM-IV Axis II comorbidities and a high rate of suicide attempts in adolescent depression. Effective treatments are available, but the treatment choices are dependent on the severity of the MDD and on the available resources. Prevention strategies targeted at high-risk groups are promising.

UPDATE ON PHARMACOLOGIC TREATMENT

Pharmacological treatment data of depression in adolescents is controversial for several reasons: 1) The accepted practices and clinical guidelines vary in different countries; 2) There are concerns about the use of antidepressant drugs in patients younger than 18 years; and 3) Some recommendations are based on consensus rather than on evidence [23]. In addition, the evidence relates to the short-term effectiveness of psychological treatments and medication. Evidence for the long-term effects of treatment on rates of recurrence and the effectiveness of non-specialist interventions is scarce [24]. A major difference between adult and adolescent depression is the response to pharmacotherapy. Tricyclic antidepressants do not appear to be clinically effective in adolescents with MDD due to the lack of a demonstrable benefit and the possibility of harmful side effects [23]. Only one pharmacologic agent, the SSRI fluoxetine, is currently approved in the United States to treat depression in patients under the age of 18 [23]. Because no current medication adequately treats adolescent depression, there is a need for new treatment modalities. A major problem in finding effective treatments for young adolescents with MDD is that the drugs used to treat depression were first developed for adults and are necessarily effective in adolescents [23]. Young adolescents are not just small adults. It is certainly possible that eventually new treatments could be found that are efficacious in adolescent depression, even if they are not useful in adult depression [23].

A recent review on treatment of adolescent MDD mainly addressed cognitive behavioral therapy and interpersonal psychotherapy [24]. Cognitive behavioral therapy has emerged as well-established treatment approach for adolescents. While the number of studies on the effectiveness of cognitive behavioral therapy for depression has increased, there is little evidence-based information indicating how or why these treatments act on [25]. Training adolescents in specific coping skills and affect regulation techniques that can be applied to thoughts and behaviors shows some initial promise. However, future trials are necessary to determine the best practices for treating this high-risk population [25]. The findings from future research of cognitive behavioral therapy may provide insight into how to use this technique to promote treatment and homework compliance, reduce the dropout risk, or suggest ways to help adolescents identify and challenge depressogenic cognitions (*e.g.*, related to peer or parental stressors) [26].

Escitalopram is a SSRI; and it is the second antidepressant to be approved for use in treating MDD in adolescent patients aged 12-17 years [27]. The other antidepressant, fluoxetine, is US Food and Drug Administration approved for acute and maintenance treatment of pediatric and adolescent depression, in patients 8 to 18 years of age [28]. Despite requests to restrict adolescent access to medications, medical management of adolescent MDD with fluoxetine, including careful monitoring for adverse events, should be made widely available, and should not be discouraged. A previous randomized controlled trial in a volunteer sample of 439 adolescents aged 12-17 years with a primary diagnosis of MDD has been reported the effects of cognitive behavioral therapy combined with fluoxetine [29]. The participant received 12 weeks of cognitive behavioral therapy alone, fluoxetine (10-40 mg/day) with cognitive behavioral therapy or fluoxetine (10-40 mg/day) alone [29]. Cognitive behavioral therapy produced greater improvements than did fluoxetine alone [29]. There were no significant differences in frequency of adverse effects between fluoxetine treatment and cognitive behavioral therapy [29]. Another randomized controlled trial examined the difference in clinical efficacy between an SSRI and an SSRI combined with cognitive behavioral therapy [30]. Adolescents with treatment resistant depression, who had previously not responded to a trial of an adequate dose of an SSRI, were randomly assigned to one of the following treatment: another SSRI, venlafaxine, venlafaxine + cognitive behavior therapy, or venlafaxine + cognitive behavioral therapy [30]. Adolescents who ended the 12-week treatment during their summer vacation were more likely to show an adequate response than those ending the treatment while being in school [31]. To examine the recovery rates and remission rates probabilities in response to fluoxetine treatment, a previous study conducted a multisite clinical trial, which randomly assigned 439 adolescents with MDD to 12 weeks of treatment with either fluoxetine, cognitive behavioral therapy, combination of luoxetine and cognitive behavioral therapy, or a placebo [32]. At week 36, the estimated remission rates for intention-to-treat cases were as follows: combination treatment, 60%; fluoxetine, 55%; cognitive–behavioral therapy, 64%; and overall, 60%. There were no significant between-treatment differences in recovery rates at week 36 [32]. Most of the depressed adolescents from all three treatment modalities achieved remission at the end of 9 months of treatment [32]. These studies indicated the SSRI was not sufficiently efficient in the treatment of adolescent MDD.

Another randomized controlled trial studied of 334 patients aged 12-18 years who had a primary diagnosis of MDD and had not responded to a 2-month initial treatment with fluoxetine or other SSRIs (*e.g.*, paroxetime, citalopram, or sertraline) [33]. In these patients, the combination of cognitive behavioral therapy and a switch to another antidepressant (venlafaxine, paroxetine, citalopram, or fluoxethine) resulted in a higher rate of clinical response compared to medication alone [33]. During treatment with medication, there was a greater increase in pulse rate (average, 6.0/min) and weight gain (average, 2.9 kg), and a higher frequently of skin problems (about 20 %) [33]. A recent 8-week, randomized, placebo-controlled trial of fluoxetine (10 mg for the first 4 weeks of the treatment and 10-20 mg for the latter 4 weeks of treatment) in a total of 34 adolescents (ages 12-17 years) with comorbid depression and substance use disorder did not detect a significant antidepressant treatment effect [30]. Subjects who had chronic MDD at baseline or no more than moderate alcohol use during the trial showed significantly greater reduced in depression symptoms in response to fluoxethine treatment when compared to placebo-assigned subgroups [30]. In this study, youth with chronic depression and no more than moderate alcohol consumption are likely to better respond to treatment with fluoxetine compared with placebo than youth with transient depression and heavy alcohol use. This study suggests that adolescents with comorbid MDD and substance abuse disorders, chronic depression at baseline, and no more than moderate alcohol consumption during therapy, will obtain greater relief from depression with fluoxetine compared with placebo [30].

A previous review article discussed the efficacy of escitalopram [27]. In a randomized, double-blind clinical trial, a daily doses of 10-20 mg escitalopram (n=154) for 8 weeks was significantly better than placebo (n=157) in improving the severity of depressive symptoms in adolescents with MDD [27]. Daily doses of 10-20 mg/day escitalopram showed better efficacy than placebo for some secondary endpoints (*e.g.,* the change in the Clinical Global Impression (CGI)-Severity score, and the CGI-Improvement response rate) but not others (*e.g.,* CDRS-R response rate, and the rate of remission) [27]. Daily doses of 10-20 mg escitalopram for 8 weeks were generally well tolerated in clinical trials in involving adolescent or pediatric patients with MDD. The incidence of suicidality-related adverse events was generally similar in patients receiving escitalopram compared to placebo [27]. In a recent clinical trial, a total of 259 with MDD aged 12-17 years were randomly assigned to 8 weeks of double-blind treatment with 10-20 mg/day escitalopram or placebo [34]. Significant improvement was observed in the escitalopram group relative to the placebo group at the endpoint of the study [34]). Adverse events occurring in at least 10% of patients treated with escitalopram were headache, menstrual cramps, insomnia, and nausea; influenza-like symptoms occurred in at least 5% of escitalopram patients, and this finding was at least twice as frequent as that observed in patients treated with placebo (7.1% *vs.* 3.2%) [34]. This study concluded that escitalopram was effective and well tolerated in the treatment of depressed adolescents.

To summarize, escitalopram as well as fluoxethine was based on a single positive behavioral cognitive therapy.

In addition, deep brain stimulation, which is a neuromodulation therapy that has been used successfully in the treatment of symptoms associated with movement disorders, has recently undergone clinical trials for individuals suffering from treatment-resistant depression. The published researches demonstrated significant reductions in depressive symptomatology and high rates of remission in a severely treatment-resistant patient group. Despite these encouraging results, the mechanisms of action are understood [35]. Propsonosed mechanisms of action include short and long-term local effects of deep brain stimulation at the neuronal level that modulate neural network activity.

ACKNOWLEDGEMENTS

None declared.

CONFLICT OF INTEREST

The authors confirm that this article content has no conflicts of interest.

REFERENCES

[1]   Goriounova MA, Mansvelder HD. Nicotine exposure during adolescence alters the rules for prefrontal cortical synaptic plasticity during adulthood. Front Synaptic Neurosci Epub 2012 Aug 2.
[2]   Kaplan, P. Adolescence. Houghton Mifflin Company, Boston; 2004.
[3]   Toa R, Calley CS, Hart J, Mayes TL, Nakanezny PA, Lu H. *et al,* Brain activity in adolescent major depressive disorder before and after fluoxetine treatment. Am J Psychiatry 2012; 168: 381-88.

[4]   Wilkinson PO, Harris C, Kelvin R, Dubicka B, Goodyer IM. Associations between adolescent depression and parent mental health, before and after treatment of adolescent depression. Eur Child Adolec Psychiatry 2012; doi:10.1007/s00787-012-0310-9.

[5]   Hazell P. Depression in children and adolescents. Clin Evid (online) 2011; Oct 21; 2011, pii:1008.

[6]   Breton JJ, Labelle R, Huynh C, Berthiaume C, St-Georges M, Guilé JM. Clinical characteristics of depressed youths in child psychiatry. J Can Acad Child Adolesc Psychiatry 2012; 21: 16-29.

[7]   Bohman H, Jonsson U, Päären A, von Knorring L, Olsson G, von Knorring AL. Prognostic significance of functional somatic symptoms in adolescence: a 15-year community-based follow-up study of adolescents with depression compared with healthy peers. BMC Psychiatry 2012; 12: 90.

[8]   Naninck EFG, Lucassen PJ, Bakker J. Sex differences in adolescent depression: do sex hormone determine vulnerability? J Neuroendocrinol 2011; 23: 383-92.

[9]   Karlsson L, Kiviruusu O, Miettunen J, Heilä H, Holi M, Ruuuttu T. *et al,* One-year course and predictors of outcome of adolescent depression: a case-control study in Finland. J Clin Psychiatry 2008; 69: 844-53.

[10]   Silk Js, Steinberg L, Morris AS. Adolescents' emotion regulation in daily life: links to depressive symptoms and problem behavior. Child Dev 2003; 74: 1869-80.

[11]   Colman I, Ataullahjan A, Naicker K, Van Lieshout RJ. Birth weight, stress, and symptoms of depression in adolescence: evidence of fetal programming in a national canadian cohort. Can J Psychiatry 2012; 57: 422-28.

[12]   Baskaran A, Millev R, McIntyre RS. The neurobiology of the EEG biomarkers as a predictor of treatment response in depression. Neuropharmacology 2012; 63: 507-13.

[13]   van Deurzen PAM, Buitelaar JK, Brunnekreef JA, Ormeal J, Minderaa RB, Hartman CA. *et al,* Response time variability and response inhibition predict affective problems in adolescent girls, not in boys: the TRAILS study. Eur Child Adolesc Psychiatry 2012; 21: 277-87.

[14]   Forbes EE, Olino TM, Ryan ND, Birmaher B, Axelson D, Moyles DL, Dahal RE. Reward-related brain function as a predictor of treatment response in adolescents with major depressive disorders. Cog Affect Behav Neurosci 2010; 10: 107-18.

[15]   Bress JN, Smith E, Foti D, Klein DN, Hajcak C. Neural response to reward and depressive symptoms in late childhood to early adolescence. Biol Psychiatry 2012; 89: 156-62.

[16]   Mitchell AM, Pöseel P. Frontal brain activity pattern predicts depression in adolescent boys. Biol Psychiatry 201: 89: 525-27.

[17]   Mingtian Z, Shuqian Y, Xiongzhao Z, Jinyao Y, Xueling Z, Xiang W. *et al,* Elevated amygdala activity to negative faces in young adults with early onset major depressive disorders. Psychiatr Res: Neuroimaging 2012; 201: 107-12.

[18]   Cullen KR. Editorial: imaging adolescent depression treatment. Am J Psychiatry 2012; 169: 384-50.

[19]   Jin C, Gao C, Chen Ce, Ma S, Netra R, Wang Y. *et al,* A preliminary study of the dysregulation of the resting networks in first-episode medication-native adolescent depression. Neurosci Let 2011; 503: 105-09.

[20]   Forbes EE, Dahl RE. Research review: altered reward function in adolescent depression: what, when and how. J Child Psychol Psychiatry 2012; 53: 3-15.

[21] Jiano Q, Ding J, Lu G, Su L, Zhang Z, Wang Z. *et al,* Increased activity imbalance in fronto-subcotrical circuits in adolescents with major depression. PlosOne 2011; 6: e25159. Doi:10:1371/journal. pone.0025159.

[22] Guerry JD, Hastings PD. In search of HPA axis dysregulation in child and adolescent depression. Clin Child Psychol Rev 2012; 14: 135-60.

[23] Bylund DB, Reed AL. Childhood and adolescent depression: why do children and adults respond differently to antidepressant drugs. Neurochem Int 32007; 51: 246-53.

[24] Thapar A, Collishaw S, Pine D, Thapar A. Depression in adolescent. Lancet 2012; 379:1060-67.

[25] Spirito A, Esposito-Smythers C, Wolff J, Uhl K. Cognitive-behavioral therapy for adolescent depresion and suicidality. Child Adolesc Psychiatr Clin N Am 2011; 20: 191-204.

[26] Webb CA, Auerbach RP, DeRubeis RL. Proceses of changes in CBT of adolescent depression: review and recommendations. J Clin Child Adolesc Psychology 2012; 41(5): 654-65.

[27] Yang LP, Scott LJ. Escitalopram: in the treatment of major depressive disorder in adolescent. Paediatr Drugs 2010; 12: 155-63.

[28] Carandang C, Jabbal R, MacBride A, Elbe D. A review of escitaloram and citaloprom in child and adolescent depression. J Can Acad Adolesc Psychiatry 2011; 20: 315-24.

[29] March J, Silva S, Petrycki S, Curry J, Wells K, Fairbank J *et al,* Fluoxetine, cognitive-behavioral therapy, and their combination for adolescents with depression. JAMA 2004: 292: 807-20.

[30] Hirschtritt ME, Pagano ME, Christian KM, McNamara NK, Stansbrey RJ, Lingler J. *et al,* Moderators of fluoxetine treatment response for children and adolescents with comorbid depression and substance use disorders. J Subst Abuse Treat 2012; 42: 366-72.

[31] Shamseddeen W, Clarke G, Wagner KD, Ryan ND, Birmaher B, Emslie G. *et al,* Treatment-resistant depressed youth showed a higher response rate if treatment ends during summer school break. J Am Acad Child Adolesc Psychiatry 2011; 50: 1140-48.

[32] Kennard BD, Silva SG, Tonev S, Rohde P, Hughes JL, Vitiello B. *et al,* Remission and recovery in the treatment for adolescents with depression study (TADS): acute and long-term outcomes. J Am Acad Child Adolesc Psychiatry 2009; 48: 186-95.

[33] Brent D, Emslie G, Clarke G, Wanger KD, Asarnow JR, Keller M. *et al,* Switching to another SSRI or to venafaxine with or without cognitive behavioral therapy for adolescents with SSRI-resistant depression. JAMS 2008; 299: 901-13.

[34] Emslie GJ, Ventura D, Korotzer A, Tourkodimitris S. Escitalopram in the treatment of adolescent depression: a randomized placebo-controlled multisite trial. J Am Acad Child Adolesc Psychiatry 2009; 48: 721-29.

[35] Anderson RJ, Frye MA, Abulseoud OA, Lee KH, McGillvrsy J, Berk M. *et al,* Deep brain stimulation for treatment-resistant depression: efficacy, safety and mechanisms of action. Neurosi Biobehav Rev 2012; 36: 1920-33.

INDEX